General Editor – Construction and Civil Engineering

C. R. Bassett, B.Sc.
Formerly Principal Lecturer in the Department of Building and Surveying, Guildford County College of Technology

Books already published in this sector of the series:

Building organisations and procedures   *G. Forster*
Construction site studies – production, administration and
    personnel   *G. Forster*
Practical construction science   *B. J. Smith*
Construction science Volume 1   *B. J. Smith*
Construction science Volume 2   *B. J. Smith*
Construction mathematics Volume 1   *M. K. Jones*
Construction mathematics Volume 2   *M. K. Jones*
Construction surveying   *G. A. Scott*
Materials and structures   *R. Whitlow*
Construction technology Volume 1   *R. Chudley*
Construction technology Volume 2   *R. Chudley*
Construction technology Volume 3   *R. Chudley*
Construction technology Volume 4   *R. Chudley*
Maintenance and adaptation of buildings   *R. Chudley*
Building services and equipment Volume 1   *F. Hall*
Building services and equipment Volume 2   *F. Hall*
Building services and equipment Volume 3   *F. Hall*
Measurement Level 2   *M. Gardner*
Structural analysis   *G. B. Vine*
Site surveying and levelling Level 2   *H. Rawlinson*
Economics for the construction industry   *R. C. Shutt*
Design procedures Level 4   *J. M. Zunde*
Environmental science   *B. J. Smith, M. E. Sweeney and
    G. M. Phillips*
Design procedures Level 2   *M. Barritt*
Design technology Level 5   *J. M. Zunde*

A. G. Smyrell
M.A., C.Eng., M.I.C.E.

Design of structural elements
Volume 1

Longman London and New York

Longman Group Limited
Longman House
Burnt Mill, Harlow, Essex CM20 2JE, England

Associated companies throughout the world

*Published in the United States of America
by Longman Inc., New York*

© Longman Group Limited 1982

First published 1982

British Library Cataloguing in Publication Data

Smyrell, A.G.
  Design of structural elements. – (Longman
  technician series. Construction and civil
  engineering)
  Vol. 1
  1. Structural design
  I. Title
  624.1'771   TA658
  ISBN 0-582-41229-3

Library of Congress Cataloguing in Publication Data

Smyrell, A.G. (Arthur Gordon)
  Design of structural elements.

  (Longman technician series. Construction
and civil engineering sector)
  Includes index.
  1. Structural design. I. Title. II. Series
TA658.S63   624.1'7   81-20836
ISBN 0-582-41229-3   AACR2

Printed in Singapore by
Huntsmen Offset Printing Pte Ltd.

Contents

Preface

This book is intended for use by technicians studying 'TEC U77/ 427–Design of Structural Elements IV' or other HND technicians studying a similar course (although it may be of too high a level for them). It will also prove useful to Building Control Officers and others wishing to implement the requirements of CP 110 and BS 449 in their working environment. It is also intended for use by first year under-graduate students, particularly those studying building and civil engineering.

Thus the text must cater for a wide range of differing expectations. Whereas the undergraduate must understand the background to the codes and should be able to derive the various formulae which are used, the technician is only interested in the implementation of these formulae.

Therefore, where theory is being presented or where difficult concepts are being discussed, that section is delineated with a vertical line on the outer edge of the text. The undergraduate should pay particular attention to these sections and become familiar with the intricacies of the argument. On the other hand the technician should only become familiar with using the formulae presented and not with the theory. Indeed the technician may be able to omit a study of them completely, concentrating on the design examples instead.

Acknowledgements

I am grateful to the following for permission to reproduce copyright material:
British Standards Institution for my Figs. 2.6, 2.8, 2.13, 2.14, 7.11, 8.3, 8.4, 8.5, 8.6, 8.7, 8.8, 8.9, 8.10, A.1, A.2, and Tables 2.2, 9.3, 9.4, 9.6, 9.7, 9.8, 9.9, 9.10 and 9.11; extracts from British Standards are reproduced by permission of the British Standards Institution.

My thanks are also due to my colleagues, Mr R. Sharpe and Mr B. Currie, for their kind and constructive comments.

Most of all my thanks are due to a very patient wife, without whose assistance this book could not have been written.

Not forgetting Dr B. Smith, Director of Studies in the School of Building at the Ulster Polytechnic, who encouraged me to devote my time to this project.

Dedication:
To God be all the Glory.

Section I

Reinforced concrete

Chapter 1

The aims of design

Section 1

Reinforced concrete

In many ways, the process of Structural Design is impossible to define, combining as it does a mixture of experience, intuition and logical thought. Experience teaches those lessons learned by previous designers in similar situations, e.g. if the client requires a house, the designer can base his design on other houses using tried and tested details. But if experience were the only tool at the designers' disposal, how would he cope with an entirely different situation? The design of a skyscraper cannot be based solely on the experience gained from the design of two or even three-storey homes.

Design must also encompass the realms of invention if it is to cope with the unfamiliar or unexpected situation. Here, intuition can play a part by extending the experience gained on the design of one structure to the development of another. However, this process can only advance in small steps over a long period, since intuition cannot make the jump from the small house to the large skyscraper in one go with any degree of safety. There must be a series of lesser buildings in-between, increasing in size and complexity over a period of time. Throughout this evolution, the processes of logical thought must be employed to learn from the mistakes that will be made along the way.

Intuition will not always lead to the best solution and may end in a dangerous one. Thus it is wisest to use intuitive solutions to design problems sparingly and only when those involved have a wide experience in the practice of design. Most designers are best advised to rely heavily on the amassed experience of the many generations of engineers who have gone before them. Yet how can this experience be passed from one designer to another?

One of the most common modes for communicating design experience is by means of the British Standard Codes of Practice, and it is with two of these that this book is largely concerned – CP 110 *The Structural Use of Concrete* and BS 449 *The Structural Use of Steel in Buildings*. These documents are not intended to provide insights into the ways in which specific design problems may be solved, but describe the building blocks on which any design can be based. They give the designer a set of rules by which he may judge the suitability of his solution, and perhaps an indication of some of the ways in which he may arrive at that solution, but they cannot teach a designer to design. Instead they present a logical approach with coverage of certain design areas as examples in their use. For example, both codes cover the various aspects to be considered in the design of beams, but arches are not specifically dealt with.

Limit-state design

When first considering the structural design process, one is apt to think only in terms of strength when assessing the suitability of a structure. Understandable as this may be, the design of structures constitutes much more than an assessment of the load at which they collapse. Imagine a road bridge constructed from a web of steel cables strong enough to carry the heaviest vehicle. Unfortunately not many drivers would be induced to use it, since it would deflect in such a way as to make any crossing difficult if not hazardous. Thus it can be seen that deflection is a primary criterion of structural 'fitness'. Nor is this the only other criterion or 'Limit-State' by which a structure must be judged.

CP 110 divides the various limit-states into three main groups:

1. The **ultimate** limit-state.
2. The **serviceability** limit-state; and
3. Other considerations.

The ultimate limit-state

This limit-state is concerned with the strength of the structure, and involves the determination of the maximum loads likely to be experienced by the structure during its lifetime. Once these loads have been calculated and their effects on the structure assessed, the individual elements must not collapse or contribute to the overall destruction of the structure while supporting them. This is often the first state to be examined, but it must not be allowed to obscure the requirements of the other states.

The serviceability limit-states

These limit-states, whilst not causing collapse of the structure, may render it unusable by virtue of their effects on the users. There are three main states mentioned in CP 110, but this must *not* be taken as an exhaustive list, as each structure will have its own unique requirements and must be assessed on its merits. The three main states applicable to all structures are:

1. *Deflection* (Cl. 2.2.3.1). There are two aspects to the limitation of deflection in a structure. First, the deflection must not be too noticeable, otherwise it will give the occupants a sense of uneasiness— an expectation of impending doom. Thus the code imposes a limit on deflection of span ÷ 250, but this may be too small or large in certain circumstances. It can only be used for the most common situations and must not be assumed in all cases without first thinking of the consequences of its imposition.

   Second, the deflection must not cause cracking or warping in any secondary structural elements such as partitions. The limits in this case are not so simple, but the main criterion is span ÷ 350 due to imposed loading only and to long-term effects such as creep.

2. *Cracking* (Cl. 2.2.3.2). Although appearance is an important factor in the determination of the allowable crack width the major influence is the effect that the width has on the extent of corrosion of the reinforcement. For this reason, a limit of 0.3 mm is imposed.

3. *Vibration* (Cl. 2.2.3.3). If the natural frequency and amplitude of vibration give an uncomfortable feeling to the users of a structure, it is deemed to be unserviceable. The analysis of vibration is rather complicated and beyond the scope of this book.

Other considerations

CP 110 mentions four other states which may need to be considered in some structures. A machine shop may cause **fatigue** loading; depending upon the degree of exposure, **durability** may be a problem; all buildings which are to be used extensively by people must be **fire resistant**; some buildings may be subject to damage by **lightning**; each structure must be judged on *all* the limit-states applicable to it.

Finally, there is one important consideration which must be made for all structures, and that is the problem of Stability. This is dealt with in the code in Cl. 3.1.2.2, and is all too often left to the end of the design process, yet as has been demonstrated frequently in the past, it can be most critical. It is suggested by some authorities that the consideration of stability be made the responsibility of one engineer who takes no other part in the design process. Whatever the organisation, it cannot be too greatly stressed that stability must be checked, especially in structures where extensive use is made of precast elements.

Loads and strengths of materials

It is a fact of life that it is not possible to predict the strength of any material or structure with absolute accuracy until it has been tested to destruction. There are many reasons for this state of affairs, and the following example will help to illustrate at least some of them.

A certain structure requires that the concrete used in its construction should have a minimum strength of $30\,N/mm^2$ after twenty-eight days from the date of casting. Over the years, as a result of thousands of experiments, the batching plant will have found that if x kg of sand, y kg of coarse aggregate and z kg of cement are mixed with u litres of water there is a fair chance that the required twenty-eight day strength will be met by the majority of the concrete. However, the plant manager cannot rely on that particular recipe to work for ever, and he must test the strength of each batch and adjust the proportions as necessary. Why should this be necessary?

For a start, there will be inaccuracies in the weights of materials actually added. Even if these inaccuracies could be eliminated the resulting concrete would not give a constant strength. There are a number of factors contributing to this variance; no two batches of cement have the same ability to give strength – here again the cement manufacturer is subject to recipe trouble. The external temperature and humidity will

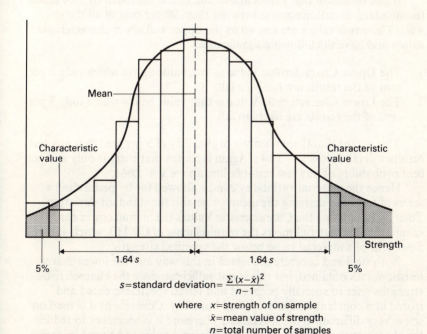

s = standard deviation = $\dfrac{\sum (x-\bar{x})^2}{n-1}$

where x = strength of on sample
$\bar{x}$ = mean value of strength
n = total number of samples

Fig. 1.1   Normal distribution curve

affect the result, while even the manner in which the contractor places the concrete can greatly alter the strength. These and many other influences ensure that the prediction will coincide with reality only very rarely. This sort of situation applies to all materials, including steel, but the effect is not so marked as is the case with concrete.

The only way to approach this problem is through statistics, but thankfully the mathematics involved are relatively simple. If the strengths of a number of randomly selected samples are measured and then plotted on a histogram of strength against frequency, a diagram such as is shown in Fig. 1 will result. For materials such as concrete and steel, their histograms can be mapped with sufficient accuracy by a curve known as 'the Normal Distribution' curve, this being the curve plotted in Fig. 1.1.

It will be noted that the 'bell-like' shape of the normal distribution curve is symmetrical about the mean or average result. This means that the number of results falling above the mean is equal to the number of results falling below the mean. If the distribution of results does not follow this pattern, the normal distribution analysis cannot be used. As well as knowing the value of the mean, it is necessary to know how widely spread the values are; in other words it is necessary to know the range of values either side of the mean which encloses a known percentage of the results. This is done by calculating the 'Standard Deviation' (most modern calculators have this facility, making the task simple).

It can be shown that values above and below the mean by 1.64 times the standard deviation enclose between them 90 per cent of all the results. These two values are known as the upper and lower **characteristic** values and have the following significance:

1. The Upper Characteristic Value is that value above which only 5 per cent of the results are likely to fall.
2. The Lower Characteristic Value is that value below which only 5 per cent of the results are likely to fall.

This means that there is only a probability of 5 per cent that any test results will fall below $\bar{x} - 1.64\,s$. Again it means that there is only as 5 per cent probability of any test result falling above $\bar{x} + 1.64\,s$.

Hence the material variability can be allowed for by performing a series of tests and deriving the mean value and the standard deviation. Then by calculating the Characteristic Values the manufacturer can ensure that his material meets the requirements of CP 110, which allows 5 per cent of a material to be below the specified strength.

Both steel and concrete are tested in this way and the lower characteristic values obtained, but this is not sufficient, since the characteristic strengths refer to specially prepared and tested specimens cured and stored at a constant temperature and humidity. Concrete as it is used on site is very different to the specimens. Therefore it is necessary to reduce the characteristic strength when determining a value to be used in design by dividing it by a partial safety factor. CP 110 advises the use of 1.5 as the partial safety factor for concrete but a smaller 1.15 for steel since the sources of variation in the manufacture of steel are less. Therefore, the design strengths are given by:

Concrete – design strength $= f_{cu}/1.5$

Steel – design strength $= f_y/1.15$

f_{cu} and f_y are the characteristic values

Although ideally the characteristic values for loading should be determined in a similar fashion, taking the upper value, this is not yet possible for dead and imposed loads. Therefore the values for the characteristic loads should be taken as the values of dead and imposed load recommended in CP 3, Chapter V, Part 1. In the case of wind loading, the pressures given in CP 3 are already expressed as characteristic values, a proper statistical analysis having been carried out.

Yet again, the characteristic values of loads do not allow for loads that for some accidental reason differ significantly from those specified. Nor do they allow for inaccuracies in assessing their effects, or for errors in construction. To account for these, loading partial safety factors must be applied. It must be fully understood that the term 'safety factor' is in some senses a misnomer. It has just been demonstrated that they are applied as a direct consequence of the mathematical treatment. The name is a carry-over from previous codes in which lack of knowledge or uncertainty concerning analytical techniques forced the use of true safety factors. There is still an element of safety in the factors used in CP 110, but it should not be supposed that it is the only ingredient. The whole purpose of using CP 110 is that it gives a better understanding of the way in which structures perform, allowing designers to dispense with an over-reliance on a guess of how safe a structure really is.

In any situation where the loads are not given in CP 3, the statistical analysis must be carried out and the characteristic values arrived at. Hopefully all loads will be derived in this fashion before too long.

When a structure is exposed to more than one type of loading (as is usual), the assessment of their effects must be carried out in a systematic form, in such a way as to account for all the possible combinations that might occur. In other words, if it is possible for wind to be blowing at gale force at the same time as some imposed load is acting then this 'Load-Case' must be investigated. However, it is not usual for this to be the case; more often the maximum, imposed load will occur at a time when the wind is not so strong. Similarly, the strongest wind load will be unlikely to occur simultaneously with the occurrence of the largest imposed load. In the ideal situation, where a proper statistical survey had been carried out, the calculation of the worst loading combination would be relatively simple. However, until such time as this is possible CP 110 allows the designer to consider certain formalised combinations. These are set out in Table 1.1. It will be noticed that the partial safety factors used for loads placed on the structure at the ultimate limit-state are different from those specified for the serviceability limit-state. This is only to be expected since the serviceability condition refers to the general or working situation.

Table 1.1 Design loads and partial safety factors

Combination	Ultimate limit-state	Serviceability limit-state
Dead + imposed loads	Maximum loads of $(1.4G_k + 1.6Q_k)$ and minimum loads of $1.0G_k$ so arranged as to give the worst effect in the structure	$1.0(G_k + Q_k)$
Dead + wind load	$0.9G_k + 1.4W_k$ or $1.4(G_k + W_k)$	$1.0(G_k + W_k)$
Dead + imposed + wind loads	$1.2(G_k + Q_k + W_k)$	$1.0G_k + 0.8(Q_k + W_k)$

where G_k: characteristic dead load (self-weight)
        Q_k: characteristic imposed load
        W_k: characteristic wind load

Chapter 2

Beams

Moments and shear forces in beams

Of all the ways in which structures may resist the loads placed upon them, bending is probably the most common. Many loaded structures such as arches, short columns, trusses or cables have elements which exhibit an almost constant stress across their section – either pure tension or pure compression. However, when bending strength is utilised, a much more complex situation develops, whereby a varying stress across the section balances moments and forces produced in the structural element by the external loads. In order to evaluate these stresses, it is first necessary to know the magnitude of the moments and forces causing them and it is to this end that Bending Moment and Shear Force Diagrams are constructed.

For indeterminate concrete structures, an 'exact' Bending moment or shear force diagram is difficult or impossible to construct since the material is not elastic and cracks when it is put into tension. The consideration of indeterminate structures is left to Volume 2 of this book, and for the moment, only statically determinate beams and cantilevers are examined. As a result, bending moment and shear force diagrams are easily constructed by examining the forces and the reactions they cause. It is assumed that the student has enough expertise in their construction for the subject to be omitted from this study. The design procedure for beams may be summarised as consisting of two stages:

1. The design moments and shears to be resisted are calculated from

the ultimate or 'design' loads.

2. A beam section is chosen and the amount of steel needed is calculated to enable the beam's strength to exceed the required capacity. It will quickly be realised that this step is one of trial and error, since there is not a unique solution for every situation. It is possible to find a number of satisfactory sections and a choice must be made between them to decide which is the best in any given case.

Stage 1 is illustrated in this next example.

Example 2.1   A simply supported beam acting as a lintel has an effective span of 3 m and carries a brick wall above it. The wall produces a triangular load of the form shown in Fig. 2.1(a), while the beam's own weight produces a Uniformly Distributed Load (UDL). Both these loads are Dead Loads, therefore the factor of safety for loads is 1.4 for both. If the lintel is 250×100 mm and the densities of concrete and brickwork are $24 \, kN/m^3$ and $19 \, kN/m^3$ respectively, calculate the magnitude of bending moment and shear force which the beam must resist.

$$\text{Characteristic self-weight of lintel} = 24 \times 3 \times 0.25 \times 0.1$$
$$= 1.8 \, kN$$

$$\text{Characteristic weight of brickwork} = 19 \times 0.1 \times 0.5 \text{ base} \times \text{height}$$
$$= 19 \times 0.1 \times 1.5 \times 3 \sin 60°$$
$$= 7.405 \, kN$$

$$\text{Total ultimate load} = 1.4(1.8 + 7.405) = 12.886 \, kN$$

$$\text{Maximum shear force} = \text{end reaction} = \frac{12.886}{2}$$
$$= 6.443 \, kN$$

$$\text{Maximum moment occurs at mid span} = M = 6.443 \times 1.5$$
$$- (1.4 \times 0.9) \times \frac{1.5}{2}$$
$$- \left(1.4 \times \frac{7.405}{2}\right) \times \frac{1.5}{3}$$
$$M = \underline{6.128 \, kNm}$$

The resulting shear force and bending moment diagrams are given in Fig. 2.1(d) and (e).

At the beginning of this example the beam's effective span was given without defining what was meant by this term, yet it is the value of span

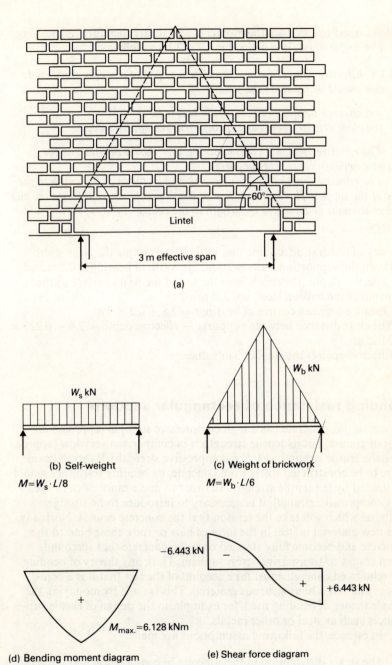

(a)

W_s kN

(b) Self-weight
$M = W_s \cdot L/8$

W_b kN

(c) Weight of brickwork
$M = W_b \cdot L/6$

(d) Bending moment diagram
$M_{max.} = 6.128$ kNm

−6.443 kN

+6.443 kN

(e) Shear force diagram

Fig. 2.1

which is used to calculate the moments and shears therefore its meaning must be understood. It is defined in Cl. 3.3.1.1 as follows:

3.3.1.1 *Effective span of beam. The effective span of a simply supported member should be taken as the smaller of*

the distance between the centres of bearings, or
the clear distance between supports plus the effective depth.

The effective span of a continuous member should be taken as the distance between centres of supports.
The effective length of a cantilever should be taken as its length to the face of the support plus half its effective depth except where it forms the end of a continuous beam where the length to the centre of the support should be used.

By way of illustration assume that in the last example the clear width between the supporting walls is 2.8 m; the width of bearing is 0.2 m and the effective depth (the depth from the top of the beam surface to the centroid of the tension steel) is 0.225 m.
Distance between centres of bearings = 2.8 + 0.2 = 3 m
The clear distance between supports + effective depth = 2.8 + 0.225 = 3.025 m
Effective span is the smaller 3 m value.

Bending resistance of rectangular sections

As a material, concrete has a high compressive strength (approaching that of granite) but its tensile strength is in comparison very low (approximately: tensile strength = 0.1 × compressive strength). If then a beam were to be constructed solely from concrete, its bending resistance would be limited by the tensile strength. In order to make more efficient use of this compressive strength it is necessary to introduce some stronger material which will take the tension that the concrete cannot. Normally, this new material is steel in the form of bars or rods; these bind to the concrete and become fully stressed once the concrete that surrounds them cracks so transferring stress to them. Thus, any theory of bending for reinforced concrete must take account of the fact that it is a composite and not a homogeneous material. This is done by modifying the simple theory of bending used, for example, in the design of elastic substances such as steel or other metals.
In essence, the following assumptions are made:

1. The strain distribution in the concrete in compression and the reinforcement (whether in tension or, as later described, in compression) is derived from the assumption that sections which were planar before bending remain so after bending. In other words the strain is

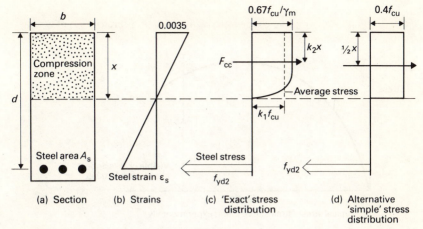

Fig. 2.2 Stresses and strains in a reinforced concrete section

    proportional to the distance from the neutral axis, i.e. it is a straight-line diagram (Fig. 2.2(b)).

2.   The tensile strength of the concrete is low enough to be ignored; this is in fact a conservative assumption, since its inclusion would increase the calculated strength of the beam.

3.   Perfect bond or adhesion is developed between the concrete and the steel so that there is no slippage (or strain difference) between the two materials. Thus the steel strain may be derived from the strain diagram shown in Fig. 2.2(b).

4.   A concrete beam fails when the maximum compressive strain in the concrete reaches a value of 0.0035.

5.   The stress in steel is constant across its section.

    Having defined these assumptions the next stage is to derive from the strain diagram the stresses in the two materials. This is empirically achieved by testing samples of concrete and steel, then plotting the results in the form of a graph of stress against strain. From these experimental curves are derived notional or approximate design curves which are simpler than the actual curves but which are accurate enough for design purposes. Figures 2.3 and 2.4 show sample curves for concrete and steel alongside the curves which have been adopted by CP 110 for use in the design process. When the design curve for concrete is applied to the strain diagram of Fig. 2.2(b) the stress block of Fig. 2.2(c) is the result, and similarly for steel, the stress may be found by reading the value from the curve in Fig. 2.4(b).

    The final step involves relating these stresses to the external moment, and once this is done the beam can be designed. Consider first the strain diagram of Fig. 2.2(b).

14

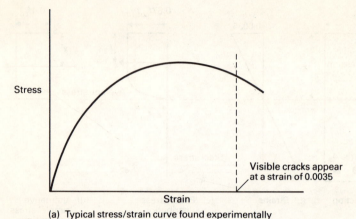

Stress

Visible cracks appear
at a strain of 0.0035

Strain

(a) Typical stress/strain curve found experimentally

*The 0.67 takes account of the ratio between the characteristic
cube strength and the bending strength in a flexural member

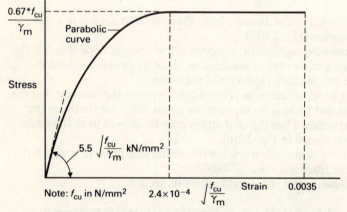

$\dfrac{0.67*f_{cu}}{\gamma_m}$

Parabolic
curve

Stress

$5.5\sqrt{\dfrac{f_{cu}}{\gamma_m}}$ kN/mm^2

Note: f_{cu} in N/mm^2 $2.4\times10^{-4}\sqrt{\dfrac{f_{cu}}{\gamma_m}}$ Strain   0.0035

(b) Short-term design stress-strain curve for normal weight concrete

Fig. 2.3 Stress/strain curves for concrete

Strain diagram

From similar triangles

$$\frac{0.0035}{x}=\frac{\epsilon_{st}}{d-x}$$

Thus $\epsilon_{st}=0.0035\left(\dfrac{d}{x}-1\right)$ [2.1]

From this it can be seen that the steel strain, and thus the steel stress
depends only on the depth of the neutral axis and not on the value of

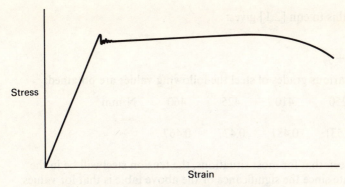

(a) Typical stress/strain curve found experimentally

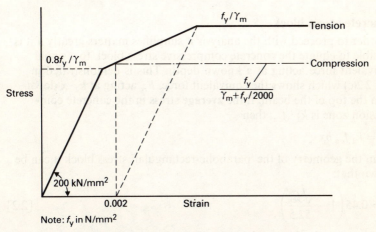

Note: f_y in N/mm²

(b) Short-term design stress/strain curve for reinforcement

Fig. 2.4 Stress/strain curves for steel

the external moment. This may seem strange since one might expect that as moment increases the stress increases. This does in fact happen, but remember that it is the ultimate state of the beam that is being examined and not some intermediate state. At failure the *maximum* stress experienced by the steel is the value being calculated and it is truly independent of the external moment.

It will be advantageous at this point to digress for a moment and ask the question – 'If the steel stress is related so closely to the value of neutral axis depth, at what value of x/d does the steel **just** yield?'

Strain in the tension steel at yield $= 0.002 + \dfrac{f_y}{1.15 \times 200\,000}$

Equating this to eqn [2.1] gives:

$$\frac{x}{d} = \frac{805}{1265 + f_y}$$

Thus for various grades of steel the following values are obtained:

f_y	250	410	425	460	N/mm²
$\dfrac{x}{d}$	0.531	0.481	0.477	0.467	

It can be seen that for most situations, the tension steel will be in the yielded state since the significance of the above table is that for values of x/d less than the ones given, the steel is yielded, and the actual stress in the steel, $f_{yd2} = f_y/1.15$.

Concrete stress block

In order to proceed with the analysis it simplifies matters greatly if it is possible to equate the concrete compressive stress block to a single equivalent force acting at a known depth. This is demonstrated in Fig. 2.2(c) which shows the equivalent force, F_{cc} acting at $k_2 \cdot x$ down from the top of the beam. If the **average** stress in the concrete compression zone is $k_1 \cdot f_{cu}$, then

$$F_{cc} = k_1 f_{cu} bx$$

From the geometry of the 'parabolic-rectangular' stress block it can be shown that:

$$k_1 = 0.45\left(1 - \frac{\sqrt{f_{cu}}}{52.5}\right) \qquad [2.2]$$

$$k_2 = \frac{\left[2 - \dfrac{\sqrt{f_{cu}}}{17.5}\right]^2 + 2}{4\left[3 - \dfrac{\sqrt{f_{cu}}}{17.5}\right]} \qquad [2.3]$$

As can be seen, eqns [2.2] and [2.3] are highly complex and Fig. 2.5 shows how they vary with f_{cu}. Due to this complexity CP 110 allows a further simplification illustrated in Fig. 2.2(d), where $k_1 = 0.4$ and $k_2 = 0.5$. This alternative will be investigated in more detail later, but for the moment it is simply noted that it is a possibility.

Equating forces and moments

In order to maintain equilibrium the **force** in the concrete compression zone must be equal to the **force** in the tension steel.

Hence $\qquad k_1 f_{cu} bx = f_{yd2} \cdot A_s \qquad [2.4]$

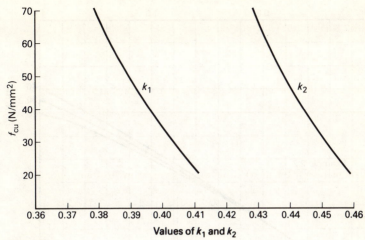

Fig. 2.5

Finally, the external moment M_u must be balanced by the internal moment. Taking moments about the tension steel:

$$M_u = k_1 f_{cu} bx(d - k_2 x) \qquad [2.5]$$

This is often rearranged to give non-dimensional terms as follows:

$$\frac{M_u}{f_{cu}bd^2} = k_1 \frac{x}{d}\left(1 - k_2 \frac{x}{d}\right) \qquad [2.5(a)]$$

Substituting [2.4] into [2.5(a)] gives:

$$\frac{M_u}{bd^2} = f_{yd2} \cdot \frac{A_s}{bd}\left(1 - \frac{k_2}{k_1} \cdot \frac{f_{yd2}}{f_{cu}} \cdot \frac{A_s}{bd}\right) \qquad [2.6]$$

since f_{yd2} may be found using eqn [2.1]; if M_u, b, d and f_{cu} are known, A_s may be found. Code of Practice 110: Part 2 has made this process easier by constructing graphs such as the one illustrated in Fig. 2.6 for steel of characteristic strength equal to 410 N/mm².

Example 2.2 A beam of width 100×300 mm effective depth is to resist a moment at the ultimate limit state of 25 kNm. If $f_y = 410$ N/mm², and $f_{cu} = 30$ N/mm² calculate the area of steel required using chart 2 of CP 110: Part 2 (Fig. 2.6).

$$\frac{M}{bd^2} = \frac{25 \times 10^6}{100 \times 300 \times 300} = 2.777 \text{ N/mm}^2$$

Therefore as shown in Fig. 2.6

$$100 A_s/bd = 0.883$$

$$A_s = 264.9 \text{ mm}^2$$

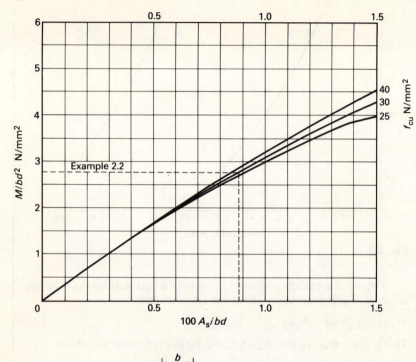

Fig. 2.6 Singly reinforced beams (Chart 2)

It is now possible to calculate the depth to the neutral axis as follows:
  Using eqn [2.5]

$$25 \times 10^6 = k_1 \times 30 \times 100x(300 - k_2 x)$$

for $f_{cu} = 30 \text{ N/mm}^2$ $k_1 = 0.403;$ $k_2 = 0.451$

     $x = 78.10 \text{ mm}$ and $f_{yd2} = 357 \text{ N/mm}^2$ (yielded)

Had the formulae been used instead of the chart then the following
would have been the procedure adopted:

1. From eqn [2.5] find x the neutral axis depth.
2. Determine whether or not the steel is yielded.
3. If yielded then calculate the stress in the steel as $f_{yd2} = f_y/1.15$ and
   use in eqn [2.4] to find the area of steel needed.

4. If the steel is not yielded then the stress in the steel must be found by one of the following three methods:
   (a) Iteratively – by trial and error
   (b) The elastic/plastic and the elastic phase of the steel curve may be given equations:

Elastic/plastic: $f_{yd2} = \left(700\dfrac{d}{x} + 900\right)\dfrac{f_y}{2300 + f_y}$ [2.7]

Elastic: $f_{yd2} = 700\left(\dfrac{d}{x} - 1\right)$ [2.8]

Thus the steel stress may be found mathematically
   (c) Graphically as follows:

Rearranging eqn [2.4] gives

$f_{yd2} = k_1 f_{cu} bx / A_s$ [2.4(a)]

Rearranging eqn [2.1] gives

$x = \dfrac{0.0035}{0.0035 + \epsilon_{st}} \cdot d$ [2.1(a)]

Combining equations [2.4(a)] and [2.1(a)] gives

$f_{yd2} = k_1 f_{cu} \cdot \dfrac{0.0035}{0.0035 + \epsilon_{st}} \cdot \dfrac{bd}{A_s}$ [2.9]

This equation can now be plotted on the same graph as the stress/strain curve for steel, and where they cross will indicate the

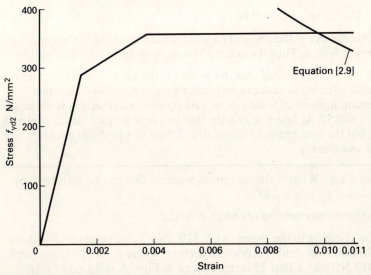

Fig. 2.7

values of stress and strain. However, eqn [2.9] depends on the value of A_s which must be known beforehand, thus this method is only useful in checking an already designed section. Figure 2.7 illustrates the method for the beam in Example 2.2.

Stipulations of CP 110 regarding singly reinforced beams

When a beam is reinforced in tension only, the depth of the neutral axis is limited by two constraints set out in Cl. 3.3.5.1 and Cl. 3.3.5.3.

1. The neutral axis depth must not exceed half the effective depth. This is to ensure that the tension steel is yielded or nearly yielded. When the failure of a beam is dictated by the crushing of the concrete before the steel has a chance to yield, the collapse is sudden, catastrophic and without warning. If the beam is designed in such a way that the steel yields before collapse, then the failure is more gradual with large cracks and deflections. This gives the users of the structure sufficient time to evacuate and thus the design is considered much safer. Hence CP 110 tries to ensure this is always the case.

2. Clause 3.3.5.3 sets out simple formulae for the design of beams and uses Z, the lever arm factor. This is given by:

$$Z = d - k_2 x \text{ (letting } k_2 = 0.5)$$

The clause then states that Z should not be allowed to exceed $0.95d$

$$0.95d > = d - 0.5x$$

Therefore $x > = 0.1d$

This ensures sufficient safety against excess crushing of the concrete before failure. Thus the neutral axis must lie between $0.1d$ and $0.5d$.

It will be appreciated that the greater is the value of x, the greater is the value of moment that can be taken by a given section. Thus the maximum moment of resistance of a singly reinforced section is obtained when $x = 0.5d$. At first it may seem that this task is again one of trial and error, but the next example demonstrates how this problem may be solved more easily.

Example 2.3   What is the maximum moment that can be withstood by the beam of Example 2.2?

Maximum moment occurs when $x = 0.5d$

On referring to the charts in CP 110: Part 2 one observes that charts 23 – 26 deal with doubly reinforced sections where $f_{cu} = 30$ N/mm^2 and $f_y = 410$ N/mm^2. Chart 23 is reproduced in Fig. 2.8 and a point A is indicated. On these charts, the depth to the neutral axis is marked in-

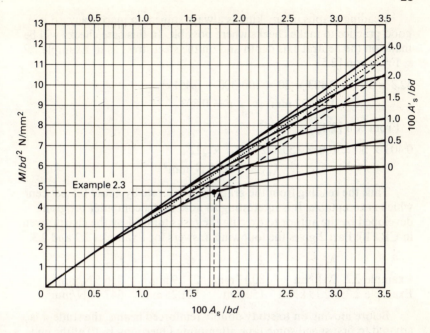

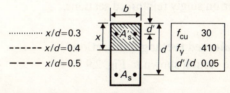

Fig. 2.8 Doubly reinforced beams (Chart 23)

cluding $x/d = 0.5$, where this cuts the lowest line is point A. The trick is to use the line on these charts for zero compression reinforcement and thus A gives both $100\,A_s/bd$ and M_u/bd^2 for the maximum moment condition.

$$\frac{M_u}{bd^2} = 4.7$$

Hence $\quad M_u = 4.7 \times 100 \times 300^2\ C = 42.3 \times 10^6\ \text{Nmm}$

$\qquad\qquad\quad = 42.3\ \text{kNm}$

$$\frac{100\,A_s}{bd} = 1.74$$

Hence $\quad A_s = \dfrac{1.74 \times 100 \times 300}{100}$

$\qquad\qquad\quad = 522\ \text{mm}^2.$

Reading graphs of any kind is always subject to inaccuracy, so it is good practice to make checks where possible. In this case checks will be made on the neutral axis depth and on the steel stress as in Example 2.2.

From eqn [2.5]

$$42.3 \times 10^6 = 0.403 \times 30 \times 100(300 - 0.451x)$$
$$x = 150.822 \text{ mm} \qquad x/d = 0.503$$

From eqn [2.4]

$$0.403 \times 30 \times 100 \times 150.822 = f_{yd2} \times 522$$
$$f_{yd2} = 349.32 \text{ N/mm}^2$$

Thus the reading from the graph is inaccurate by about 0.5 per cent which is acceptable. It is interesting to note that a computer program developed by the author to execute the tasks fulfilled by the charts given in CP 110: Part 2 gives the following answers for Examples 2.2 and 2.3.

	M_u	x	A_s	f_{yd2}
Example 2.2:	25 kNm	80.17 mm	271.9 mm²	356.5 N/mm²
Example 2.3:	42.13 kNm	150 mm	521.25 mm²	347.97 N/mm²

Before moving on to study doubly reinforced beams, the student is advised to first spend some time attempting Questions 1–10 at the end of this chapter on singly reinforced sections.

Doubly reinforced rectangular sections

Consider the section shown in Fig. 2.9

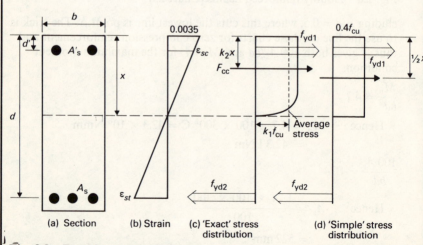

Fig. 2.9 Doubly reinforced beams

Strain diagram

Equation [2.1] still holds for the tension steel

$$\epsilon_{sc} = 0.0035\left(l - \frac{d'}{x}\right)$$ [2.10]

Equating forces and moments

At this point it is quite possible to continue as for singly reinforced beams and obtain the following formulae:

$$k_1 f_{cu} b x + f_{yd1} \cdot A'_s = f_{yd2} \cdot A_s$$ [2.11]

(similar to eqn [2.4])

and taking moments about the tension steel

$$M_u = k_1 f_{cu} b x (d - k_2 x) + f_{yd1} \cdot A'_s (d - d')$$ [2.12(a)]

or taking moments about the compression steel

$$M_u = k_1 f_{cu} b x (d' - k_2 x) + f_{yd2} \cdot A_s (d - d')$$ [2.12(b)]

Substituting [2.11] into [2.12(a)] or [2.12(b)] gives

$$\frac{M_u}{bd^2} = \rho f_{yd2} - \rho' f_{yd1} \cdot \frac{d'}{d} - \frac{k_2}{k_1} (\rho f_{yd2} - \rho' f_{yd1})^2 / f_{cu}$$ [2.13]

In eqn [2.13] there are six independent variables:

1. M_u/bd^2.
2. $\rho = A_s/bd$.
3. $\rho' = A'_s/bd$.
4. f_y since both f_{yd1} and f_{yd2} depend on this value.
5. f_{cu} since both k_1 and k_2 depend on this value.
6. d'/d.

Therefore Charts 7–42 in CP 110: Part 2 need to have separate charts for different values of f_{cu}, f_y and d'/d. The other three variables are then plotted on a single sheet by having curves for different values of A'_s/bd. It is also to be noted that the values of x/d are indicated on these charts.

However, there is another approach which can be taken as illustrated in Fig. 2.10. As this has a number of advantages and as it is the method which shall be used when developing the formulae for the simplified concrete stress block of Fig. 2.9(d), it shall be dealt with in some detail.

The complete, doubly reinforced section may be thought of as being the sum of two constituent sections. First is the section shown in Fig. 2.10(a). This consists of a concrete compression zone and the tension steel A_{ss} required to balance F_{cc}.

Thus $$f_{yd2} \cdot A_{ss} = k_1 f_{cu} b x$$ [2.14]

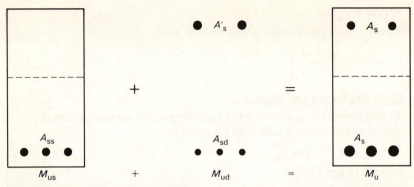

Fig. 2.10 Doubly reinforced beams

The second, imaginary section consists of steel only. The compression steel A_s' is balanced by the other part of the tension steel A_{sd}.

Thus $\qquad f_{yd1} \cdot A_s' = f_{yd2} \cdot A_{sd}$ [2.15]

Similarly, the moment is broken into two parts:

$$M_u = M_{us} + M_{ud}$$ [2.16]

where $\qquad M_{us} = k_1 f_{cu} bx(d - k_2 x)$ [2.17]

and $\qquad M_{ud} = f_{yd1} \cdot A_s'(d - d')$ [2.18]

It will be remembered that the tension steel stress is related to the neutral axis depth by eqns [2.7] and [2.8]. Similarly, the compression steel stress may be related to the neutral axis by eqns [2.19] and [2.20].

Compression steel equations:

Elastic: $\qquad f_{yd1} = 700\left(1 - \dfrac{d'}{d} \cdot \dfrac{d}{x}\right)$ [2.19]

Elastic/plastic: $f_{yd1} = \left(2300 - 700\,\dfrac{d'}{d} \cdot \dfrac{d}{x}\right) \cdot \dfrac{f_y}{2300 + f_y}$ [2.20]

The compression steel is plastic when

$$\frac{x}{d} > = \frac{7}{3} \cdot \frac{d'}{d}$$ [2.21]

The design procedure when using this approach is slightly different, but it has the advantage that x, f_{yd1} and f_{yd2} are automatically calculated.

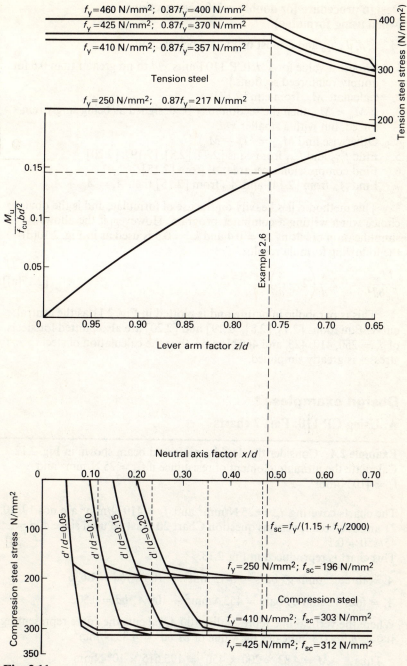

Fig. 2.11

**Design procedure for doubly reinforced sections
(when using formulae)**

M_u, b, d, d', f_{cu} and f_y must be known beforehand

1. Choose a value for x/d. (CP 110 limits x/d to no greater than 0.6 for doubly reinforced sections.)
2. Calculate M_{us} from eqn [2.17].
3. If $M_u < M_{us}$ then the section may be designed as being singly reinforced, but with a smaller x/d.
4. Otherwise find $M_{ud} = M_u - M_{us}$.
5. Find f_{yd2} and f_{yd1} from eqns [2.7], [2.8], [2.19], [2.20].
6. Find compression steel area, A'_s from [2.18].
7. Find A_{ss} from [2.14] and A_{sd} from [2.15] then $A_s = A_{ss} + A_{sd}$.

This method relies heavily on the use of formulae, and is the obvious choice when writing a computer program. However, if the allowed simplification of letting $k_1 = 0.4$ and $k_2 = 0.5$ is used as in Fig. 2.10(d) the following formula results:

$$\frac{M_{us}}{f_{cu}bd^2} = 0.2\,\frac{x}{d}\left(2 - \frac{x}{d}\right) \qquad\qquad [2.17(a)]$$

This is parabolic in nature and is plotted in Fig. 2.11 as the central curve. Equations [2.7], [2.8], [2.19] and [2.20] are also plotted for steels of $f_y = 250, 410, 425,$ and $460\,\text{N/mm}^2$, thus the calculation of steel stresses is greatly simplified.

Design examples

A. Using CP 110: Part 2 charts

Example 2.4   Consider the doubly reinforced beam shown in Fig. 2.12. Calculate the ultimate moment of resistance if $f_{cu} = 25\,\text{N/mm}^2$ and $f_y = 410\,\text{N/mm}^2$.

The charts covering $f_{cu} = 25\,\text{N/mm}^2$ and $f_y = 410\,\text{N/mm}^2$ are nos 19, 20, 21 and 22. To answer this question Chart 20 must be used since $d'/d = 45/450 = 0.1$
This chart is reproduced in Fig. 2.13.

A_s = three 25-mm $\varnothing$ bars = $1473\,\text{mm}^2$ $- 100\,A_s/bd = 1.636$

A'_s = four 12-mm $\varnothing$ bars = $452.4\,\text{mm}^2$ $- 100\,A'_s/bd = 0.502$

Where the vertical line representing $100\,A_s/bd$ cuts the curve representing $100\,A'_s/bd = 0.5$ indicates the value of $M/bd^2 = 4.83\,\text{N/mm}^2$

Thus $M = 4.83 \times 200 \times 450^2 = 195.615 \times 10^6\,\text{Nmm}$

$\underline{\qquad = 195.615\,\text{kNm}}$

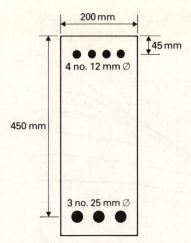

200 mm

45 mm

4 no. 12 mm ∅

450 mm

3 no. 25 mm ∅

Fig. 2.12

The position of the neutral axis may be calculated in the following manner:

Stage I: Estimate the value from Chart 20 as $x/d = 0.42$.

Stage II: Using this value for x/d calculate the values for the steel stresses, either from Fig. 2.11 or from eqns [2.7], [2.8], [2.19] and [2.20].

        Thus $f_{yd1} = 302.58 \text{ N/mm}^2$ (yielded)
        and $f_{yd2} = 356.52 \text{ N/mm}^2$ (yielded)

Stage III: Using eqn [2.11] recalculate x.

$$0.407 \times 25 \times 200x \quad + 302.58 \times 452.4 = 356.52 \times 1473$$

$$x = 190.73 \text{ mm}$$

and

$$x/d = 0.424.$$

Note: In this case the problem was relatively simple since for a large range of values of x/d the stresses in both tension and compression steel were yielded. In general, the problem is much more complex, involving a number of repeats of Stages II and III before a better estimate is obtained.

When the designer has access to a computer, the problem is more quickly and easily solved, but this example serves to illustrate the difference between checking a beam having only tension reinforcement and checking one with compression reinforcement also. The presence of compression steel makes the problem much more complicated. However, it is usually sufficient to estimate the depth to the neutral axis from the charts as CP 110 is not too concerned with the precise value.

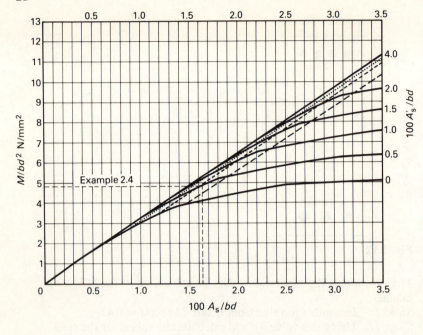

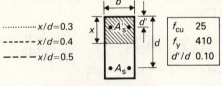

Fig. 2.13 Doubly reinforced beams (Chart 20)

Example 2.5 illustrates another difference between singly and doubly reinforced beams, this time in the design procedure.

Example 2.5 A beam is required to resist a moment of 800 kNm; the effective depth is 600 mm and the width is 300 mm. Determine the reinforcement needed if $d'/d = 0.1$, $f_{cu} = 25 \, \text{N/mm}^2$, and $f_y = 425 \, \text{N/mm}^2$.

$$M/bd^2 = 800 \times 10^6/300 \times 600^2 = 7.407 \, \text{N/mm}^2$$

This line has been drawn on Chart 32, reproduced in Fig. 2.14. It will be observed that the line crosses a number of different compression steel ratios. This shows that there is no one solution when a beam contains compression steel.

For example, if a neutral axis depth of 300 were chosen i.e. $x/d = 0.5$, then $100 \, A'_s/bd$ would be about 1.25. If x were chosen to be nearer 240 $(x/d = 0.4)$ then $100 \, A'_s/bd$ would be 1.5 for the same moment. Hence it

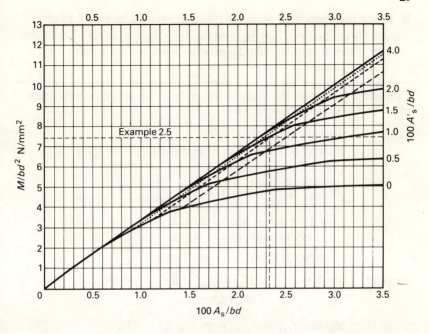

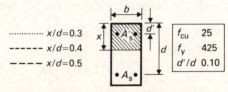

Fig. 2.14 Doubly reinforced beams (Chart 32)

can be seen that one can play about with the areas of steel to a certain extent.

It can be shown that the minimum area of steel will occur when the tension steel has just yielded. Therefore it is best to keep the neutral axis depth between 0.4 and 0.5 times the effective depth. In this example, a neutral axis depth of 0.4 (approx.) is chosen.

Thus $A'_s = 1.5 \times 300 \times 600/100 = 2700 \text{ mm}^2$

and $A_s = 2.315 \times 300 \times 600/100 = 416 \text{ mm}^2$

The computer program developed by the author actually shows that the minimum area of steel occurs when

$A'_s = 2311.9 \text{ mm}^2$ and $A_s = 4319.8 \text{ mm}^2$ and

$$x = 0.477 \times 600$$
$$= 286.2 \text{ mm}$$

B. Using the simplified design method illustrated in Fig. 2.11

Note: This method can only be used for designing beams. If it is required to check a section, then the procedure set out in Example 2.4 should be used.

Example 2.6 Design the beam in Example 2.5 for a neutral axis depth of 0.477 times the effective depth.

Stage I: Calculate M_{us}
This may be done either from the diagram, or from eqn [2.17(a)].
From the equation

$$M_{us} = 25 \times 300 \times 600^2 \times 0.2 \times 0.477 \times (2 - 0.477)$$
$$= 25 \times 300 \times 600^2 \times 0.1453$$
$$= 392.29 \times 10^6 \, \text{Nmm}$$
$$= 392.29 \, \text{kNm}.$$

Stage II: Calculate $M_{ud} = M_u - M_{us} = 800 - 392.29$
$$= 407.7 \, \text{kNm}.$$

Stage III: Calculate f_{yd1} and f_{yd2} from Fig. 2.11
$$f_{yd1} = 311.93 \, \text{N/mm}^2 \qquad f_{yd2} = 369.565 \, \text{N/mm}^2.$$

Stage IV: Calculate A'_s from eqn [2.18]

$$A'_s = \frac{M_{ud}}{f_{yd1} \cdot (d - d')} = \frac{407.7 \times 10^6}{311.93 \times (600 - 60)}$$
$$= 2420.4 \, \text{mm}^2.$$

Stage V: Calculate A_{ss} and A_{sd} from eqns [2.14] and [2.15] $(k_1 = 0.4)$
$$A_{ss} = 0.4 \times 25 \times 300 \times 286.2/369.565 = 2323.3 \, \text{mm}^2$$
$$A_{sd} = 311.93 \times 2420.2/369.565 \qquad = \underline{2042.7} \, \text{mm}^2$$
$$A_s = 4366.0 \, \text{mm}^2.$$

Comparing this with Example 2.5:
Total steel used in Example 2.5 = 2700 + 4167 = 6867 mm²
Total steel used in Example 2.6 = 2420.4 + 4366 = 6786.4 mm²
Economy in steel = 80.6
Thus it can be seen that sufficient accuracy is achieved by keeping the neutral axis between 0.4 and 0.5.

However, the computer gives a total area of 6631.7 mm², which is a saving of 229.3 mm² over Example 2.5 and a saving of 154.5 mm² over the simplified method. Thus the computer works out the best solution of all, as is only to be expected.

Practical considerations

The areas of steel which have been calculated in all the foregoing examples must be converted into actual bars of steel, but it is not possible simply to specify a bar of 16.095 mm diameter so that the area calculated and the area of bars match. Instead, the reinforcement manufacturer makes a range of sizes of bars, and the designer must come as close to his calculations as possible with these standard sizes.

The designer's choice is further limited as follows:

1. Theoretically it is true to say that four 8-mm $\varnothing$ bars are equivalent to one 16-mm $\varnothing$ bar. But this is not always practical. For instance, suppose that in a certain design either twenty 8-mm bars or five 16-mm $\varnothing$ bars are needed to give the required area of steel. Obviously, the designer will use the lesser number of bars, firstly because it will be cheaper to place the larger bars, but secondly, because the larger bars will take up less room in the beam.

2. It is good practice to so arrange bars in a section that they lie symmetrically about the vertical axis of the cross-section. In other words, if the bars used in a beam at one level consist of two 20-mm $\varnothing$ and one 25-mm $\varnothing$ bars, it is always best to put the odd bar in the centre. This has the consequence that the following bar arrangement would not be possible:
Three 20-mm $\varnothing$ and one 25-mm $\varnothing$ bars – it is impossible to place the 25-mm $\varnothing$ bar in a central position.
The rule to abide by is 'If mixing sizes of bars at a level, make sure that they can be arranged to be symmetrical about the beam's vertical axis.' In general it is also true that one should never mix more than two different bar sizes at any one level.

The following bar arrangements would be sufficient for the three examples already considered:

Example 2.2: $A_s = 264.9 \, mm^2$ use two, 10 mm + one, 12 mm bars $= 270.2 \, mm^2$

Example 2.3: $A_s = 522.0 \, mm^2$ use four 12 mm + one, 10 mm bars $= 530.9 \, mm^2$

Example 2.5: $A_s' = 2311.9 \, mm^2$ use three, 32 mm bars $= 2410 \, mm^2$
$A_s = 4319.8 \, mm^2$ use four 25 mm + three 32 mm bars $= 4376.2 \, mm^2$
or three 40 mm + two 20 mm bars $= 4398.0 \, mm^2$
(using less bars this fits across the section more easily)

3. Table 3 of CP 110 restricts the use of high-strength bars to certain diameters and does not allow the use of bars higher than 16 mm in diameter for steel of strength greater than or equal to $460 \, N/mm^2$.

Table 3 Strength of reinforcement

Designation	Nominal sizes	Specified characteristic strength f_y
	mm	N/mm^2
Hot rolled mild steel (BS 4449)	All sizes	250
Hot rolled high yield (BS 4449)	All sizes	410
Cold worked high yield (BS 4461)	Up to and including 16	460
	Over 16	425
Hard drawn steel wire (BS 4482)	Up to and including 12	485

Bending resistance of flanged sections

In a rectangular section, the concrete below the neutral axis is cracked and serves only to contain the tension steel. As a result, for large beams, the extra weight can be economically disastrous, it being much more efficient to eliminate some of the concrete in this region. Thus the T-section illustrated in Fig. 2.15(a) is very often encountered by the designer, and he must be aware of the problems involved with its design. This type of section is characterised by a top flange, sometimes containing compression reinforcement, taking most of the compressive force and a slimmer web housing both the tension steel and the vertical shear reinforcement.

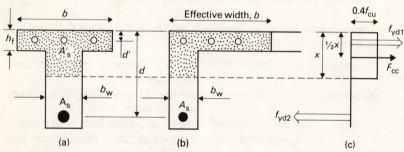

Fig. 2.15

Besides the isolated T-beam, this sort of section is used in structural design when slabs and their supporting beams are made to act compositely as shown in Fig. 2.15(b). When the beam is on the edge of a slab as in Fig. 2.15(b), an L-section results whereas, if the beam is at the boundary between two slabs, the T-shape must be analysed. The obvious question in this situation is what width of slab can be used when analysing the T- or L-section, and this is defined in Cl. 3.3.1.2. of CP 110.

3.3.1.2. *Effective width of flanged beam.* *In the absence of any more accurate determination, the effective flange width for a T-beam should not exceed the lesser of:*

> *the width of the web plus one-fifth of the distance between the points of zero moment, or the actual width of the flange*

and for an L-beam, the effective flange should not exceed the lesser of:

> *the width of the web plus one-tenth of the distance between points of zero moment, or the actual width of the flange.*

> *For a continuous beam the distance between the points of zero moment may be taken as 0.7 times the effective span.*

This effective width also applies to the isolated T-beam which is not part of a slab since the factors influencing effective width depend, not on whether it is part of a slab or not, but upon the way in which the stress in the flange varies from the outer edge to the centre of the web. Although it does vary because of a phenomenon known as 'shear-lag', in analysis it is assumed that the stress at any level above the neutral axis is constant. In other words, if the neutral axis lies within the flange:

$$F_{cc} = k_1 f_{cu} bx$$

as before and the beam may be analysed as if it was a rectangular section of width 'b'.

However, if the neutral axis is within the web, a more complicated situation arises and new formulae must be derived. Although it can be achieved using the 'exact', parabolic-rectangular concrete stress block, it becomes highly involved, therefore the design method presented is based on the rectangular stress block and the concept of M_{us} and M_{ud} illustrated in Fig. 2.10 for rectangular beams.

The Strain relationships developed for rectangular beams still hold. Therefore Fig. 2.11 may still be used to obtain the values of steel stress from the value of neutral axis depth.

Equating moments and forces

If $x < = h_f$

$$F_{cc} = 0.4 f_{cu} bx \tag{2.22}$$

If $x > h_f$

$$F_{cc} = 0.4 f_{cu}(b - b_w)h_f + 0.4 f_{cu} b_w x \tag{2.23}$$

Taking moments about the tension steel:

If $x < = h_f$

$$M_{us} = 0.4 f_{cu} bx(d - 0.5x) \tag{2.24}$$

(This is a rearrangement of [2.17(a)])

If $x > h_f$

$$M_{us} = 0.4 f_{cu}[(b - b_w)h_f(d - 0.5h_f) + b_w x(d - 0.5x)] \qquad [2.25]$$

In either case:

$$M_{ud} = A'_s f_{yd1}(d - d') \qquad [2.18]$$

$$f_{yd2} A_{ss} = F_{cc} \qquad [2.26]$$

$$f_{yd2} A_{sd} = f_{yd1} A'_s \qquad [2.15]$$

Thus the design procedure follows closely that presented on p. 26 for rectangular beams:

Knowing M_u, b, d, d', f_{cu} and f_y:

1. Choose a value for x/d.
2. Calculate M_{us} from eqns [2.24] or [2.25] as appropriate.
3. If $M_u < M_{us}$ then the section may be designed as being singly reinforced, but with a smaller x/d.
4. Otherwise find $M_{ud} = M_u - M_{us}$.
5. Find f_{yd1} and f_{yd2} from eqns [2.7], [2.8], [2.19], [2.20] or from Fig. 2.11.
6. Find the compression steel area A'_s from eqn [2.18].
7. Find A_{ss} from eqn [2.26] and A_{sd} from eqn [2.15]
   thus $A_s = A_{ss} + A_{sd}$.

If it is necessary to check an already designed section the procedure is one of trial and error unless x is less than h_f. Therefore it is advisable to first assume that x *is* less than h_f and treat the section as rectangular. Then x is calculated to check if this assumption was correct. If it proves to be so then the check of the beam is completed, but if x proves to be greater than the depth of the flange then at least it can be said that the neutral axis does not lie within the flange. The calculation then proceeds by trial of a value for x and adjustment to a new value until an answer of sufficient accuracy is found.

The author has again written a computer program for the design and checking of flanged beams and this enables all these calculations to be carried out much more quickly and more accurately. The student will be well advised at some time in his career to become familiar with the techniques of computer programming, and he may like to try to produce a program to design reinforced concrete beams.

Example 2.7 Design the isolated T-beam of Fig. 2.16 to carry a bending moment at the ultimate limit-state of 350 kNm, if $f_{cu} = 30 \, \text{N/mm}^2$, $f_y = 425 \, \text{N/mm}^2$ and the span is 10 m.

Stage I: Find the effective width.

$$\text{Width of web} + \text{span}/5 = 200 + \frac{10\,000}{5} = 2200 \, \text{mm}$$

Actual width $= 400 \, \text{mm} = $ effective width.

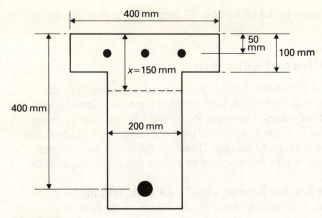

Fig. 2.16

Stage II: Let $x = 150$ mm.
M_{us} is found from eqn [2.25]

$$M_{us} = 0.4 \times 30 \times \{(400 - 200) \times 100 \times (400 - 50) \\ + 200 \times 150(400 - 75)\}$$

$$= 201 \times 10^6 \text{ Nmm} = 201 \text{ kNm}$$

$$M_{ud} = 350 - 201 = 149 \text{ kNm}.$$

Stage III: Find f_{yd1} and f_{yd2}.
From Fig. 2.11 both are yielded

$$f_{yd1} = 312 \text{ N/mm}^2; \qquad f_{yd2} = 370 \text{ N/mm}^2.$$

Stage IV: $A'_s = \dfrac{M_{ud}}{f_{yd1}(d - d')} = \dfrac{149 \times 10^6}{312 \times 350} = 1364.5 \text{ mm}^2.$

Stage V: From eqn [2.23].
$$A_{ss}f_{yd2} = F_{cc} = 0.4f_{cu}(b - b_w)h_f + 0.4f_{cu}b_w x$$

$$A_{ss} = \frac{0.4 \times 30 \times \{(400 - 200) \times 100 + 200 \times 150\}}{370}$$

$$= 1621 \text{ mm}^2$$

$$A_{sd} = f_{yd1}A'_s/f_{yd2} = 312 \times 1364.5/370$$

$$= 1150.6 \text{ mm}^2$$

$$A_s = \underline{2772.2 \text{ mm.}^2}$$

The main problem with all flanged beams is one of fitting the tension steel into such a narrow web. In this case, the actual bars which must be used are most closely matched by two, 32-mm $\emptyset$ bars, plus one 40-mm $\emptyset$ bar.

Shear in reinforced concrete

Designing a beam in bending so that its capacity is greater than the imposed ultimate moment will not ensure that the beam does not collapse by some other means. However, it is desirable that when a beam does fail it does so in flexure, since this type of collapse is a gradual one giving the occupants time to escape. Therefore designers must prevent the other modes of failure from occurring before flexural collapse takes place.

Shear failure is a case in point; usually if a beam fails in shear, the resulting collapse is sudden, without much chance of the occupants escaping. This means that in order to prevent a beam collapsing in shear one must first understand how a beam carries a shear load. Consider those beams which are reinforced in tension only; compression reinforcement does not materially affect the value of the shear resistance.

The mechanism of shear failure

Much research effort has been put into the way in which reinforced concrete beams react to shear forces. A good general review of the various themes put forward may be found in *Shear Strength of Reinforced Concrete Beams* published by the Institution of Structural Engineers in 1969. The mechanism is quite complex and depends on the distance between the support and the nearest concentrated load. The worst conditions of shear occur when a beam is loaded by concentrated loading and the distance between load and support is known as the 'shear span' – a_v in Fig. 2.17.

Case I: $6 > \dfrac{a_v}{d} > 2.5$ (approx.) (Fig. 2.17(a)).

The first cracks to appear as the load increases are those resulting from bending. This means that they are vertical, extending from the underside or soffit of the beam to the neutral axis at that section. However, as the load is further increased, the crack ceases to be vertical and turns to about 45° and progresses to about mid depth – point c in Fig. 2.17(a). The next stage in this process depends on whether a_v is closer to 6 or closer to 2.5 times the effective depth.

1. a_v/d nearer to 6.   Suddenly and without warning the crack shoots to e and from b back to g and h. This catastrophic failure is called **diagonal tension failure** and the load at which it occurs is the **diagonal cracking** load.
2. a_v/d nearer to 2.5.   The crack stops at c and debonding of the tension reinforcement occurs along g and h. This is called **shear tension failure.**

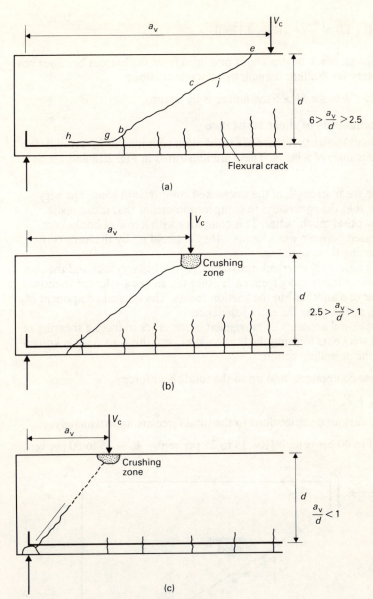

Fig. 2.17 Mechanisms of shear failure

Case II: $2.5 > \dfrac{a_v}{d} > 1.0$ (Fig. 2.17(b)).

Here the diagonal crack is independent of the flexural cracks. Ultimate failure occurs by crushing of the concrete. This failure is known as **shear compression failure** and the failure load is about twice the Diagonal Tension Failure Load.

Case III: $1.0 > \dfrac{a_v}{d} > 0$ (Fig. 2.17(c)).

The diagonal crack begins about one third from the bottom but does not extend very far. Failure is again by concrete crushing.

Case IV: When $a_v/d > 6$ the failure is by flexure.

The mechanisms of shear resistance

Experiment shows that there are three main components involved in the shear resistance of a beam. These are illustrated in Fig. 2.18 and are as follows:

V_{cz} The shear strength of the uncracked compression zone. The very fact that the concrete is in compression means that it can resist shear that much better. This compares with a row of books compressed between one's hands – they are held up by the force compressing them together.

V_a The aggregate interlock resistance. As the beam cracks and the two sections move away from each other the stones are forced to move over one another but the friction resists. The vertical component of this action gives the extra resistance.

V_d The dowel action. The movement of the crack induces a shearing of the steel bars forming the tension steel, and this resistance is known as the dowelling action.

These components add up to the total shear force:

$$V_c = V_{cz} + V_a + V_d$$

The various contributions to the total force are approximately:

V_{cz} – 20 to 40 per cent; V_a – 15 to 25 per cent; V_d – 35 to 50 per cent.

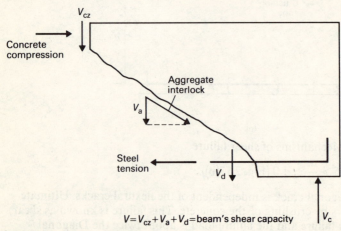

Fig. 2.18 Shear resistance

Thus it can be seen that the steel plays a very important part in the shear resistance. This it does in two ways: 1. by virtue of its dowelling action; and 2. as the proportion of steel is increased the depth to the neutral axis is increased thus the value of V_{cz} is increased. This fact is reflected in Table 5 of CP 110 where the value of the Ultimate Shear Resistance of beams reinforced in tension only can be calculated. The value of shear stress that can be taken by a beam is related to the strength of the concrete as is to be expected, but is also more strongly related to the amount of steel present in the tension zone.

Table 5   Ultimate shear stress in beams

$\dfrac{100 A_s}{bd}$	Concrete grade			
	20	**25**	**30**	**40 or more**
	N/mm^2	N/mm^2	N/mm^2	N/mm^2
0.25	0.35	0.35	0.35	0.35
0.50	0.45	0.50	0.55	0.55
1.00	0.60	0.65	0.70	0.75
2.00	0.80	0.85	0.90	0.95
3.00	0.85	0.90	0.95	1.00

The term A_s in Table 5 is that area of longitudinal tension reinforcement which continues at least an effective depth beyond the section being considered except at supports where the full area of tension reinforcement may be used in all cases provided the requirements of **3.11.7** are met.

Example 2.8   Consider the beam of Example 2.5. Determine from CP 110, Cl. 3.3.6.1 and Table 5 the value of shear capacity for this beam.

The ultimate shear capacity of a beam reinforced in tension only is given by

$$V_c = v_c bd \qquad\qquad [2.27]$$

v_c, the allowable stress is derived from Table 5 of CP 110

$100 A_s/bd = 100 \times 4319.8/(300 \times 600) = 2.4$

$f_{cu} = 25\,\text{N/mm}^2$

Thus from Table 5: $v_c = 0.87\,\text{N/mm}^2$ (interpolated)

Thus ultimate shear capacity $= V_c = 0.87 \times 300 \times 600 = 156.6\,\text{kN}$

If the actual maximum shear force experienced by the beam exceeds this value shear reinforcement must be provided to resist the balance that the concrete cannot resist.

Shear reinforcement

When the concrete area with longitudinal reinforcement only has sufficient strength to resist a part of the imposed shear force at the

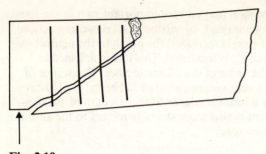

Fig. 2.19

ultimate limit state, additional resistance to shear may be provided by web reinforcement. This may take the form of stirrups (hoops of steel encircling the tension and compression steels) inclined or vertical and anchored in the compression zone or it may be partly given by some of the tension steel being bent up. In the latter case bars can be bent up only when they are no longer required at that section to carry bending moment. In other words, towards the ends of a beam it is very often the case that the moment to be resisted is smaller than at the centre of the beam. Thus those bars no longer required to be there can be used to give extra shear capacity. Figure 2.19 illustrates how this can be the case. As the shear crack opens it puts the web steel into tension, and this steel force prevents the beam from losing any strength due to failure in shear. Although the addition of the web steel affects the values of the other various factors involved in the mechanism of shear resistance, increasing them to a marked degree, it is conservatively assumed by CP 110 that they remain the same. Thus the total resistance of the beam with web steel is given by:

$$V = V_{cz} + V_a + V_d + V_s = V_c + V_s$$

where V_s is the contribution of the web steel

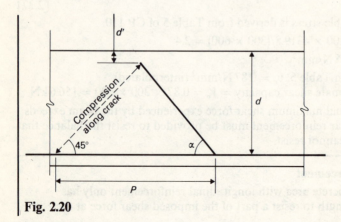

Fig. 2.20

The shear reinforcement is designed by the so-called 'Truss Analogy', in which the tension in the shear reinforcement is opposed by imaginary compression 'members' formed of the concrete along the line of the crack as shown in Fig. 2.20. This figure shows an isolated bar used as shear reinforcement, inclined at an angle of α to the horizontal. The vertical component of the force in the bar contributes to the overall value of V_s, the shear resistance of the shear reinforcement. P is the effective pitch of each 'truss' segment, i.e. the distance of the beam under the strengthening protection of the inclined bar. If S_v is the horizontal spacing of these stirrups or inclined bars, then:

Number of stirrups cutting every shear crack

$$= P/S_v$$
$$= (1 + \cot \alpha) \cdot (d - d')/S_v$$
$$= (1 + \cot \alpha) \frac{d}{S_v} \text{ [if } d' \text{ is small]}$$

Vertical force in each bar $= f_{yv} A_{sv} \sin \alpha / 1.15$

(1.15 is the partial safety factor for steel)

Therefore, resistance of stirrups to shear:

V_s = vertical force in each bar $\times$ number of bars

$$= 0.87 f_{yv} A_{sv} \cdot \sin \alpha (1 + \cot \alpha) \frac{d}{S_v}$$
$$= V - V_c$$

Thus, rearranging: letting $V = vbd$ and $V_c = v_c bd$

$$\frac{A_{sv}}{S_v} = \frac{b(v - v_c)}{0.87 f_{yv} (\sin \alpha + \cos \alpha)} \qquad \text{[2.28]}$$

For vertical shear reinforcement:

$\alpha = 90°$

$$\frac{A_{sv}}{S_v} = \frac{b(v - v_c)}{0.87 f_{yv}} \qquad \text{[2.28(a)]}$$

Note:
1. A_{sv} must take account of **all** the shear reinforcement. If there are two legs to the stirrup both must be included in A_{sv}.
2. f_{yv} is the characteristic strength of the shear reinforcement. It is usually equal to f_y, except that it must not exceed a value of 425 N/mm^2.
3. v is the shear stress calculated from the actual loads. v_c is the capacity of the beam to resist shear stress and is found from Table 5 in CP 110.

Example 2.9   Design the section shown in Fig. 2.21 to resist a shear force of 250 kN. $f_{cu} = 30$ N/mm²; $f_{yv} = 425$ N/mm²

$$A_s = 3 \times \pi \times 16^2 = 2412.7 \text{ mm}^2$$

$$100 A_s/bd = 1.6$$

Thus from Table 5 $v_c = 0.82$ N/mm²

'Actual' shear stress $= \dfrac{250\,000}{300 \times 500} = 1.667$ N/mm²

Assuming vertical links:
From eqn. [2.28]

$$\frac{A_{sv}}{S_v} = \frac{300 \times (1.667 - 0.82)}{0.87 \times 425} = 0.69$$

Assuming links of 10 mm $\varnothing$ bars: $A_{sv} = 2 \times 25\pi = 157$ mm²

Therefore S_v must be less than $157/0.69 = 227.5$ mm
let $S_v = 200$ mm

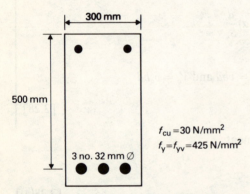

Fig. 2.21

However, the design does not end there since there are number of restrictions placed upon A_{sv} and S_v by CP 110 as follows:

1.   Even if the calculations had shown that no shear reinforcement was needed it is always good practice to include some unless the element is of minor importance such as a lintel or if

$$v < 0.5v_c$$

(Cf. Cl. 3.11.4.3 of CP 110)

In all other cases the minimum areas of shear steel are:

$$\frac{A_{sv}}{S_v} = 0.0012b_t \qquad \text{for high yield links}$$

$$\frac{A_{sv}}{S_v} = 0.002b_t \qquad \text{for mild steel links}$$

where b_t is the breadth of the beam at the level of the steel.

2. Referring again to Fig. 2.20 it can be seen that if the links were to be placed at more than P apart it would be possible to have a crack occurring between them. This would mean that no steel would cross that crack, leaving the beam to fail at that section. The smallest value for P occurs when $\alpha = 90°$; $P = d$. Therefore in Cl. 3.11.4.3 of CP 110 the code states that stirrups or links shall not be placed at more than $0.75d$ apart.

3. It has already been stated that shear can be taken by some of the tension bars being bent up at the ends of the beam. However, CP 110 allows only 50 per cent of the shear steel to be of this form. Thus up to $0.5 V_s$ may be taken by bent-up bars.

4. To avoid web crushing CP 110 imposes a maximum value on the permissible shear stress. This is given in Table 6 of CP 110. If the calculated shear stress V/bd exceeds this value, the dimensions of the section should be increased.

Table 6 Maximum value of shear stress in beams

Concrete grade			
20	25	30	40 or more
N/mm²	N/mm²	N/mm²	N/mm²
3.35	3.75	4.10	4.75

5. Clause 3.11.6.4 of CP 110 calls for the shear reinforcement to be adequately anchored in the compression zone as illustrated in Fig. 2.22.

3.11.6.4 *Anchorage of links. A link may be considered to be fully anchored if it passes round another bar of at least its own size through an angle of 90° and continues beyond for a minimum length of eight times its own size or, through 180° and continues for a minimum length of four times its own size. In no case should the radius of any bend in the link be less than twice the radius of a test bend guaranteed by the manufacturer of the bar.*

Summary of design procedure

1. Calculate the capacity of the section – $V_c = v_c bd$.
2. If $0.5V_c < V < V_c$ then use nominal links unless the member is of minor importance such as a lintel.

44

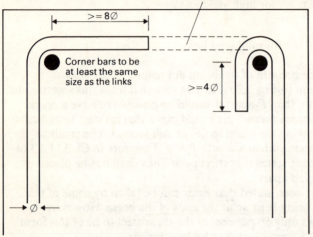

Fig. 2.22 Anchorage of shear steel

3. If $V > V_{max}$ redesign the section.
4. If $V < V_{max}$ calculate $V_s = V - V_c$.
5. Design links and bent-up bars as preferred, remembering $S_v < 0.75d$.

Example 2.10 Design the beam shown in Fig. 2.23 for shear, if $f_{cu} = 30\,\text{N/mm}^2$ and $f_y = 460\,\text{N/mm}^2$. Neglect self-weight.

Stage I: Calculate $V_c = v_c bd$.

$A_s = 3004\,\text{mm}^2 \qquad 100\,A_s/bd = 2.0$

From Table 5 of CP 110 $\quad v_c = 0.9\,\text{N/mm}^2$

$V_c = 0.9 \times 300 \times 500 = 135\,000\,\text{N} = 135\,\text{kN}$

Stage II: The design must be divided into three parts according to the three sections of the shear force diagram.

L.H. section: $v = V/bd = 200\,000/(300 \times 500) = 1.333\,\text{N/mm}^2$ this is less than the maximum allowed.

$$\frac{A_{sv}}{S_v} = \frac{300 \times (1.333 - 0.9)}{0.87 \times 425}$$

$$= 0.352$$

However, minimum shear steel is given by:

$$\frac{A_{sv}}{S_v} = 0.0012b_t = 0.36$$

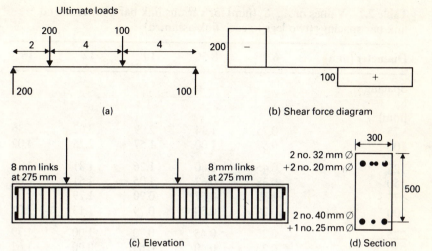

Fig. 2.23

Therefore use minimum steel using 10 mm links
$A_{sv} = 157.1\,\text{mm}^2$

$S_v = 436.4\,\text{mm}$ (say 400 mm)

R.H. section: $v = 0.667\,\text{N/mm}^2$
Therefore use nominal steel as for the L.H. section.

Central section: Shear force $= 0$,
Therefore no shear steel is needed.

Stage III: Check that S_v is less than $0.75d = 375\,\text{mm}$.
Thus alter the size of link used if possible. Try 8-mm links
$A_{sv} = 100.5\,\text{mm}^2 \qquad S_v = 100.5/0.36 = 279.3\,\text{mm}$
(say $S_v = 275\,\text{mm}$)
Thus the final design is as shown in Fig. 2.23(c).

Table 2.1 will be found useful in designing shear reinforcement. It tabulates values of A_{sv}/S_v for various link bar diameters and link spacings.

Beams where the shear span is small $(a_v/d < 2.0)$
Refer to Cl. 3.3.6.2 of CP 110.

When a concentrated load is close to the support the shear resistance is greater than predicted by eqn. [2.27]. Therefore CP 110 allows it to be modified to give:

$$V_c = \frac{2d}{a_v} \cdot v_c bd \qquad\qquad\qquad [2.27(\text{a})]$$

Table 2.1 Values of A_{sv}/S_v (mm) for various link bar diameters and link bar spacings (two legs to each link assumed)

Diameter (mm)	6	8	10	12	16
Spacing S_v (mm)					
75	0.75	1.34	2.09	3.02	5.36
100	0.57	1.00	1.57	2.26	4.02
125	0.45	0.80	1.26	1.81	3.22
150	0.38	0.67	1.05	1.51	2.68
175	0.32	0.57	0.90	1.29	2.30
200	0.28	0.50	0.79	1.13	2.01
225	0.25	0.45	0.70	1.00	1.79
250	0.23	0.40	0.63	0.90	1.61
275	0.21	0.37	0.57	0.82	1.46
300	0.19	0.34	0.52	0.75	1.34
325	0.17	0.31	0.48	0.70	1.24
350	0.16	0.29	0.45	0.65	1.15
375	0.15	0.27	0.42	0.60	1.07

Thus eqn. [2.28(a)] becomes:

$$\frac{A_{sv}}{S_v} = \frac{b(v - 2dv_c/a_v)}{0.87 f_{yv}} \qquad [2.28(b)]$$

Shear resistance of flanged beams

Flanged sections are usually designed to give as large a moment as possible at mid span with as little concrete deadweight as necessary. However, the shear force is also usually large, but the beam has less with which to resist compared to an equivalent rectangular section. In fact CP 110 insists that only the web be taken into account in determining the shear capacity of a flanged beam. This is conservative and means that flanged beams are heavily reinforced in shear.

The shear steel is designed in exactly the same way as for rectangular sections. However, the addition of shear steel will often dictate the way in which the main tension steel is arranged since it further restricts the width available. In addition to the web steel used to resist the shear force, it is good practice to include links in the flange to enclose all the compression steel. This is done more by way of making the job of placing the steel easier than to provide strength, though it may be necessary to prevent the compression steel from buckling in some situations.

Bond in reinforced concrete

As well as ensuring that a beam does not fail in shear, the designer must also prevent bond failure. In the design of beams for bending it was tacitly assumed that the stresses which were calculated as being developed in the tension steel could in fact be obtained in reality. However, when it is understood how the stress can be developed it is seen that there is a distinct possibility of its not occurring. As its name suggests, bond failure is characterised by the steel and concrete parting company, and usually consists of the concrete surrounding the bar splitting and the steel being withdrawn from it.

Bond stress consists of three main ingredients:

1. *Adhesion.* When wet concrete is cast around the steel it binds or adheres to the surface of the steel rather like a glue. This must be broken in order for bond failure to occur and is the largest component of bond strength.
2. *Friction.* Even the smoothest surface is not quite what it seems as, when viewed under a microscope it appears undulated and pitted. A reinforcing bar is far from smooth and this results in large frictional forces being set up when a beam is bent.
3. *Bearing.* In certain types of bar, particularly the high-yield deformed bars, the surface is covered with numerous protrusions at right angles to the bar centre line. The concrete between these tends to act in bearing when resisting bond forces.

Bond stresses are set up in response to the way in which stress in the steel bar varies. This variance is a direct result of the way in which the moment varies along the beam, and also as a consequence of the fact that it is not possible to have a fully stressed bar without there being a distance (known as the anchorage length) over which the steel stress rises from zero to maximum. The first kind of stress is known as 'Local Bond' while the second is known as 'Anchorage Bond'.

Local bond stress

Consider the short length of beam shown in Fig. 2.24. At one side of the section, the moment is M while the moment increases to $M + dM$ at the other side. Thus the force in the tension steel increases also:

$$dT = \frac{M + dM}{z} - \frac{M}{z}$$

where z is the lever arm

$$dT = dM/z$$

This change in steel stress is resisted by the bond force acting on the surface area of the bars at that section. If the perimeter of these bars

48

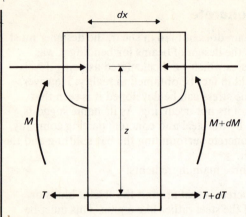

Fig. 2.24

is equal to ΣU_s, then the surface area is given by:

$$dA = dx \sum U_s$$

and the bond force by:

$$dT = f_{bs} \cdot dx \sum U_s$$

Thus $\dfrac{dT}{dx} = f_{bs} \sum U_s = \dfrac{1}{z} \cdot \dfrac{dM}{dx} = \dfrac{V}{z}$ since $V = dM/dx$

For ease of use this formula is modified in CP 110 by changing z to d, the effective depth, and the allowable bond stresses altered accordingly.

Thus the local bond stress is given by:

$$f_{bs} = \frac{V}{\sum U_s \cdot d} \qquad [2.29]$$

Table 21   Ultimate local bond stresses

Bar type	Concrete grade			
	20	25	30	40 or more
	N/mm²	N/mm²	N/mm²	N/mm²
Plain bars	1.7	2.0	2.2	2.7
Deformed bars	2.1	2.5	2.8	3.4

Where there would be an advantage, and the deformed reinforcement to be used is Type 2, as defined in **E.1** of Appendix E of CP 110, the values of bond stress for deformed bars may be increased by 20 per cent.

The largest local bond stresses will occur where the shear forces are largest, or where the number of bars has been decreased, in order to bend them up or because they are no longer needed to resist the bending moment. In practice these positions are one and the same for simply supported beams such as have been considered in this chapter. In other words the shear force is greatest and the number of bars least at the beam's support.

The allowable values of the ultimate local bond stress are to be found in Table 21 in CP 110.

Anchorage bond stress

Consider the stressed bar in Fig. 2.25. The assumption is made that the bond stress is uniformly distributed along the embedded length of the bar and is equal to f_{ba}.

Thus force in bar = Bond Force

$$f_u \cdot A_s = \pi \varnothing\, L \cdot f_{ba}$$

$$f_{ba} = \frac{\varnothing f_u}{4L} \quad \text{since } A_s = \pi \varnothing^2/4 \qquad [2.30]$$

Rearranging $\quad L = \tfrac{1}{4} \varnothing\, f_u/f_{ba}$

where $\quad f_u$ is the actual bar stress
$\qquad\quad f_{ba}$ is the bond stress

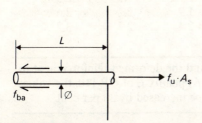

Fig. 2.25

Anchorage bond stress is usually checked where a bar is curtailed either when it is no longer required to resist bending or at the ends of a member. The allowable anchorage bond stresses are given in Table 22 of CP 110.

It should be noted that in both Tables 21 and 22 the stresses are not characteristic but design stresses, viz. they already contain an allowance for the partial safety factor for materials, γ_m.

It will have been noted that stresses are given for three types of bar: plain; deformed type 1; and deformed type 2. Types 1 and 2 are illustrated in Figure 2.26. It will also have been noticed that for anchorage

50

bond stresses a different value is given for a bar in tension to that given for a bar in compression. There are two main reasons for this:

1. A bar in tension tends to reduce in diameter thus decreasing the frictional and adhesional components of bond, while the reverse is the case for a bar in compression.
2. In compression there is a tendency for the end of the bar to be in bearing against the concrete.

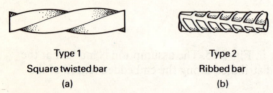

Type 1   Type 2
Square twisted bar   Ribbed bar
(a)   (b)

Fig. 2.26   Deformed bars

Table 22   Ultimate anchorage bond stresses

Bar type	Concrete grade			
	20	**25**	**30**	**40 or more**
	N/mm²	N/mm²	N/mm²	N/mm²
Plain bar in tension	1.2	1.4	1.5	1.9
Plain bar in compression	1.5	1.7	1.9	2.3
Deformed bar in tension	1.7	1.9	2.2	2.6
Deformed bar in compression	2.1	2.4	2.7	3.2

Where there would be an advantage, and the deformed reinforcement to be used is Type 2, as defined in **E.1** of Appendix E of CP 110, the values of bond stress for deformed bars may be increased by 30 per cent.

In practical bar arrangements it is often necessary to have some of the longitudinal bars touching one another. This obviously affects the bond strength and this is reflected in Table 23 and Cl. 3.11.6.3 of CP 110 where an effective perimeter is defined. In other words, when calculating bond stress, instead of using the actual bar diameters, the designer must use the effective perimeter when working out the surface area in contact with the concrete.

3.11.6.3 *Effective perimeter of a bar or group of bars. The effective perimeter of a single bar may be taken as 3.14 times its nominal size. The effective perimeter of a group of bars* (see **3.11.3.1**) *should be taken as the sum of the effective perimeters of the individual bars multiplied by the appropriate reduction factor given in Table 23.*

Table 23 Reduction factor for effective perimeter of a group of bars

No. of bars in a group	Reduction factor
2	0.8
3	0.6
4	0.4

Example 2.11 The cantilever shown in Fig. 2.27 carries a point load of 250 kN at its end. Calculate the ultimate local bond stress and the length of anchorage needed at the support, if $f_y = 425 \text{ N/mm}^2$ and bars of deformed type (2) are being used. $f_{cu} = 30 \text{ N/mm}^2$.

Local bond

Only the tension steel is checked for local bond stress

$$\sum U_s = (0.6 \times 6 \times \pi \times 32) + (2 \times \pi \times 20) = 487.6 \text{ mm}$$

$$f_{bs} = \frac{V}{d \sum U_s} = \frac{250\ 000}{500 \times 487.6} = 1.025 \text{ N/mm}^2 < 1.2 \times 2.8 \text{ allowed}$$

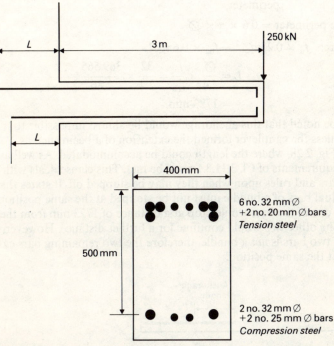

6 no. 32 mm ∅
+2 no. 20 mm ∅ bars
Tension steel

2 no. 32 mm ∅
+2 no. 25 mm ∅ bars
Compression steel

Fig. 2.27

Anchorage bond

The section at the support must be checked to obtain the steel stresses. When this is done it is found that both tension and compression steels are yielded.

Compression
$$f_u = 311.93 \text{ N/mm}^2$$
$$f_{ba} = 1.3 \times 2.7 = 3.51 \text{ N/mm}^2$$

25-mm $\varnothing$ bars:
$$L_{25} = 0.25 \times 25 \times 311.93/3.51$$
$$= 555.4 \text{ mm}$$

32-mm $\varnothing$ bars:
$$L_{32} = 0.25 \times 32 \times 311.93/3.51$$
$$= 711.0 \text{ mm}$$

Tension
$$f_u = 369.565 \text{ N/mm}^2$$
$$f_{ba} = 1.3 \times 2.2 = 2.86 \text{ N/mm}^2$$

20-mm $\varnothing$ bars:
$$L_{20} = 0.25 \times 20 \times 369.565/2.86$$
$$= 646.1 \text{ mm}$$

32-mm $\varnothing$ bars: The usual formula must be amended to take account of the bundling of the bars. Thus the effective perimeter of the bars is equal to 0.6 times the actual perimeter.

Effective perimeter $= 0.6 \times \pi \times \varnothing$

Therefore: $f_u \times 0.25\pi\varnothing^2 = f_{ba} \times 0.6\pi\varnothing L$
$$L = \frac{\varnothing}{2.4} \times \frac{f_u}{f_{ba}} = \frac{32}{2.4} \times \frac{369.565}{2.86}$$
$$= 1723 \text{ mm}.$$

It should be noted that this anchorage would be almost impossible to provide unless the cantilever formed the extension of a beam such as shown in Fig. 2.28, where the length could be accommodated. As well as this, the requirements of Cl. 3.11.3.1 must be met. This clause deals with bundled bars and rules upon when they may be stopped off. It states that the individual bars of a bundle must not be stopped at the same position. Thus only one of the bars would stop at a distance of 1723 mm from the support. The other two would continue for a further distance. However, a group of two bars is *not* a bundle, therefore the two remaining bars can be ended at the same position.

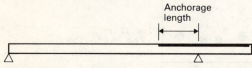

Anchorage length

Fig. 2.28

Even then, anchorage is not the only consideration involved in stopping off a bar. These extra considerations will be dealt with in Volume 2 of this textbook, but suffice it to say that the section of the code which deals with this subject is 3.11.7.

The provision of anchorage

When a bar is in compression there is only one way in which to provide the required anchorage length. This is by extending the bar by a distance equal to the anchorage length beyond the point at which anchorage was considered, **on both sides of the section**.

When a bar is in tension, however, although the straight bar extension is still a possibility, there are two other options. These consist of bending the bar through either 90° or 180° – known respectively as a bend or a hook. The equivalent value of anchorage afforded by such a device is set in Cl. 3.11.6.7 of CP 110.

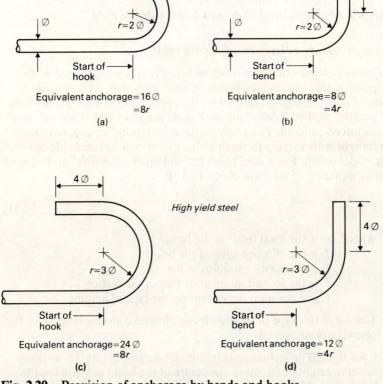

Fig. 2.29 Provision of anchorage by bends and hooks

3.11.6.7. *Hooks and bends.* *Hooks, bends and other reinforcement anchorages should be of such form, dimension and arrangement as to avoid overstressing the concrete. Hooks, which should be used only to meet specific design requirements, should be of U- or L-type, as specified in BS 4466.*

The effective anchorage length of a hook or bend should be measured from the start of the bend to a point four times the bar size beyond the end of the bend, and may be taken as the lesser of twenty-four times the bar size or

1. *for a hook, eight times the internal radius of the hook;*
2. *for a 90° bend, four times the internal radius of the bend.*

In no case should the radius of any bend be less than twice the radius of the test bend guaranteed by the manufacturer of the bar and in addition it should be sufficient to ensure that the bearing stress at the mid point of the curve does not exceed the value given in Cl. 3.11.6.8.

When a hooked bar is used at a support, the beginning of the hook should be at least four times the bar size inside the face of the support.

Figure 2.29 illustrates the two types of anchorage; the top two represent mild steel and the bottom two high yield steel, as most manufacturers will be able to bend to at least half the shown radii.

Deflection in reinforced concrete

The subject of deflection in a concrete beam is very complicated when compared with deflections in beams of other materials such as steel. Whereas in the design of steel beams the application of a simple formula will give the predicted deflection with good accuracy this is not the case for reinforced concrete. For concrete the calculations become tedious and fraught with varying factors resulting in at best a reasonable estimate of deflection. For a steel beam the mid-span deflection can be found from an equation of the form of eqn [2.31].

$$a = C \cdot \frac{wL^3}{EI} \qquad [2.31]$$

where w is the total load on the beam
L is the effective span of the beam
E is the elastic modulus of the material
I is the second moment of area of the section
C is a constant dependent on the type of loading

The use of this type of formula for reinforced concrete is complex for the following reasons:

1. Even though deflection calculations are carried out at the working or serviceability limit-state, the centre of the beam will still tend to be cracked. This means that the value for I will vary along the length

of the beam. Although this variance is predictable it results in extra and difficult calculations.

2. The elastic modulus of steel is usually taken as being $200 \, kN/mm^2$. However the value of E for concrete is not so simple to determine. It depends firstly on the characteristic strength of the concrete, increasing with increase in concrete strength. Secondly, due to the phenomenon of Creep, the value of E will be different when considering deflections in the short-term from that used for long-term deflections. Creep is the time-dependent strain increase which takes place under the application of a constant stress. Thus over a period of time the beam will increase its deflection if the load is constant. However, thankfully most ultimate loads are only present for short times, but even so the value of E used must be given careful thought.

3. The eqn [2.31] is based upon the assumption that the material's response is elastic (or largely so) up to failure. Obviously this is not the case and any calculations of deflection in reinforced concrete must take account of this fact.

Returning again to the elastic beam represented in eqn [2.31], if the section is rectangular, the maximum moment which may be taken by the beam is:

$$M = fbh^2/6$$

where f is the allowed stress
        b is the width of section
        h is the depth of section

The actual moment caused by the external load is given by:

$$M = B \cdot wL$$

where B is another constant dependent on the loading

Thus $wL = fbh^2/6B$

Substituting this into eqn [2.31] gives:

$$\frac{a}{L} = C'\frac{L}{h} \qquad C' = 2fC/EB \quad (I = bh^3/12)$$

CP 110 states that deflection shall be controlled by making it a certain fraction of the span length, viz. a/L must be made a constant. This can be done if the ratio L/h is kept below a certain value.

Although this argument has been developed for a homogeneous elastic material, the principle can be shown to work for a concrete beam as well. Thus CP 110 uses this to control deflection in reinforced concrete beams. Depending on the type of the beam being dealt with (cantilever or simply supported – continuous beams will be dealt with in the second volume of this book) so long as the span/effective depth ratio is kept below the tabulated values, the deflections are deemed to meet CP 110's requirements.

However, even this seemingly simple procedure is further complicated by the need to allow for varying amounts of steel in the beam and for the requirement of Cl. 2.2.3.1(2) that an absolute deflection limit of 20 mm be imposed. The second of these is met by altering the values of span/effective depth for beams over 10 m in span. The first adjustment for percentage of steel is done by multiplying the 'Basic' ratios by factors obtained in Tables 10 and 11. All the relevant tables are reproduced below.

Table 8 Basic span/effective depth ratios
for rectangular beams

Support conditions	Ratio
Cantilever	7
Simply supported	20
Continuous	26

Table 8 should only be used for spans greater than 10 m if the engineer is satisfied that a deflection of span/250 is acceptable. When it is necessary further to restrict the deflection, to avoid damage to finishes or partitions, Table 9 should be used for spans exceeding 10 m.

Table 9 Special span/effective depth ratios for rectangular beams

Span	Cantilever	Simply supported	Continuous
m			
10		20	26
12	Value to be	18	23
14	justified by	16	21
16	calculation	14	18
18		12	16
20		10	13

Deflection is influenced by the amount of tension reinforcement and its stress and therefore the span/effective depth ratios should be modified according to the area of reinforcement provided and its service stress at the centre of the span (or at the support in the case of a cantilever). Values of span/effective depth ratio obtained from Table 8 or 9 should therefore be multiplied by the appropriate factor obtained from Table 10.

Table 10 Modification factor for tension reinforcement

Service stress (f_s)	$\dfrac{100 A_s}{bd}$							
	0.25	0.50	0.75	1.00	1.50	2.00	2.50	$\geqslant 3.0$
N/mm^2								
145 $(f_y = 250)$	2.0	1.98	1.62	1.44	1.24	1.13	1.06	1.01
150	2.0	1.91	1.58	1.41	1.22	1.11	1.04	0.99
200	2.0	1.46	1.26	1.15	1.02	0.94	0.89	0.85
238 $(f_y = 410)$	1.60	1.23	1.09	1.00	0.90	0.84	0.80	0.77
246 $(f_y = 425)$	1.55	1.20	1.06	0.98	0.88	0.83	0.79	0.76
250	1.52	1.18	1.05	0.97	0.87	0.82	0.78	0.75
267 $(f_y = 460)$	1.41	1.11	0.99	0.92	0.84	0.78	0.75	0.72
290 $(f_y = 500)$	1.27	1.03	0.92	0.86	0.79	0.74	0.71	0.68
300	1.22	0.99	0.90	0.84	0.77	0.72	0.69	0.67

The service stress may be estimated from the equation

$$f_s = 0.58 \, \frac{f_y A_{s,req}}{A_{s,prov}} \times \frac{1}{\beta_b}$$

where f_s is the estimated service stress in the tension reinforcement,

         f_y is the characteristic strength of the reinforcement,

         $A_{s,req}$ is the area of tension reinforcement required at mid-span to resist the moment due to ultimate loads (at support for a cantilever),

         $A_{s,prov}$ is the area of tension reinforcement provided at mid-span (at support for a cantilever),

         β_b is the ratio of the resistance moment at mid span obtained from the redistributed maximum moments diagram to that obtained from the maximum moments diagram before redistribution

Moment redistribution will be covered in Volume 2 of this book, and in any case, for determinate structures such as cantilevers and simply supported beams, no redistribution is possible. Therefore, β_b is always 1.0.

Example 2.12   Check the cantilever of Example 2.11 for deflection.

Stage I: Obtain the 'basic' span/effective depth ratio from Table 8 (since the span is less than 10 m).

$$\frac{L}{d} = 7.$$

Stage II: Obtain the modification factor for tension reinforcement from Table 10.

Table 11 Modification factor for compression reinforcement

$\dfrac{100\,A'_s}{bd}$	Factor
0.25	1.07
0.50	1.14
0.75	1.20
1.0	1.25
1.5	1.33
2.0	1.40
$\geqslant 3.0$	1.50

Intermediate values may be interpolated.

The area of compression reinforcement A'_s used in Table 11 may include all bars in the compression zone even those not effectively tied with links.

$$f_s = 0.58\,\frac{f_y A_{s\,req}}{A_{s\,prov}} \qquad \frac{100\,A_s}{bd} = 2.73$$

$$= 0.58\,\frac{425 \times 4806}{5454} \qquad \text{(4806 obtained by checking the section)}$$

$$= 217.2\,\text{N/mm}^2$$

Thus from Table 10: $M_{ten} = 0.87$ (double interpolation).

Stage III: Obtain the modification factor for compression reinforcement from Table 11.

$$\frac{100\,A'_s}{bd} = 1.295$$

Thus from Table 11: $M_{com} = 1.30$.

Stage IV: Obtain modified span/effective depth ratio.

$$\frac{L}{d} = 7 \times 0.87 \times 1.3 = 7.9$$

Minimum effective depth $= 3000/7.9 = 380\,\text{mm}$ O.K.

Example 2.13 a fully designed example

Design the beam shown in Fig. 2.30(a) $f_{cu} = 30\,\text{N/mm}^2$; $f_y = 425\,\text{N/mm}^2$

Stage I: Obtain the section dimensions.

This is largely a matter of experience and will involve some informed guesswork. The dimensions chosen are as shown in Fig. 2.30(b)

Stage II: Obtain the bending moment and shear force diagrams. These are given in Figs. 2.30(c) and 2.30(d). They include an allowance for the self-weight of the beam as follows:

Assume density = $24\,\text{kN/m}^3$

Self-weight UDL = $24 \times 0.5 \times 0.75 \times 1.4 = 12.6\,\text{kN/m}$

Note: the loads in Fig. 2.30(a) are already in their design form

Stage III: Obtain the areas of steel necessary in tension and compression.

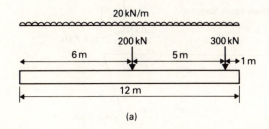

(a)

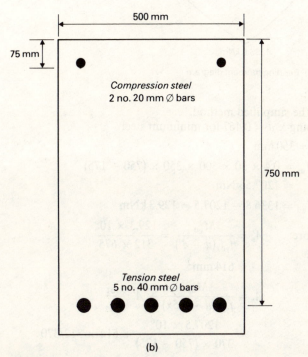

(b)

Fig. 2.30

60

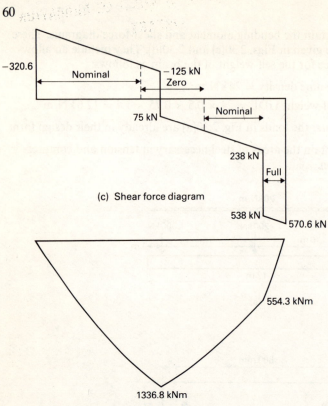

-320.6

Nominal

-125 kN
Zero

75 kN

Nominal

238 kN

Full

538 kN

570.6 kN

(c) Shear force diagram

554.3 kNm

1336.8 kNm

(d) Bending moment diagram

Fig. 2.30 (*Cont.*)

Using the simplified method:
And using $x/d = 0.467$ for minimum steel

$$x = 350\,\text{mm}$$

$$M_{us} = 0.4 \times 30 \times 500 \times 350 \times (750 - 175)$$
$$= 1207.5\,\text{kNm}$$

$$M_{ud} = 1336.8 - 1207.5 = 129.3\,\text{kNm}$$

Therefore:
$$A'_s = \frac{M_{ud}}{f_{yd1}(d - d')} = \frac{129.3 \times 10^6}{312 \times 675}$$
$$= 614\,\text{mm}^2$$

$$A_s = \frac{M_{us}}{f_{yd2}(d - 0.5x)} + A'_s \cdot \frac{f_{yd1}}{f_{yd2}}$$
$$= \frac{1207.5 \times 10^6}{370 \times (750 - 175)} + 614 \times 312/370$$
$$= 6193.4\,\text{mm}^2$$

Stage IV: Choose bars to give the required areas of steel. This is best
done by consulting Tables 2.2 (a), (b), (c) and (d).

Table 2.2(a) Weights of groups of bar (kg/m run)
     (b) Weight in kg/m^2 for various bar spacings
     (c) Sectional areas of groups of bar (mm^2)
     (d) Sectional areas per m width for various bar spacings (mm^2)

(a)   Weights of groups of bars (kg/m run)

Bar size (mm)	No. of bars									
	1	**2**	**3**	**4**	**5**	**6**	**7**	**8**	**9**	**10**
6	0.222	0.444	0.666	0.888	1.110	1.332	1.554	1.776	1.998	2.220
8	0.395	0.790	1.185	1.580	1.975	2.370	2.765	3.160	3.555	3.950
10	0.616	1.232	1.848	2.464	3.080	3.696	4.312	4.928	5.544	6.160
12	0.888	1.776	2.664	3.552	4.440	5.328	6.216	7.104	7.992	8.880
16	1.579	3.158	4.737	6.316	7.895	9.474	11.053	12.632	14.211	15.790
20	2.466	4.932	7.398	9.864	12.330	14.796	17.262	19.728	22.194	24.660
25	3.854	7.708	11.562	15.416	19.270	23.124	26.978	30.832	34.686	38.540
32	6.313	12.626	18.939	25.252	31.565	37.878	44.191	50.504	56.817	63.130
40	9.864	19.728	29.592	39.456	49.320	59.184	69.048	78.912	88.776	98.640

(b)   Weight in kg/m^2 for various bar spacings

Bar size (mm)	Spacing of bars (mm)								
	50	**75**	**100**	**125**	**150**	**175**	**200**	**250**	**300**
6	4.440	2.960	2.220	1.776	1.480	1.269	1.110	0.888	0.740
8	7.900	5.267	3.950	3.160	2.633	2.257	1.975	1.580	1.317
10	12.320	8.213	6.160	4.928	4.107	3.520	3.080	2.464	2.053
12	17.760	11.840	8.880	7.104	5.920	5.074	4.440	3.552	2.960
16	31.580	21.053	15.790	12.632	10.527	9.023	7.895	6.316	5.263
20	49.320	32.880	24.660	19.728	16.440	14.091	12.330	9.864	8.220
25	77.080	51.387	38.540	30.832	25.693	22.023	19.270	15.416	12.847
32	126.260	84.173	63.130	50.504	42.087	36.074	31.565	25.252	21.043
40	197.280	131.520	98.640	78.912	65.760	56.366	49.320	39.456	32.880

Compression steel = two 20-mm $\emptyset$ bars
               = 628 mm^2
Tension steel = five 40-mm $\emptyset$ bars
             = 6280 mm^2

It is unusual to be able to get so close to the required values.

(c) Sectional areas of groups of bars (mm^2)

Bar size (mm)	No. of bars									
	1	2	3	4	5	6	7	8	9	10
6	28.3	56.6	84.9	113	142	170	198	226	255	283
8	50.3	101	151	201	252	302	352	402	453	503
10	78.5	157	236	314	393	471	550	628	707	785
12	113	226	339	452	566	679	792	905	1 020	1 130
16	201	402	603	804	1 010	1 210	1 410	1 610	1 810	2 010
20	314	628	943	1 260	1 570	1 890	2 200	2 510	2 830	3 140
25	491	982	1 470	1 960	2 450	2 950	3 440	3 930	4 420	4 910
32	804	1 610	2 410	3 220	4 020	4 830	5 630	6 430	7 240	8 040
40	1 260	2 510	3 770	5 030	6 280	7 540	8 800	10 100	11 300	12 600

(d) Sectional areas per m width for various bar spacings (mm^2)

Bar size (mm)	Spacing of bars (mm)								
	50	75	100	125	150	175	200	250	300
6	566	377	283	226	189	162	142	113	94.3
8	1 010	671	503	402	335	287	252	201	168
10	1 570	1 050	785	628	523	449	393	314	262
12	2 260	1 510	1 130	905	754	646	566	452	377
16	4 020	2 680	2 010	1 610	1 340	1 150	1 010	804	670
20	6 280	4 190	3 140	2 510	2 090	1 800	1 570	1 260	1 050
25	9 820	6 550	4 910	3 930	3 270	2 810	2 450	1 960	1 640
32	16 100	10 700	8 040	6 430	5 360	4 600	4 020	3 220	2 680
40	25 100	16 800	12 600	10 100	8 380	7 180	6 280	5 030	4 190

Stage V: Design the shear reinforcement.
Obtain v_c and V_c

$$100 A_s/bd = 1.675 \qquad v_c = 0.835 \, \text{N/mm}^2$$

$$V_c = 0.835 \times 500 \times 750 = 313\,100 \, \text{N}$$
$$= 313.1 \, \text{kN}$$

If nominal links are provided the capacity of the beam rises in such a fashion that v_c may be represented by $v_c + 0.43$ (cf. Question 24 at the end of this chapter) Thus nominal links will be used where the shear force is less than

$$V = (0.835 + 0.43) \times 500 \times 750$$
$$= 474.35 \, \text{kN}$$

This extends from the L.H. support to the 300 kN point load. Part of the central section needs no shear reinforcement since $V < 0.5V_c = 151.1$ kN.
From the 300 kN point load to the R.H. support requires properly designed shear reinforcement, remembering that the load is less than $2d$ from the support.

From 300 kN point load to support

$$V = 570.6 \text{ kN} \qquad v = V/bd = 570\,600/500 \times 750$$
$$= 1.521 \text{ N/mm}^2$$

From eqn [2.28(b)]

$$\frac{A_{sv}}{S_v} = \frac{500 \times (1.521 - 1500 \times 0.835/1000)}{0.87 \times 425}$$

$$= 0.363$$

Therefore use 8 mm bars at 250 mm (250 mm chosen to fit into 1000 easily)

Nominal shear steel

$$\frac{A_{sv}}{S_v} = 0.0012b_t = 0.0012 \times 500 = 0.6$$

Thus the surprising result arises that the nominal steel is more severe than that for the high shear zone. This is due to the closeness of the point load to the support

Therefore, use nominal shear steel throughout (except where no shear steel is needed)

Nominal steel: use 10 mm bars at 250 mm spacing or 12 mm bars at 375 mm spacing.

Stage VI: Check bond (Assume deformed type (2) bars).
    (a) **Local bond** – tension steel only

$$f_{bs} = \frac{V}{d \sum U_s}$$

The greatest value of local bond stress will occur where the shear force is largest.
Thus $V = 570.6$ kN
No reduction for grouping of bars, therefore use the actual bar diameters.

$$\sum U_s = 5\pi 40 = 628.32 \text{ mm}$$

$$f_{bs} = \frac{570\,600}{750 \times 628.32} = 1.211 \text{ N/mm}^2$$

Allowed $f_{bs} = 1.2 \times 2.8 = 3.36 \text{ N/mm}^2$

Therefore the beam is safe against local bond failure

(b) **Anchorage bond** – at position of maximum moment.

(i) Tension steel: $\qquad f_u = 370 \text{ N/mm}^2$

$$f_{ba} = 1.3 \times 2.2 = 2.86 \text{ N/mm}^2$$

40-mm bars: $\qquad L_{40} = 0.25 \times 40 \times 370/2.86$

$$= 1293.7 \text{ mm}$$

Satisfactory

(ii) Compression steel $\qquad f_u = 312 \text{ N/mm}^2$

$$f_{ba} = 1.3 \times 2.7 = 3.51 \text{ N/mm}^2$$

20-mm bars: $\qquad L_{20} = 0.25 \times 20 \times 312/3.51$

$$= 444.4 \text{ mm}$$

Satisfactory.

Stage VII: Check deflection.

From Table 9: $\qquad \dfrac{L}{d} = 18$

From Table 10: $\qquad$ as $A_{s \cdot \text{req}} \approx A_{s \cdot \text{prov}}$

$f_s = 0.58 \times 425 = 246.5 \text{ N/mm}^2$

$100\, A_s/bd = 1.675$

Modification factor for tension reinforcement

$M_{\text{ten.}} = 0.863$

From Table 11: $\qquad 100\, A_s'/bd = 0.1675$

Modification factor for compression reinforcement

$M_{\text{com}} = 1.047$

Modified $\dfrac{L}{d} = 18 \times 0.863 \times 1.047 = 16.264$

Therefore $\qquad d_{\text{min.}} = \dfrac{12\,000}{16.264} = 737.8 \text{ mm}$

Thus the design is complete (apart from a possible check for cracking).

Questions

1. (a) Using the charts in CP 110: Part 2 calculate the area of tension steel required to resist an ultimate design moment of 25 kNm, in a beam of width 100 mm, if $f_{cu} = 30\,N/mm^2$ and $f_y = 250\,N/mm^2$ and effective depth is 300 mm.

   (b) What is the maximum moment that this beam will carry with only tension steel? What is the area of steel and its stress for this case? (At all times remember to check the neutral axis depth and the steel stress.)

2. A singly reinforced concrete beam in which $f_{cu} = 40\,N/mm^2$ and $f_y = 500\,N/mm^2$ contains 1400 mm² of tension steel. If the width = 200 mm and the effective depth = 500 mm calculate the ultimate moment of resistance. Find the depth to the neutral axis and the stress and strain in the steel.

3. Calculate the area of tension steel required by a reinforced concrete beam of width 150 mm and effective depth 450 mm if it is to resist a moment of 150 kNm. Assume that $f_{cu} = 30\,N/mm^2$ and $f_y = 425\,N/mm^2$. Does the section meet the neutral axis requirements of CP 110?

4. By examining Chart 6 in CP 110: Part 2 is it possible to determine at what value of moment and steel ratio the steel yields?

5. When a beam of width 100 mm and effective depth 400 mm is designed, a neutral axis depth of 160 mm is obtained. What moment was it designed for, and what area of steel was used? Assume $f_{cu} = 25\,N/mm^2$; $f_y = 410\,N/mm^2$.

6. A beam has an effective span of 3 m and carries a design point load of 100 kN. If the width of the beam is limited to 100 mm, find the minimum effective depth and the corresponding area of steel needed to carry the load. Assume $f_{cu} = 40\,N/mm^2$; $f_y = 410\,N/mm^2$. (*Hint:* maximum moment occurs when neutral axis is deepest.)

7. Design the following beam so that the tension steel has just yielded. Effective depth = 500 mm; $f_{cu} = 40\,N/mm^2$; $f_y = 410\,N/mm^2$; $M_u = 150$ kNm. What is the width of the section and what area of steel is required?

8. A beam is required to resist 150 kNm. If concrete of $f_{cu} = 40\,N/mm^2$ and steel of $f_y = 425\,N/mm^2$ is used design three sections with width 100, 125 and 150 mm each of effective depth 500 mm. If concrete costs £40/m³ and steel costs £350/t, which beam is cheapest?

9. A beam 100 mm wide by 300 mm to the centre of the tension steel is to resist a moment of 27 kNm. If $f_y = 425\,N/mm^2$ compare the areas of steel needed if concrete of $f_{cu} = 25\,N/mm^2$ and concrete of $f_{cu} = 40/mm^2$ is tried.

10. What is the moment of resistance, depth to the neutral axis and stress in the steel when a beam of $f_{cu} = 40\,N/mm^2$ and $f_y = 425\,N/mm^2$; width = 250 mm and effective depth of 600 mm has a tension steel area of 3000 mm².

11. A reinforced concrete beam of section 400 mm wide by 1000 mm effective depth is to carry a bending moment of 1500 kNm, where the depth to the neutral axis must not exceed 0.35 times the effective depth. If $f_{cu} = 25 \, \text{N/mm}^2$ and $f_y = 425 \, \text{N/mm}^2$, use the 'exact' formulae to determine the areas of steel needed and the stresses in them if the depth to the compression steel is 100 mm.

12. A reinforced concrete beam 300×700 mm effective depth is made from concrete of $f_{cu} = 30 \, \text{N/mm}^2$ and steel of $f_y = 425 \, \text{N/mm}^2$. If the depth to the compression steel is 70 mm and the steel is arranged as three bars at 20-mm $\emptyset$ in compression and three bars at 32-mm $\emptyset$ in tension, find the ultimate moment of resistance, the depth to the neutral axis and the stresses in the steel using the charts in CP 110: Part 2.

13. A beam is required to resist a bending moment of 800 kNm, the effective depth is 600 mm and the breadth is 300 mm. Determine the reinforcement needed if $f_{cu} = 25 \, \text{N/mm}^2$, $f_y = 425 \, \text{N/mm}^2$, $d'/d = 0.1$ and $x/d = 0.5$. Use the simplified method as illustrated in Fig. 2.11.

14. Design the following beam for minimum steel. Breadth is 200 mm and the effective depth is 500 mm, with the depth to the compression steel being 50 mm. If $f_{cu} = 30 \, \text{N/mm}^2$, and $f_y = 425 \, \text{N/mm}^2$ use the 'exact' formulae to calculate the areas of steel if $M_u = 250 \, \text{kNm}$.

15. Find the ultimate moment of resistance for the following beam. Breadth = 200 mm, effective depth = 600 mm, depth to the compression steel = 60 mm, $f_{cu} = 30 \, \text{N/mm}^2$, $f_y = 425 \, \text{N/mm}^2$, $A'_s = 250 \, \text{mm}^2$ and $A_s = 2500 \, \text{mm}^2$. What is the depth to the neutral axis and the stresses in the steel?

16. Design a beam of effective depth 500 mm and width 300 mm to carry a moment of 300 kNm if $f_{cu} = 25 \, \text{N/mm}^2$ and $f_y = 460 \, \text{N/mm}^2$ and $d'/d = 0.15$. The minimum area of steel should be found first from the 'exact' formulae then from the simplified method in Fig. 2.11. What are the stresses in the steel?

17. Examine Chart 31 in CP 110: Part 2. Shade the following areas: 1. that region of the chart in which the tension steel is in its elastic phase; 2. that region of the chart in which the tension steel is plastic (yielded); 3. those regions where it is impossible to meet the requirements of CP 110 for concrete of $f_{cu} = 25 \, \text{N/mm}^2$ and steel of $f_y = 425 \, \text{N/mm}^2$.

18. Design on a minimum steel basis, reinforced concrete beams of widths 200 mm, 250 mm, 260 mm and 300 mm all with an effective depth of 800 mm and $d'/d = 0.1$ to resist a bending moment of 1000 k Nm. Calculate the areas of steel needed and price the beams on the basis of concrete costing £30/m^3 and steel costing £350/t. Assume steel density = 7843 kg/m^3, $f_{cu} = 40 \, \text{N/mm}^2$ and $f_y = 410 \, \text{N/mm}^2$. Can you see a principle concerning the cost of beams?

19. (a) Consider Fig. 2.9. By taking moments about the centroid of the concrete compression zone, prove that:

$$\frac{M}{bd^2} = \frac{A_s}{bd} f_{yd2}\left(1 - k_2 \cdot \frac{x}{d}\right) + \frac{A_s'}{bd} \cdot f_{yd1}\left(k_2 \cdot \frac{x}{d} - \frac{d'}{d}\right)$$

(b) Let $A_s'/bd = 0.01$; $k_2 = 0.451$ $(f_{cu} = 30)$; $d'/d = 0.1$ and $f_y = 410 \, \text{N/mm}^2$

Remembering that when steel is yielded in compression $f_{yd1} = 2000 f_y/(2300 + f_y)$, and when in tension $f_{yd2} = f_y/1.15$ and using eqns [2.7], [2.8], [2.19], [2.20] construct the M/bd^2 V $100 \, A_s/bd$ curve. Check your answer against the one given in Chart 24 of CP 110: Part 2.

20. Check a reinforced concrete beam of width 100 mm, effective depth 250 mm and depth to the compression steel of 50 mm. The tension steel consists of two, 25-mm $\varnothing$ bars and the compression steel of two 16-mm $\varnothing$ bars. If $f_{cu} = 40 \, \text{N/mm}^2$ and $f_y = 425 \, \text{N/mm}^2$ find M_u and the stresses in the steel.

21. Design a simply supported beam which is on the boundary between two slabs and spans 5 m, supporting a UDL of 200 kN/m. The slab is 100 mm thick, the depth to tension steel from the slab surface is 500 mm and the beams width is 200 mm. Assume the depth to neutral axis is 125 mm, $f_{cu} = 25 \, \text{N/mm}^2$, $f_y = 460 \, \text{N/mm}^2$ and $d' = 50 \, \text{mm}$.

22. Design the section shown in Fig. 2.31 for a moment of 500 kNm if $f_y = 425 \, \text{N/mm}^2$, $f_{cu} = 30 \, \text{N/mm}^2$ and depth to neutral axis = 120 mm.

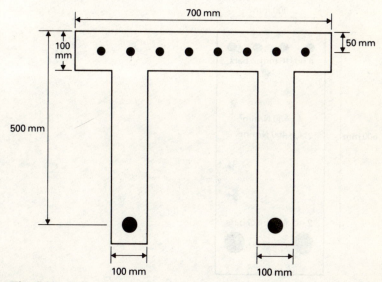

Fig. 2.31

23. A beam spanning 5 m carries a uniformly distributed load of 100 kN/m. The beam's width is 200 mm and the effective depth is 500 mm with tension steel comprising two 32-mm ∅ and one 25-mm ∅ bars. If $f_{cu} = 30$ N/mm², $f_y = f_{yv} = 410$ N/mm², 1. over what portion of the beam is no shear reinforcement needed; 2. where is only nominal shear reinforcement needed; and 3. what shear steel is needed near to the supports?

24. Show that for a beam with nominal links to CP 110, the permissible nominal shear stress may be taken as $v_c + 0.43$ N/mm².

25. A beam of width 350 mm and effective depth 550 mm contains four 32-mm ∅ bars as tension steel and spans 6.4 m carrying a UDL of 125 kN/m. Near the beam supports two of the tension bars are to be bent up as shear reinforcement. Calculate: (a) where these bars may be bent up, and (b) the amount of shear links needed to give the full shear resistance. Assume that $f_{cu} = 25$ N/mm² and $f_{yv} = 425$ N/mm² and the bars are bent up at an angle of 45°.

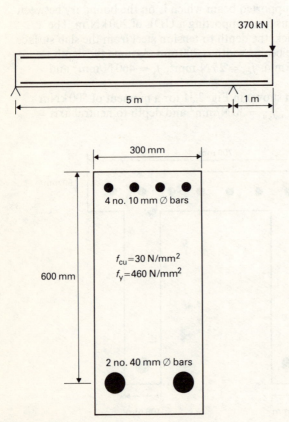

Fig. 2.32

26. The beam in Question 22 spans 10 m carrying a point load of 250 kN at a distance of 2.764 m from one end. Design the shear reinforcement for this beam.

27. A rectangular beam of width 300 mm and effective depth 600 mm spans 6 m carrying a point load 1 m from the support of 240 kN. Design the shear reinforcement if the tension steel consists of two 25-mm $\emptyset$ bars, $f_{cu} = 30 \, \text{N/mm}^2$, and $f_{yv} = 425 \, \text{N/mm}^2$.

28. Check the beam in Question 27 for local and anchorage bond stresses if the bars are of deformed type (2).

29. Check the beam in Fig. 2.32 for local and anchorage bond. Suggest remedies for any deficiencies.

30. Redesign the beam of Example 2.13 using a section of width 475 mm and effective depth 775 mm.

31. Redesign the beam of Example 2.13 using the original section dimensions but using the exact formulae.

Chapter 3

Slabs

Moments and shear forces in slabs

A slab is characterised by having a width which is very much greater than its thickness and of a comparable dimension to its length. They are most often used in buildings to provide floors or flat roofs, but they are also used in bridges and car parks, both horizontally and sloped as ramps. In common with beams, they support load by bending, but they differ from beams in that they can bend in two directions at right angles – along the span and perpendicular to the span. This latter effect is of most importance in the support of concentrated loads, as it gives the slab the ability to spread the load from the point of application, sideways to other unloaded parts of the slab.

The principle is illustrated in Fig. 3.1. In Fig. 3.1(a) a floor is shown as consisting of a series of unconnected beam-like strips. On one of these strips a weight is being supported, and as can be seen, only that strip deflects. On the other hand, Fig. 3.1(b) shows the weight supported first on a cross-strip spanning at right angles to the main span, and then on the main strips. In this case the weight is more evenly shared between the strips, but the main strip directly under the load still bears the major contribution. This is a rather simplified model of the load sharing action of a slab, but it serves the purpose of illustration. Although, theoretically, the whole slab plays a part in the support of the load, the design to account for this is rather complicated and therefore there are a number of simpler, less exact ways of dealing with the situation. One of these methods will be dealt with later.

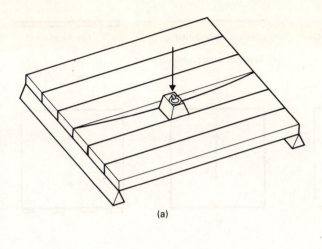

(a)

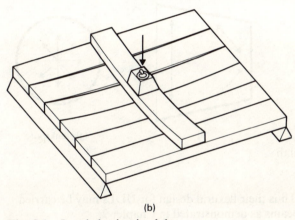

(b)

Fig. 3.1 Load sharing in slabs

In Fig. 3.1 it was tacitly assumed that the slab was simply supported only along two of its edges and that those edges were parallel to one another. Although this chapter is indeed limited to the study of these types of slab, it must be realised that slabs may be supported in countless ways. Figure 3.2 illustrates just some of the types of slab encountered in structural design, from the rectangular slab simply supported on all four sides, to a non-regular slab with fixed and free edges. The type shown in Fig. 3.2(b), with which this study is concerned, is known as a slab spanning in only one direction, even though this is not the case as was demonstrated in Fig. 3.1. However, the essence of the name is that the slab has only one principal span direction. Also, under the influence of purely uniformly distributed loads, the slab **does** act as a beam, without any

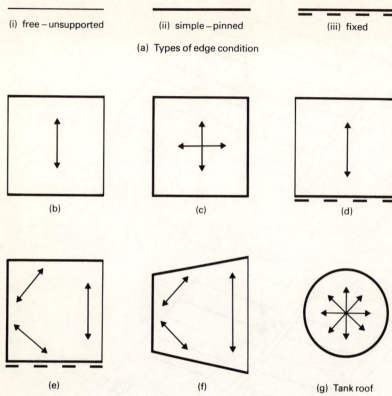

(a) Types of edge condition

(i) free – unsupported   (ii) simple – pinned   (iii) fixed

(b)   (c)   (d)

(e)   (f)   (g) Tank roof

Fig. 3.2 Types of slab

spreading of loads. Thus their flexural design for UDLs may be carried out as if they were beams as demonstrated in Chapter 2.

Concentrated loads on slabs

The simplest way of designing a slab for a concentrated load is by assigning a strip of slab to carry the whole load to the supports. Then the problem reduces to obtaining the effective width of that strip, since the assumption is made that the strip acts as a beam, with constant stress across the section as with a beam.

As will be realised from a commonsense approach to the problem, the closer a concentrated load is to a support, the less will be the sharing action of the rest of the slab. Therefore, it is important that whatever value of effective width is chosen, it reflects this fact. In other words, the nearer to a support a load is, the less should be the value of the effective width.

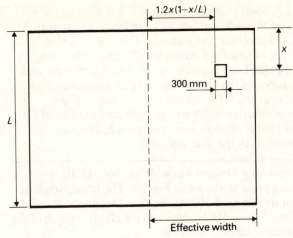

Fig. 3.3

This is accomplished as illustrated in Fig. 3.3. The effective width of the slab strip is given by:

$$b_e = a_y + 2.4x(1 - x/L)$$

where a_y is the loaded width
x is the distance to the support from the centre of the loaded area
L is the span

Effectively this states that on each side of the loaded area a width of $1.2x(1 - x/L)$ is considered to act as a beam in supporting the load. Thus if the load is too close to the edge of a slab for this width to be available, then the actual width from load to slab edge should be used. The following example demonstrates the application of this equation.

Example 3.1 A solid, one-way spanning slab spans 5 m carrying a concentrated load of 4.5 kN over a small area of 300 × 300 mm. The load is placed 1.2 m from the support and 1.0 m from an unsupported edge of the 6-m wide slab (these distances measured from the centre of the loaded area). This load is the value that CP 3 recommends for a library or public room with storage.

$$1.2x(1 - x/L) = 1.2 \times 1.2 \times (1 - 1.2/5) = 1.094 \text{ m}$$

Therefore, the load is just too close to the edge for the full effective width to be used
Effective width, $b_e = 1.094 + 0.3 + 0.85 = 2.244$ m
The strip would now be designed as if it were a beam. (The value 0.85 derives from the fact that the edge distance was measured from the centre of the loaded area.)

74

However, the 4.5 kN point load specified in CP 3 is allowed to be placed at any position on the slab, not just at 1.2 m from the support. As well as a point load, CP 3 specifies that instead a UDL of 7.5 kN/m² may be applied. This type of situation occurs for all types of building – a UDL or an alternative **Point Load** is given, either of which may be used, the more severe being taken as the Design Load. In all cases the point load will not be as severe as the UDL for the flexural calculations.

Therefore, the use of effective width will be restricted to cases of loading where a fixed point load is present. For example a heavy piece of machinery may be present as in the next example.

Example 3.2 A slab spanning 5 m and 6 m wide carries a UDL of 7.5 kN/m² and a printing press as shown in Fig. 3.4. The press, weighing 150 kN is supported on six feet each 100 mm square. Design the slab for bending if $f_{cu} = 30$ N/mm²; $f_y = 410$ N/mm² and the effective depth of the slab $= d = 200$ mm. Overall depth $= 230$ mm.

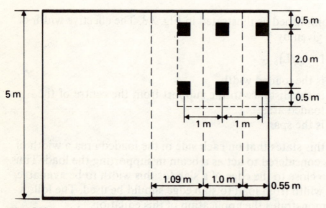

Fig. 3.4

Design Dead Load $= 1.4 \times 0.23 \times 24 = 7.728$ kN/m²

Design Live Load $= 1.6 \times 7.5 = 12$ kN/m²

Design Point Load $= 1.6 \times 150/6 = 40$ kN

As can be seen from Fig. 3.4 the effective widths for the point loads overlap, thus the actual distances between the legs of the press must be used, except towards the middle of the slab. Here the smaller effective width of the load nearest the support is used as this is conservative.

$1.2x(1 - x/L) = 1.2 \times 0.5 \times (1 - 0.5/5) = 0.54$ m

Figure 3.5 gives the shear force and bending moment diagrams.

$M_u = 289.204$ kN m

Thus $M/bd^2 = \dfrac{289.204 \times 10^6}{2640 \times 40\,000}$

$\qquad\qquad = 2.739$

Referring to Chart 2 in CP 110: Part 2 gives

$100\,A_s/bd = 0.866$

$\qquad A_s = 0.866 \times 2640 \times 200/100$

$\qquad\quad = 4572.5\,\text{mm}^2$

Use twenty-three 16-mm bars at approximately 115 mm spacing, total area $= 4624.4\,\text{mm}^2$.

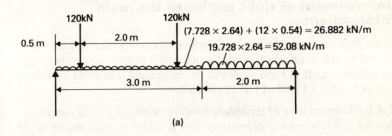

(a)

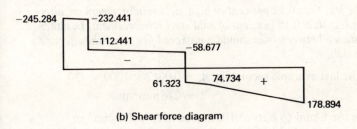

(b) Shear force diagram

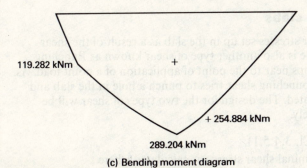

(c) Bending moment diagram

Fig. 3.5

Outside the line of the press the UDL is still present, therefore the outer left half of the slab must be designed separately.
Consider a 1 m width.

$$M_u = w L^2/8 = 19.73 \times 5^2/8 = 61.65 \, \text{kN m}$$

$$M/bd^2 = 1.54 \qquad 100 \, A_s/bd = 0.46$$

$$A_s = 0.46 \times 1000 \times 200/100$$
$$= 920 \, \text{mm}^2$$

Use 12-mm $\varnothing$ bars at 120 mm spacing = 942 mm²/m.

Reinforcement at right angles to the main reinforcement

In order that the slab may be able to spread the concentrated loads as predicted by use of an effective width analysis, it is necessary to provide some reinforcement at right angles to the main steel. The rules for this are set out in Cl. 3.11.4.2 of CP 110 as follows:

3.11.4.2 *Minimum area of secondary reinforcement. In a solid concrete suspended slab, the amount of reinforcement provided at right angles to the main reinforcement, expressed as a percentage of the gross cross-section, should not be less than 0.12 per cent of high yield reinforcement or, alternatively, not less than 0.15 per cent of mild steel reinforcement. In either case, the distance between bars should not exceed five times the effective depth of the slab.*

Thus in the last example secondary $A_s = 0.0012 \times 1000 \times 230$
$$= 276 \, \text{mm}^2/\text{m}$$

Therefore use 6-mm $\varnothing$ bars at 100 mm centres = 283 mm²/m. These 6-mm bars would be laid directly on top of the main steel.

Shear in solid slabs

As well as the shear stresses set up in the slab as a result of the shear force diagram, there is also another type of shear known as Punching Shear which develops near to the point of application of a point load. As its name suggests, punching shear tries to punch a hole in the slab and this must be prevented. The design for the two types of shear will be considered separately.

Ordinary shear (Cl. 3.4.5.1)

As for beams, a nominal shear stress v is calculated where

$$v = V/bd$$

where v is the shear force due to ultimate loads
         b is the width of slab under consideration (usually 1 m)
         d is the effective depth

Again, as for flexure, due to load spreading the shear resistance of a slab is comparatively greater, therefore the nominal shear stress which is calculated may be higher than allowed for beams. This is reflected in Table 14 of CP 110 which gives a modification factor ξ_s which, when multiplied on v_c obtained from Table 5 of CP 110 gives the allowed shear stress for slabs.

Table 14   Values of ξ_s

Overall slab depth	ξ_s
mm	
300 or more	1.00
275	1.05
250	1.10
225	1.15
200	1.20
175	1.25
150 or less	1.30

1. No shear reinforcement is needed if $v < \xi_s v_c$, not even nominal shear steel as for beams.
2. For slabs of overall depth < 200 mm v must not exceed $\xi_s v_c$ and these slabs no shear reinforcement can be allowed.
3. For slabs of overall depth > 200 mm v must not exceed half the value obtained from Table 6 of CP 110. For shear stress between $\epsilon_s v_c$ and half the value from Table 6, shear reinforcement must be provided. However, if shear steel is used then there must be some compression steel for the links to be anchored around. These compression bars must be of a diameter at least equal to the diameter of the links. If shear steel is provided, then eqn [9] of CP 110 must be modified to

$$\frac{A_{sv}}{S_v} = b(v - \epsilon_s v_c)/0.87 f_{yv} \qquad [3.1]$$

Example 3.3   Calculate the ultimate shear stress for the slab of Example 3.2

$V_{max} = 254.284\,\text{kN}$

$$v = \frac{254\,284}{2640 \times 200} = 0.465\,\text{N/mm}^2$$

78

From Table 14 of CP 110: overall depth = 230 mm

$$\xi_s = 1.13$$

From Table 5 of CP 110:

$$v_c = 0.660 \, \text{N/mm}^2$$
$$100 \, A_s/bd = 0.866$$

Therefore, no shear steel is needed.

Punching shear (Cl. 3.4.5.2)

When an over-large concentrated load punches a hole in a solid slab, the hole tends to be of the shape and size shown in Fig. 3.6. As can be seen the hole is not quite as small as might be first thought, but has a perimeter which is about 1.5 times the thickness of the slab greater than the loaded area. Thus CP 110 states that the nominal shear stress shall be calculated using this effective perimeter u_{crit}

$$u_{\text{crit}} = 3\pi h + 2(a + b)$$

where h is the overall thickness of the slab

$$v = V/u_{\text{crit}} \cdot d.$$

1. If $v < \xi_s v_c$ then no shear reinforcement under the load is needed. To calculate v_c, the value of A_s used should be the average of the tension steel areas in the two directions at right angles.
2. No shear reinforcement is allowed in slabs of overall thickness 200 mm or less. In other words, v must not exceed $\xi_s v_c$.
3. For slabs of overall thickness > 200 mm
    If $v > \xi_s v_c$ but < half the maximum value given in Table 6, then shear reinforcement must be provided. This steel should satisfy eqn [3.2] below. A similar area of steel should be provided, not only

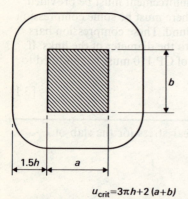

$$u_{\text{crit}} = 3\pi h + 2 \, (a+b)$$

Fig. 3.6   Effective perimeter for punching shear

at the critical perimeter, but also halfway between it and the edge of the loaded area.

$$0.4 < = \frac{\sum A_{sv}(0.87 f_{yv})}{u_{crit} \cdot d} > = v - \xi_s v_c \qquad [3.2]$$

where $\sum A_{sv}$ is the total area of shear reinforcement viz. all the legs of all the links added together

f_{yv} is the characteristic strength of the reinforcement, which should not be taken as greater than $425 \, N/mm^2$

u_{crit} is the effective perimeter for punching shear

4. As well as adding the shear reinforcement in 3. above it is then also necessary to make further checks on punching shear for, first, a critical perimeter which is at $1.5 + 0.75$ times the overall depth out from the loaded area and then for critical perimeters which are successive 0.75 times the overall depth even further out. This is done until the value of v is less than $\xi_s v_c$. At each of the other effective perimeters, shear steel must be added. Thus it can be seen that the procedure for punching shear is more complex than for ordinary shear.

Example 3.4 Check the Example 3.2 for punching shear.

Loaded area $= 100 \times 100$

Effective perimeter $= 3\pi \times 235 + 400$

$\qquad\qquad\qquad\quad = 2614.8 \, mm$

Ultimate load per press foot $= 40 \, kN$

$$v = \frac{40\,000}{2614.8 \times 200} = 0.0765 \, N/mm^2$$

Thus punching shear is no problem.

Example 3.5 Check a slab supporting a concentrated load of $900 \, kN$ over a loaded area of $100 \times 100 \, mm$, if the slab depth of $200 \, mm$ contains $5000 \, mm^2$ of main steel per metre width and $500 \, mm^2$ of secondary steel per metre of span. Assume that the overall depth is $250 \, mm$.

$f_{cu} = 30 \, N/mm^2$

$\qquad u_{crit} = 3\pi \times 250 + 400 = 2756.2 \, mm$

$$v = \frac{900\,000}{2756.2 \times 200} = 1.633 \, N/mm^2$$

Average $A_s = (5000 + 500)/2 = 2750 \, mm^2$

$100 \, A_s/bd = 1.375$

From Table 5 of CP 110: $\quad v_c = 0.775 \, \text{N/mm}^2$

From Table 14 of CP 110: $\quad \xi_s = 1.2$

$$\xi_s v_c = 0.93 \, \text{N/mm}^2$$

$$v - \xi_s v_c = 0.703 \, \text{N/mm}^2$$

Thus $\sum A_{sv} = (v - \xi_s v_c) \times u_{crit} \cdot d / 0.87 f_{yv}$

$$= 0.703 \times 2765.2 \times 200/356.52 = 1087 \, \text{mm}^2$$

This is divided equally around the effective perimeter as shown in Fig. 3.7. Note that the spacing of the links must not exceed $0.75 \, h = 176.25 \, \text{mm}$.

Using sixteen 8-mm links gives $\sum A_{sv} = 1605 \, \text{mm}^2$

$$\text{spacing} = 172 \, \text{mm Outer}$$

$$= 94.2 \, \text{mm Inner}$$

New critical perimeter at $2.25 \, h$ from the loaded area

$$u_{crit} = 4.5\pi \times 235 + 400 = 3722.2 \, \text{mm}$$

$$v = \frac{900\,000}{3722.2 \times 200} = 1.21$$

$$v - \xi_s v_c = 0.28$$

Therefore shear steel must be provided here as well.

$$\sum A_{sv} = 0.4 \times u_{crit} \cdot d / (0.87 f_{yv})$$

$$= 0.4 \times 3722.2 \times 200/356.52 = 835.2 \, \text{mm}^2$$

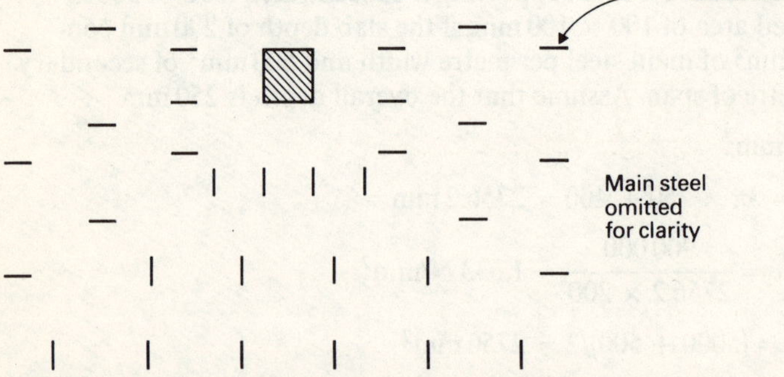

Main steel omitted for clarity

Fig. 3.7 Shear reinforcement for Example 3.5

Using twenty-two 6-mm links gives $\sum A_{sv} = 1244 \, \text{mm}^2$

$$\text{spacing} = 170 \, \text{mm}.$$

New critical perimeter at 3 h from loaded area

$u_{\text{crit}} = 6\pi \times 235 + 400 = 4829.6$

$$v = \frac{900\,000}{4829.6 \times 200} = 0.93 \, \text{N/mm}^2$$

Therefore no more shear steel is needed.

As Fig. 3.7 shows quite clearly that whilst it is possible to use links in this fashion it is very untidy. Figure 3.8 shows two alternative methods which are much easier to place and make a tidier cage. Note also that there must be some compression steel to anchor the shear reinforcement.

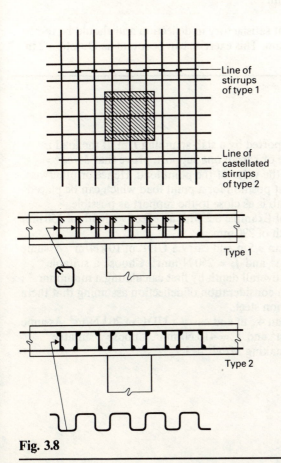

Fig. 3.8

Deflection in solid slabs

Deflection in solid slabs is dealt with in exactly the same way as for beams. The only point to watch is that the area of tension steel is confined to the steel in the direction of the span.

Example 3.6 Check the slab of Example 3.2 for deflection.

From Table 8 of CP 110: basic $L/d = 20$
From Table 10 of CP 110: Modification factor for tension steel

$100 A_s/bd = 0.866$

  Therefore $M_{ten} = 1.048$

  Modified $L/d = 20 \times 1.048 = 20.96$

$$d_{min} = \frac{5000}{20.96} = 238.55 \, mm$$

Therefore, the slab is not satisfactory in deflection and should be redesigned with $d = 250 \, mm$. This exercise can be found as Question 2 in this chapter.

Questions

1. A point load is supported by a slab spanning L m. If the load is placed at a distance of x from one support, prove that $M/bd^2 = P/2.4d^2$ where P is the value of the point load. Hence or otherwise prove that the worst position for a point load which can be placed anywhere on the slab is as close to the support as possible.

2. Redesign the slab of Example 3.2 with an effective depth of 250 mm and an overall depth of 290 mm.

3. Design a slab to span 6.5 m and carry a UDL of 10 kN/m². Assume that $f_{cu} = 25 \, N/mm^2$ and $f_y = 250 \, N/mm^2$. Choose a suitable effective depth and overall depth by first calculating a minimum effective depth from consideration of deflection assuming that there is 1 per cent of tension steel.

4. Design a slab to span 4.5 m and carry a UDL of 20 kN/m². Assume that $f_{cu} = 25 \, N/mm^2$ and $f_y = 410 \, N/mm^2$. Choose a suitable effective depth by making $M/bd^2 \approx 1.7$.

Chapter 4

Columns

The art of structural design may be thought of as providing a path by which loads may be transferred to the earth from the point at which they must be supported or at which they are applied. The path by which this transferal takes place is the structure and it may take many forms, but the one most prevalent in reinforced concrete design is that in which the load is first transferred to beams or slabs. Thence the beams transfer the load to the columns which support them, and these in turn transmit their load to other columns and finally to the foundations. To this point in the book, only the first of these transferals, the beam or the slab has been considered, and it has been assumed that the elements supporting them can be provided. However, the study must now turn to the design of columns, and at the outset it must be understood that the subject of column design is highly complex and many of the issues at stake are far beyond the scope of this book. This chapter serves only as a brief introduction and has *very* limited application in the world of real structures. Therefore the student must be careful to understand the limitations of the design procedures here set out and must not assume that he knows all there is to know about reinforced concrete columns when he has studied this chapter. This study is restricted to 'Short Braced' Columns.

Column failure

Most students will be familiar with the derivation of the Euler buckling formula:

$$P_c = \frac{\pi^2 EI}{(kL)^2}$$

where P_c is the critical axially applied load
E is Young's modulus
I is the second moment of area of the strut
L is the actual length
k is a factor depending on the restraint at the supports of the strut

The term 'kL' is sometimes referred to as the 'effective length' of the strut.

Figure 4.1 illustrates the way in which k varies with the end restraint conditions, e.g. the pin-ended strut has a value of $k = 1.0$, whereas the fully restrained strut has a value of $k = 0.5$. As shown, the Euler formula predicts column failure by sideways buckling. In other words, the failure occurs as a result of lateral instability, not because of over-stressing.

For a number of reasons this formula cannot be used to design any real column. The most important is illustrated in Fig. 4.2, where a graph of P_c against kL has been plotted for a real column. It can be seen that for cases where L is large, the Euler formula predicts the failure load quite accurately, but for small values of L the value of P_c is much less than predicted by the formula. This type of failure is characterised by the crushing of the concrete and local buckling of the longitudinal reinforcement as shown in Fig. 4.2(b). Thus the design of short columns may be

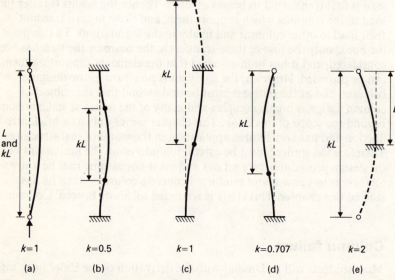

Fig. 4.1 Effective column lengths

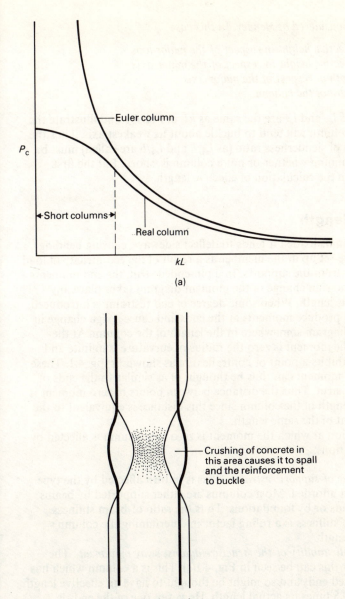

Euler column

P_c

Short columns

Real column

kL

(a)

Crushing of concrete in this area causes it to spall and the reinforcement to buckle

(b) Behaviour of a short column

Fig. 4.2

carried out without reference to buckling theory, and CP 110 recognises this fact in Cl. 3.5.1.2 where the definition of a short column is given.

3.5.1.2 *Short and slender columns: definitions.* *A column may be considered as short when both the ratios l_{ex}/h and l_{ey}/b are less than 12. It should* *otherwise be considered as slender. In this case*

> l_{ex} *is the effective height in respect of the major axis*
> l_{ey} *is the effective height in respect of the minor axis*
> h *is the depth in respect of the major axis*
> b *is the width of the column.*

The terms l_{ex} and l_{ey} are the same as kL in Fig. 4.2 and illustrate the point that a column will tend to buckle about its weakest axis. Thus the greatest value of slenderness ratio (as l_{ex}/h and l_{ey}/b are called) must be used in determining whether or not a column is 'short'. But the first problem lies in the calculation of effective length.

Effective length

When a column is loaded it tends to deflect sideways, causing bending moments to be set up in the member as a result of the eccentricity of load at sections far from the supports. In a pin-ended strut, the end moments are zero and no sign change of the moment diagram takes place any-where along its length. When some degree of end restraint is introduced, the effect is to produce moments at the ends and cause a sign change in the moment diagram somewhere in the length of the column. At the point where the moment is zero the radius of curvature is infinite and the column exhibits a point of contraflexure as shown in Fig. 4.1. These points of zero moment can thus be thought of as similar to the ends of the pin-ended strut. Thus the distance between points of zero moment is the effective length of the column since this distance is equivalent to the pin-ended strut of the same length.

The position at which the moment is zero in a column is affected by two considerations:

1. *The degree of support restraint.* This is in turn affected by the type of support afforded. Most columns are either supported by beams at their ends or by foundations. Thus the ratio of beam stiffness to column stiffness is a ruling factor in determining the column's effective length.

2. *The overall stability of the structure against sway movement.* The reason for this can be seen in Fig. 4.1(c). This is a column which has rigidly-fixed ends and so might be thought to have an effective length equal to 0.5 times its actual length. However, one of the ends is allowed to move sideways. This increases the eccentricity of the load and reduces the capacity of the strut. Columns in which this type of movement is eliminated or restricted are known as **braced** columns. Thus CP 110 defines braced columns as:

3.5.1.3 *Braced and unbraced columns: definitions. A column may be considered braced in a given plane if lateral stability to the structure as a* *whole is provided by walls or bracing designed to resist all lateral forces in that plane. It should otherwise be considered as unbraced.*

This study is restricted to short braced columns.

Example 4.1 Find the effective lengths in the two directions for the column shown in Fig. 4.3, and thus determine the minimum section dimensions needed to make the column **short** as defined in CP 110. The column may be regarded as braced in both directions.

CP 110 allows the use of Table 15 in determining effective length, but it is not accurate enough for the type of situation illustrated by this example. Thus the formulae set out below it must be used instead.

Table 15. Effective column height

Type of column	Effective column height
Braced column properly restrained in direction at both ends	$0.75l_o$
Braced column imperfectly restrained in direction at one or both ends	A value intermediate between $0.75l_o$ and l_o depending upon the efficiency of the directional restraint
Unbraced or partially braced column, properly restrained in direction at one end but imperfectly restrained in direction at the other end	A value intermediate between l_o and $2l_o$ depending upon the efficiency of the directional restraint and bracing

For a framed structure, effective heights may alternatively be obtained from:
for a braced column, the lesser of

$$l_e = l_o[0.7 + 0.05(\alpha_{c1} + \alpha_{c2})] \leqslant l_o \qquad [20]$$

$$l_e = l_o(0.85 + 0.05\alpha_{c\,min}) \leqslant l_o \qquad [21]$$

for an unbraced column, the lesser of

$$l_e = l_o[1.0 + 0.15(\alpha_{c1} + \alpha_{c2})] \qquad [22]$$

$$l_e = l_o(2.0 + 0.3\alpha_{c\,min}) \qquad [23]$$

where l_o is the clear height between end restraints
α_{c1} is the ratio of the sum of the column stiffnesses to the sum of the beam stiffnesses at one end of the column

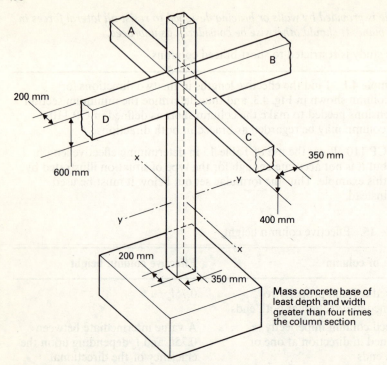

200 mm

600 mm

200 mm

350 mm

350 mm

400 mm

Mass concrete base of least depth and width greater than four times the column section

Fig. 4.3

α_{c2} is the ratio of the sum of the column stiffnesses to the sum of the beam stiffness at the other end of the column

$\alpha_{c\,min}$ is the minimum of α_{c1} and α_{c2}.

When calculating α_c, only members properly framed into the end of the column in the appropriate plane of bending should be considered. The stiffness of each member should be obtained by dividing the second moment of area of its concrete section by its actual length. For flat slab construction, an equivalent beam should be taken as having the width and thickness of the slab forming the column strip. When connection between a column and its base is not designed to resist other than nominal moment, or when the beams framing into a column are designed as simply supported, α_c at such positions should be taken as 10. If a base is designed to resist the column moment, α_c may be taken as 1.0.

Span of beam A = 4 m
Span of beam B = 6 m
Span of beam C = 6 m
Span of beam D = 3 m

Actual height of column from top of base to centre of highest beam = 5 m
Actual height of upper column from beam centres = 4 m.

Consider x–x axis of column first

$\alpha_{\text{base}} = 0$ Since according to the *Handbook on the Unified Code for Structural Concrete* the base is massive enough to be considered fully fixed

$I_{\text{column}} = 200 \times 350^3/12$

Lower column stiffness = I/actual length $= \dfrac{200 \times 350^3}{12 \times 5000} = 142916.7$

Upper column stiffness $= \dfrac{200 \times 350^3}{12 \times 4000} = 178\,645.8$

$I_{\text{beam}} = 200 \times 600^3/12$

Stiffness of beam D $= I_{\text{beam}}/3000 = \dfrac{200 \times 600^3}{12 \times 3000} = 1.2 \times 10^6$

Stiffness of beam B $= I_{\text{beam}}/6000 = 0.6 \times 10^6$

$\alpha_{\text{top}} = \dfrac{\sum \text{column stiffness}}{\text{sum of beam stiffnesses}} = \dfrac{321\,562.5}{1.8 \times 10^6} = 0.18 \approx 0$

Using eqn [20] of Table 15 of CP 110 $l_e = l_o(0.7 + 0.05 \times 0.18) = 0.709 l_o$

Since α_{top} and α_{base} were so close to zero it would have sufficed to assume that both top and bottom were fully fixed. From Fig. 4.1(b) it might have been expected that $k = 0.5$ instead of the 0.7 obtained. The difference is the result of the way in which a real reinforced concrete column behaves. Similarly it will be found that the value for k used in Fig. 4.1(d) is 0.85 in CP 110 as opposed to the 0.7 shown.

Consider y–y axis of column

$\alpha_{\text{base}} = 0$

$I_{\text{column}} = 350 \times 200^3/12$

$\sum \text{Column stiffness} = \dfrac{350 \times 200^3}{12 \times 5000} + \dfrac{350 \times 200^3}{12 \times 4000} = 105\,000$

$I_{\text{beam}} = 350 \times 400^3/12$

Stiffness of beam A $= \dfrac{350 \times 400^3}{12 \times 4000} = 466.67 \times 10^3$

Stiffness of beam C $= \dfrac{350 \times 400^3}{12 \times 6000} = 311.11 \times 10^3$

$\alpha_{\text{top}} = 0.13$

Again $l_e = l_o \times 0.7$

Minimum dimensions for a short column

$$l_{ex} = 0.7 \times 5000 = 3500 = l_{ey}$$

$$\frac{l_{ex}}{h} < = 12$$

Therefore $\quad h > = \dfrac{l_{ex}}{12} = 292\,\text{mm}$

Similarly $b > = 292\,\text{mm}$

Hence the dimensions are satisfactory in the x–x axis but must be increased to (say) 300 mm in the y–y axis.

Strength of axially loaded sections

In contrast to beams where only a part of the cross-section is utilised in resisting load, when a column is axially loaded, all of the concrete will be in compression to the full ultimate stress of $0.67f_{cu}/\gamma_m = 0.45f_{cu}$. The steel also will be in the yielded compression state with a stress of $f_y/(\gamma_m + f_y/2000)$ which is approximately equal to $0.75f_y$. It will be realised that the steel should be so distributed as not to cause an out of balance moment as well as an axial stress; therefore, in a rectangular column half the steel is placed near one side and half near the other. Also, since it must be evenly distributed in both axes, it is usual to use four bars or groups of bars at the corners. For a circular column, therefore, the steel is placed around the section in a circle consisting of at least six bars all of the same size. Thus the theoretical compressive strength of a column is easily found by adding the separate concrete and steel forces:

$$N = 0.45f_{cu}A_c + 0.75f_y \cdot A_{sc} \qquad [4.1]$$

where $\quad A_c$ is the area of the concrete section
A_{sc} is the total area of compression steel remembering that it must be evenly distributed

However, no real column will be truly axially loaded since construction tolerances will give rise to loads being applied eccentrically and even beams designed to act as being simply supported will not be quite free of end moment. Luckily, many columns may be classified as being nearly axially loaded, viz. the magnitude of the moments will be much smaller than the axial load to be resisted. Code of Practice 110 classifies these types of column into two categories: A. those supporting a rigid structure or beams which are deep in comparison to the dimensions of the column supporting them (say about twice at least), and B. those supporting an approximately symmetrical arrangement of beams. In this last category the spans of the beams should not differ from the span of the longest by more than 15 per cent. Therefore the unhappy column of Example 4.1 does not quite fit into either category. Both categories of column must be part of a structure which may be considered as being braced, thus this study is limited to these cases.

Category A is covered in CP 110 by Cl. 3.5.3 short braced axially loaded columns while Category B is covered in Cl. 3.5.4, short braced columns supporting an approximately symmetrical arrangement of beams. In both cases the formula used to design them is arrived at by reducing the theoretical value obtained from eqn [4.1] to allow for inevitable eccentricity of load.

3.5.3 *Short braced axially loaded columns.* *To allow for eccentricity due to construction tolerances the ultimate axial load for a short column which by the nature of the structure cannot be subjected to significant moments should not exceed the value N given by:*

$$N = 0.4 f_{cu} A_c + 0.67 A_{sc} f_y$$

where f_{cu} *is the characteristic strength of the concrete*
           A_c *is the area of concrete*
           A_{sc} *is the area of longitudinal reinforcement*
           f_y *is the characteristic strength of the compression reinforcement.*

3.5.4 *Short braced columns supporting an approximately symmetrical arrangement of beams.* *The ultimate axial load for a short column of this type where*

1. *the beams are designed for uniformly distributed imposed loads and*
2. *the beams spans do not differ by more than 15 per cent of the longer*

should not exceed the value N given by:

$$N = 0.35 f_{cu} A_c + 0.60 A_{sc} f_y$$

where the symbols have the same meaning as in eqn [25] of CP 110.

Example 4.2   Design a short braced axially loaded column to support a load of 800 kN, assuming $f_{cu} = 30 \, \text{N/mm}^2$ and $f_y = 250 \, \text{N/mm}^2$.

Try a square section 200 × 200 mm

     Thus $N = 800\,000 = 0.4 \times 30 \times 200^2 + 0.67 \times A_{sc} \times 250$

$A_{sc} = 1910 \, \text{mm}^2$

Use four 25-mm bars at four corners = 1964 mm²

$100 \, A_{sc}/A_c = 4.9$

Example 4.3   Design a short braced column supporting an approximately symmetrical arrangement of beams to carry a load of 2800 kN assuming that $f_{cu} = 30 \, \text{N/mm}^2$ and $f_y = 410 \, \text{N/mm}^2$.

Try a circular section of diameter 450 mm:

$N = 2800\,000 = 0.35 \times 30 \times \pi \times 225^2 + 0.6 \times A_{sc} \times 410$

$A_{sc} = 4594 \, \text{mm}^2$

Use six 32-mm bars evenly spread around the section = 4825 mm^2

$100 A_{sc}/A_c = 3.03$.

Restrictions on reinforcement

3.5.1.1 *Size and reinforcement of columns.* The size of a column and the position of the reinforcement in it may be affected by the requirements for durability and fire resistance, and these should be considered before the design is commenced.

The minimum cross-sectional area of longitudinal reinforcement provided should not normally be less then 1 per cent. However, for lightly loaded members, it will be sufficient to provide a minimum area A_{sc} such that

$$A_{sc}f_y \geqslant 0.15 N$$

where f_y is the characteristic strength of the reinforcement
N is the ultimate axial load.

A column which is intended to act as a very short length of plain concrete wall may be designed in accordance with 5.5.

Links

Links, or ties as they are sometimes called in columns since they serve to tie the longitudinal steel against buckling, should be at least a quarter the size of the largest compression bar at a spacing no greater than twelve times the size of the smallest compression bar. Links should be so arranged that every corner and alternate bar or group in an outer layer of reinforcement is supported by a link passing round the bar having an included angle of not more than 135°. All other bars or groups within a compression zone should be within 150 mm of a restrained bar. For circular columns, where the longitudinal reinforcement is located round the periphery of a circle, adequate lateral support is provided by a circular tie passing round the bars or groups.

In the last two examples the following will suffice:

Example 4.2: 8 mm links at 300 mm spacing,
Example 4.3: 8 mm links at 375 mm spacing.

Questions

1. Determine the effective lengths of the upper and lower sections of the column shown in Fig. 4.4.
2. Determine whether or not the column sections in Question 1 are classified as 'short' by CP 110. How should the section be modified to make the column short?

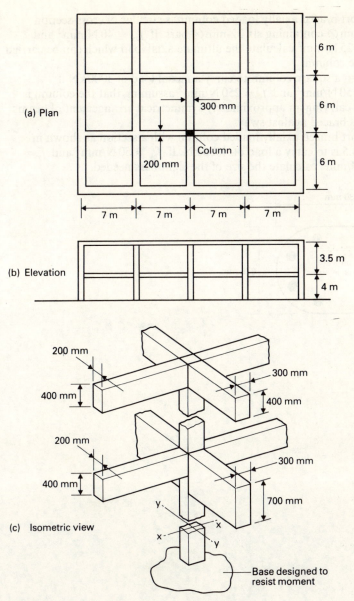

(a) Plan

300 mm

6 m

6 m

6 m

Column

200 mm

7 m   7 m   7 m   7 m

(b) Elevation

3.5 m

4 m

(c) Isometric view

200 mm

400 mm

300 mm

400 mm

200 mm

400 mm

300 mm

700 mm

y

x

x

y

Base designed to
resist moment

Fig. 4.4

3. Using Cl. 3.5.4 design the modified column above for the following
characteristic imposed loads from the beams: upper storey – 190 kN;
lower storey – 500 kN. The **characteristic** dead loads are: upper
storey – 125 kN; lower storey – 160 kN. Assume that $f_{cu} = 30\,\text{N/mm}^2$
and $f_y = 425\,\text{N/mm}^2$.

4. A short braced axially loaded column is circular of cross-section 300 mm $\varnothing$ containing six 32-mm $\varnothing$ bars. If $f_{cu} = 40\,\text{N/mm}^2$ and f_y is $425\,\text{N/mm}^2$ calculate the ultimate axial load which can be carried by the column.

5. Design a square column to carry an axial load of 3000 kN if $f_{cu} = 50\,\text{N/mm}^2$ and $f_y = 250\,\text{N/mm}^2$ assuming that the column is short carrying an approximately symmetrical arrangement of beams and is braced against sway.

6. A short braced axially loaded column with a section as shown in Fig. 4.5 is to carry a load of 4000 kN. If $f_{cu} = 50\,\text{N/mm}^2$ and $f_y = 460\,\text{N/mm}^2$ calculate the size of the eight bars needed.

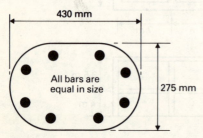

Fig. 4.5

Chapter 5

Bases and foundations

The need to make special consideration of the design of foundations may readily be understood in the light of the many failures of recent years involving the soil on which the buildings have been constructed. All too often the main structural frame is designed with great thoroughness but the foundations are given only little thought, yet they can materially affect the performance of the whole building. The main difficulty arises from the nature of the soil itself; firstly, it is not a strong material, being capable of withstanding stresses in the range of 0.1 to 1 N/mm² compared to 50 that concrete can be called to take. Secondly, it is not homogeneous, and thirdly it tends to settle unevenly causing additional stress in the structure. These and other factors make foundation design of paramount importance.

This chapter will consider two of the many types of foundation: 1. the isolated, reinforced pad used as the base of an individual column; and 2. the strip foundation which may be used to support a wall as in the domestic house, or a line of columns where the provision of individual pads would prove uneconomic. Of the many other types of foundation, the student should also be familiar with the mass concrete pad which is usually much deeper than its reinforced counterpart and the raft which consists of a flat slab covering the whole area of the superstructure and serves to tie the structure together as well as spreading the loads placed on it.

No matter which type is employed, they all have one purpose – to reduce the stresses in the structure's columns to stresses which are more acceptable to the soil underneath. In other words, a column of, say,

300 mm square stressed to 10 N/mm^2 could not stand directly on the soil if it can only support a stress of 0.4 N/mm^2. To support the same load, the soil requires an area of 1500 mm square, and thus a base of at least this size must be provided. But in so doing, this will set up large bending and shearing forces in the base which must be allowed for by introducing adequate reinforcement. To do this the following assumptions are made:

1. The soil behaves elastically and the base acts as a rigid body. Thus if stress in the soil varies to accommodate moment or eccentrically applied loading, it does so linearly. Where it does vary it is some-times sensible to limit the ratio of maximum to minimum stress to about 3 or 4 since particularly in soft cohesive soils the difference in stress will result in tilting of the base due to differential settlement.

2. The critical sections are:
   Bending –   at the face of the column
   Flexural shear – at a vertical section 1.5 times the effective depth from the face of the column
   Punching shear – around the periphery of a line 1.5 times the total depth from the faces of the column as was the case for slabs as considered in Chapter 3.

Thus the first problem must be to determine the stresses induced in the soil by the loads and moments placed on the base or strip.

Stresses under foundations

I. Pad footings

Consider the pad footing illustrated in Fig. 5.1, in which it is loaded by an eccentric load W acting at a distance from the centre of the section. Note that there is no eccentricity about the y–y axis; this will be con-sidered later.

The eccentricity of loading causes a moment about the x–x axis equal to $W \cdot e$, thus the stresses f_1 and f_2 are given by:

$$f_1 = \frac{W}{BL} + \frac{We}{Z} \qquad\qquad [5.1(a)]$$

$$f_2 = \frac{W}{BL} - \frac{We}{Z} \qquad\qquad [5.1(b)]$$

$$Z = BL^2/6.$$

Thus as the value of e is increased the difference between f_1 and f_2 becomes greater until in the limit $f_2 = 0$. This occurs when $e = L/6$ and obviously if the load was placed on the other side of the base f_1 would equal zero when $e = L/6$ on the other side.

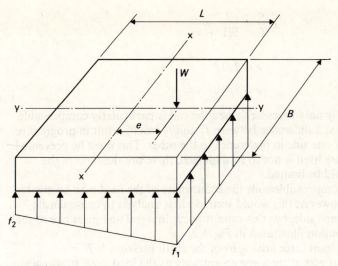

Fig. 5.1

If it is necessary to ensure that the **ratio** between f_1 and f_2 is less than a given value the following is the result:

$$f_1/f_2 = \frac{(1 + 6e/L)}{(1 - 6e/L)} = \alpha$$

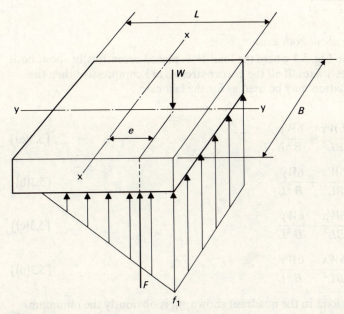

Fig. 5.2

Thus
$$\frac{e}{L} = \frac{\alpha - 1}{6(1 + \alpha)}$$

For $\alpha = 3$: $e = L/12$

For $\alpha = 4$: $e = L/10$

This is usually only necessary when the soil is particularly compressible since too great a difference between f_1 and f_2 would result in progressive settlement of one side in preference to the other. This must be prevented if the structure itself is not to be damaged, therefore the ratio of the stresses should be limited.

In less compressible soils the eccentricity of the load may be greater than $L/6$. However, this would seem on first analysis to cause tensile stress along one side, but this cannot occur in a soil and gives rise to the stress distribution illustrated in Fig. 5.2.

The resultant force arising from the earth pressure $= F$
This resultant acts at the same eccentricity as the load itself; it also acts through the centroid of the stress block, i.e. one third from the maximum stress to zero stress. Thus:

$$F = W = 0.5f_1 \times \text{stressed length of base} \times B$$
$$= 0.5f_1 \times 3(0.5L - e) \times B$$
$$f_1 = \frac{2W}{3B(0.5L - e)} \qquad [5.2]$$

Eccentricity about both axes

Consider the Fig. 5.3 where the load W is placed eccentrically about both the x and the y axes. If all the corner stresses are compressive then the bending equation may be used as for the last case.

Therefore:

$$f_A = \frac{W}{BL} - \frac{6Wx}{BL^2} + \frac{6Wy}{B^2L} \qquad [5.3(a)]$$

$$f_B = \frac{W}{BL} + \frac{6Wx}{BL^2} + \frac{6Wy}{B^2L} \qquad [5.3(b)]$$

$$f_C = \frac{W}{BL} + \frac{6Wx}{BL^2} - \frac{6Wy}{B^2L} \qquad [5.3(c)]$$

$$f_D = \frac{W}{BL} - \frac{6Wx}{BL^2} - \frac{6Wy}{B^2L} \qquad [5.3(d)]$$

For the load in the quadrant shown, f_D is obviously the minimum stress, while the maximum stress is given by f_B. Thus, if f_D is to be

99

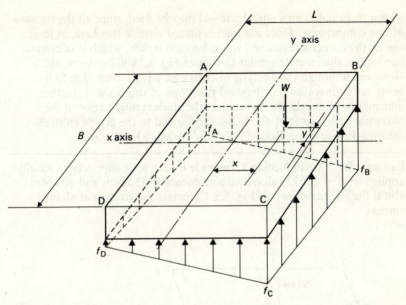

Fig. 5.3

greater than zero, then in the limit, letting $f_D = 0$

$$f_D = 0 = \frac{W}{BL}(1 - 6x/L - 6y/B)$$

Thus $\quad 6(x/L + y/B) = 1$ [5.4]

Performing this for all the quadrants results in the edge of the shaded area of Fig. 5.4 being defined. In other words, so long as the load is

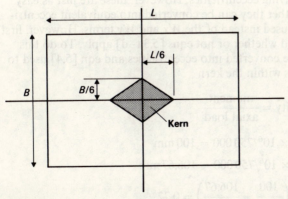

Fig. 5.4

100

within this shaded area eqns [5.3a–d] may be used, since all the stresses will be compressive. Once the load is placed outside this **kern**, at least one of the corner stresses will try to become tensile, which is of course impossible, thus a state similar to that of Fig. 5.2 will be the result. However, although the analysis of such a case is possible, due to its being rather involved, it is beyond the scope of this book. If further information is sought on this subject the student might refer to the November 1977 issue of *Civil Engineering* and to the article entitled 'Bearing Pressures on Bridge Footings' by Mike Hackman.

Example 5.1 A pad footing 1.5 m wide by 2 m long supports an axially applied load of 750 kN along with moments of 75 kN m and 80 kN m about the axes as shown in Fig. 5.5. Calculate the stresses at all four corners.

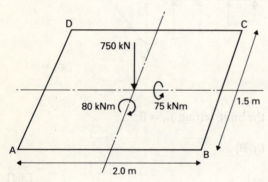

Fig. 5.5

Unlike the theory set out above, the question specifies moments about the axes instead of giving eccentricities. However, these are just as easy to deal with, since either they can be converted into equivalent eccentricities or they can be used instead of the Wx and Wy terms. However, first it must be determined whether or not eqns [5.3a–d] apply. To do this the moments must be converted into eccentricities and eqn [5.4] used to check that the load is within the kern

$$\text{Effective eccentricity} = \frac{\text{moment}}{\text{axial load}}$$

Therefore $x = 75 \times 10^6/750\,000 = 100\,\text{mm}$

$$y = 80 \times 10^6/750\,000 = 106.67\,\text{mm}$$

$$6(x/L + y/B) = 6 \times \left(\frac{100}{1500} + \frac{106.67}{2000}\right) = 0.72$$

Therefore the load is within the kern and eqns [5.3a–d] do apply

$$f_A = \frac{750\,000}{1500 \times 2000} + \frac{6 \times 75 \times 10^6}{2000 \times 2\,250\,000} - \frac{6 \times 80 \times 10^6}{1500 \times 4\,000\,000}$$

$$= 0.25 \qquad\qquad + 0.1 \qquad\qquad - 0.08$$

$$= 0.27\,\text{N/mm}^2$$

$$f_B = 0.25 + 0.1 + 0.08 = 0.43\,\text{N/mm}^2$$

$$f_C = 0.25 - 0.1 + 0.08 = 0.23\,\text{N/mm}^2$$

$$f_D = 0.25 - 0.1 - 0.08 = 0.07\,\text{N/mm}^2.$$

II. Strip footings

Consider the strip footing shown in Fig. 5.6, where it is supporting a row of columns at an equal spacing of S m. Each column imparts to the strip an equal load of P kN, and this is resisted by a uniformly distributed load along the whole of the strip footing of u kN/m. One of the inter-column 'spans' has been isolated in Fig. 5.6(b) and it is from this that the following equations are derived.

It is, of course, possible to design this foundation beam as a continuous beam with moments being experienced at the positions of the columns as well as in the centre of the span. However, as the subject of continuous beam design is left to Volume 2 of this book the simpler and conservative assumption of simply supported action is here assumed. The consequence of this assumption results in a greater area of steel being

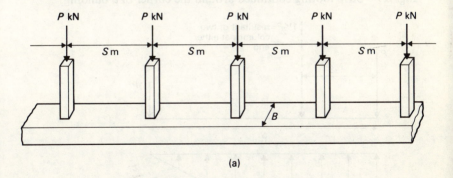

(a)

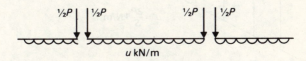

(b) Schematic force diagram

Fig. 5.6 Strip footing

used in the centre of the span than is really necessary and in the formation of large cracks on the underside of the strip underneath the columns, unless some steel is placed there. This problem will be considered later. Taking the isolated span, the total downward force from the columns is equal to P since only half the column load is taken by the span, the other half being taken by the neighbouring span.

Therefore:

$$u \times S = P \qquad [5.5]$$

And the bearing pressure under the strip

$$f_b = u/B \qquad [5.6]$$

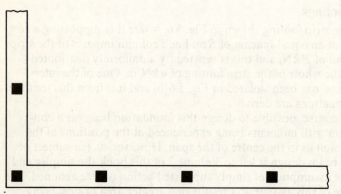

Fig. 5.7 Strip footing continued around the corner of a building

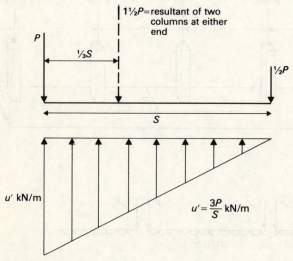

Fig. 5.8 Condition existing when the strip is not continued around the corner of the building

One might expect that this analysis would not apply to the end span of a strip footing since all of the load P must be taken by the end span. However, this problem is usually relieved by continuing the strip around the corner as shown in Fig. 5.7. Then the other half P is taken by the continued span and the above analysis applies. However, if this is not the case then the stress distribution of Fig. 5.8 will give satisfactory results.

This type of situation does not occur unless the base of the column is designed to have no moment acting, when the load from the column

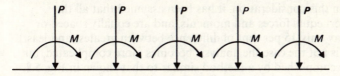

(a) Moment about axes perpendicular to the line of the columns

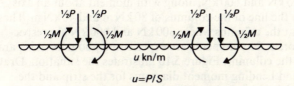

u kn/m

$u = P/S$

(b) Moments, forces and stresses induced by (a) above. Note that the stress in the soil is assumed to be the same as when only axial load acts. This is a conservative assumption

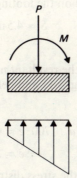

(c) Moments about an axis parallel to the line of the columns showing the resultant stresses in the soil

Fig. 5.9 Moments in strip footings

is truly axial. Moment acting at the base can be one of two sorts as demonstrated in Fig. 5.9 – it can produce moments in the strip beam which can be assumed not to alter the stress in the soil; or it can produce stresses in the soil which react to produce moments in the strip but at right angles to the span moments. If these latter moments occur, then it may prove more economic to design the foundation as a series of isolated pad footings since the width of strip necessary will approach that of the pad in any case. But for small values the strip may still prove viable.

So far in this consideration, it has been assumed that all the columns exert equal forces and moments, and are equally spaced or nearly so (say only 15 per cent of difference between greatest and least). Where this is not the case the analysis becomes more complicated, but it may be accomplished by a method similar to that shown in Fig. 5.8, so long as the difference is not too great.

Example 5.2 A strip foundation is required to support a line of columns at an equal spacing of 4.5 m. The columns carry alternate axial loads of 800 kN and 600 kN, along with moment about an axis perpendicular to the line of the columns of 80 kN m and 40 kN m. These moments occur at the columns where 800 kN and 600 kN are respectively the axial forces. No moments are experienced about an axis parallel to the line of the columns. Figure 5.10 illustrates the situation. Draw the shear force and bending moment diagrams for the strip and the stress distribution in the soil. Also find the width of strip needed to keep the stresses in the soil less than $0.2 \, \text{N/mm}^2$ $(200 \, \text{kN/m}^2)$.

Stage I: Calculate the stresses in the soil.

To do this the moments can be neglected as they are assumed not to affect the stresses in the soil.

First, the sum of the soil stresses must be equal to the sum of the vertical loads placed on the footing.

$$0.5(u_1 + u_2) \cdot S = 700 \, \text{kN} \qquad S = 4.5 \, \text{m}$$

$$u_1 + u_2 = 311.11 \, \text{kN/m} \qquad\qquad\qquad\qquad [a]$$

Second, the position of the resultant may be found in two ways as illustrated in Fig. 5.11 and below.

From the forces

Taking moments about u_1 $\quad 700 \cdot e = 400 \times 4.5$

$$e = 2.571 \, \text{m}$$

Thus the resultant of the stress distribution must act at 2.571 m from u_1. Referring to Fig. 5.11:

$$e = \frac{u_1 + 2u_2}{3(u_1 + u_2)} \cdot S = (u_1 + 2u_2) \times \frac{4.5}{3 \times 311.11} = 2.571$$

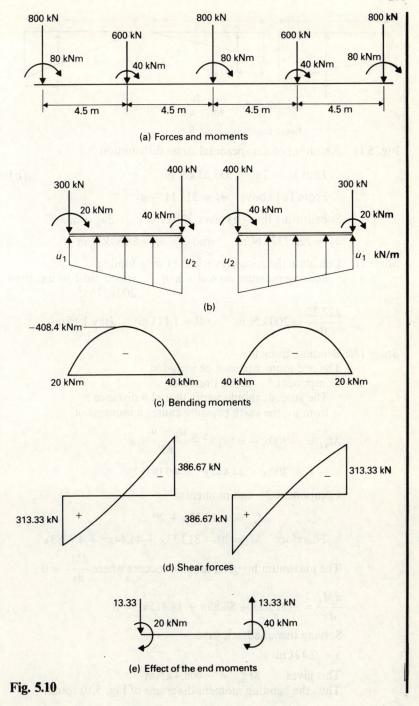

(a) Forces and moments

(b)

(c) Bending moments

(d) Shear forces

(e) Effect of the end moments

Fig. 5.10

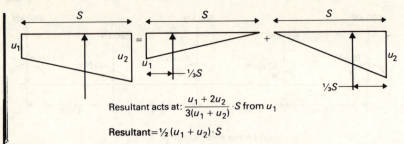

Resultant acts at: $\dfrac{u_1 + 2u_2}{3(u_1 + u_2)} \cdot S$ from u_1

Resultant $= \frac{1}{2}(u_1 + u_2) \cdot S$

Fig. 5.11   Resultant of a trapezoidal stress distribution

Thus $u_1 + 2u_2 = 533.33\,\text{kN/m}$ [b]

From [a] above $u_1 = 311.11 - u_2$

Substituting into [b] gives $311.11 - u_2 + 2u_2 = 533.33$

$u_2 = 222.222\,\text{kN/m}$ and $u_1 = 88.89\,\text{kN/m}$

Stage II: Calculate the necessary width of strip footing.
Bearing pressure on soil $= u/B$ which must be less than $200\,\text{kN/m}^2$

$\dfrac{222.22}{B} = 200\,\text{kN/m}^2$ $B = 1.111\,\text{m}$ (say $1.15\,\text{m}$).

Stage III: Bending moments.
The end moments *must* be included.
Component 1 – Earth Pressure:
The student should verify that at a distance x
from u_1, the earth pressure causes a moment of

$$M_{x1} = -300x + 0.5u_1 x^2 + \frac{u_2 - u_1}{6S} \cdot x^3$$

$$= -300x + 44.444x^2 + 4.9383x^3.$$

Component 2 – end moments:

$$M_{x2} = -13.33x + 20$$

Therefore $M_x = 20 - 313.33x + 44.44x^2 + 4.9383x^3$

The maximum hogging moment occurs where $\dfrac{dM_x}{dx} = 0$

$$\frac{dM_x}{dx} = -313.33 + 88.88x + 14.815x^2 = 0$$

Solving this quadratic gives

$x = 2.491\,\text{m}$

This gives $M_{max} = -408.4\,\text{kNm}$
Thus the bending moment diagrams of Fig. 5.10 result.

Stage IV: Shear forces.
Again the end moments must be included.
The shear force is related to the bending moment by the familiar equation:

$$V = -\frac{dM_x}{dx}$$

Thus $V = 313.33 - 88.88x - 14.815x^2$
This gives the shear force diagrams in Fig. 5.10(d)
The maximum values are: 313.33 kN and 386.67 kN.

Design of foundation reinforcement

I. Pad footings

The tendency of a pad footing is to act as a slab. In other words the design must cater for bending in two directions, shear and punching shear. Also of major importance is anchorage bond where (as is often the case) the pad is cast first and the column cast directly on top of it. In order to secure the column to the pad it is necessary to allow some of the column steel to be cast into the pad before the column is cast. This is illustrated in Fig. 5.12 and the bars are usually referred to as starter bars, since they start the column steel. This form of construction means that the main column steel must be lapped with the starter bars if a proper moment connection is to be formed. However, this may have the effect of placing the compression reinforcement at the column/pad interface in full stress, therefore the depth of the pad must be such as to allow full anchorage in compression.

(a) Bending in the base

In order to simplify the following discussion, only uni-axial bending of the pad will be considered. The same general rules will still apply to a column exerting bi-axial bending on the pad and their design may be

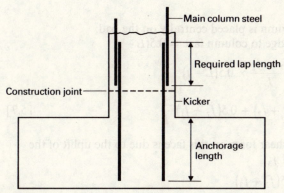

Fig. 5.12

inferred from the theory here developed for moment about only one axis.

Consider the pad footing of Fig. 5.13. At first thought it might be imagined that as the position of the maximum moment can readily be found, the design should be carried out at that position. However, this will always occur underneath the column, therefore the design is carried out at the face of the column on the most heavily loaded side. Firstly, the stress f_3 must be found from f_1 and f_2, the edges stresses.

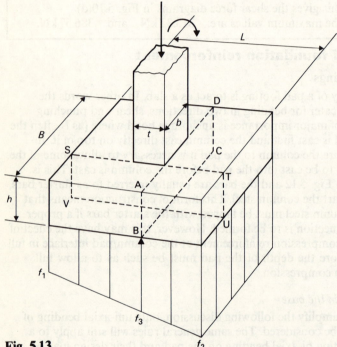

Fig. 5.13

Assuming that the column is placed centrally on the pad:

Distance from the edge to column face $= 0.5(L - t)$

Hence $f_3 = f_2 - \dfrac{f_2 - f_1}{L} \times 0.5(L - t)$

$$f_3 = 0.5(f_1 + f_2) + 0.5(f_2 - f_1) \cdot \frac{t}{L} \qquad [5.7]$$

Section ABCD The shear force on this face is due to the uplift of the earth pressures f_2 and f_3:

Average stress $= 0.5(f_2 + f_3)$

Shear force $= V_L = 0.5(f_2 + f_3) \times B \times 0.5(L - t) \qquad [5.8]$

Distance of the centroid of the stress block from the face is given by:

$$C = \frac{f_3 + 2f_2}{3(f_3 + f_2)} \times 0.5(L - t) \qquad\qquad [5.9]$$

(cf. Fig. 5.11)

Moment at ABCD $= M_L = V_L \times C$

$$= \frac{B}{24}(f_3 + 2f_2)(L - t)^2 \qquad\qquad [5.10]$$

Section STUV

Average stress $= 0.5(f_1 + f_2)$

Distance from edge to column face $= 0.5(B - b)$

$$V_B = 0.5(f_1 + f_2) \times L \times 0.5(B - b) \qquad\qquad [5.11]$$

Distance to centroid $= 0.25(B - b)$

$$M_B = \frac{L}{16}(f_1 + f_2)(B - b)^2 \qquad\qquad [5.12]$$

Longitudinal reinforcement. The total areas of steel needed to resist M_L and M_B may now be found by using the methods set out in Chapter 2. However, the way in which this steel is distributed depends on the ratio B/L. If $B < L$ then section ABCD will act as a beam and the steel is distributed evenly across it. The remote edges of the other section STUV will be more lightly stressed than might be expected, thus the majority of the steel must be placed in the central region. Alternately, if $L < B$, STUV will act as the beam while the edges of ABCD will be more lightly stressed. Thus Cl. 3.10.4.1 of CP 110 defines a central band along the

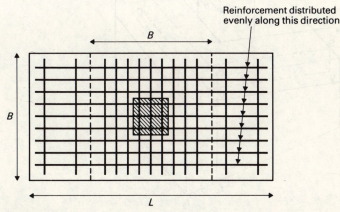

Fig. 5.14

long sides of the base in which most of the reinforcement for the short span is placed.

3.10.4.1 *Resistance to bending. Bases should be considered as beams or solid slabs as appropriate. In general, for pad footings, the reinforcement area thus determined should be distributed evenly across the section considered. Any reinforcement required to resist the moment across the short span of a rectangular base, however, should be placed as follows:*

$\dfrac{2}{\beta_1 + 1}$ *times total area of reinforcement is spread over a band centred on the column or support and of width equal to the short side dimension of the base.*

The remainder is spread evenly over the outer parts of the section.
In the above β_1 is the ratio of the longer to the shorter side.

(b) Shear

Flexural shear in the base is not considered at either sections ABCD or STUV in Fig. 5.13 since no shear crack is ever likely to form there. Instead, as shown in Fig. 5.15, a parallel section 1.5 times the effective depth from the face of the column is considered. However, unlike a slab or a beam, due to the practical difficulties involved, shear reinforcement is almost never used. Therefore, the shear stress calculated must not exceed v_c from Table 5 of CP 110.

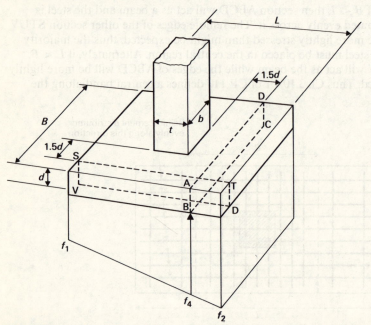

Fig. 5.15

Consider Fig. 5.15.

Section ABCD

$$f_4 = 0.5(f_1 + f_2) + 0.5(f_2 - f_1)(t + 3d)/L \qquad [5.13]$$
$$V = 0.25B(f_2 + f_4)(L - t - 3d) \qquad [5.14]$$
$$v = V/bd = 0.25(f_2 + f_4)(L - t - 3d)/d \qquad [5.15]$$

Section STUV

$$V = 0.5(f_1 + f_2) \times L \times 0.5(B - b - 3d) \qquad [5.16]$$
$$v = 0.25(f_1 + f_2)(B - b - 3d)/d \qquad [5.17]$$

(c) Punching shear

As with slabs, there is a danger that the column will punch through the
base to the earth beneath. The effective perimeter which is considered to
resist this action is, as before, taken to be 1.5 times the overall depth out
from the faces of the column.

$$\text{Perimeter} = 2(b + t) + 3\pi h \qquad [5.18]$$

Area effective in resistance = pd (note the use of the effective depth)

$$= 2(b + t)d + 3\pi dh$$

$$v = \frac{\text{column load} - \text{the earth pressure on the enclosed area}}{pd}$$

The base is then checked as for slabs.

(d) Local bond stresses

The critical sections for local bond are those for which the moment
design is carried out, i.e. ABCD and STUV (Fig. 5.13). Therefore, V_L and

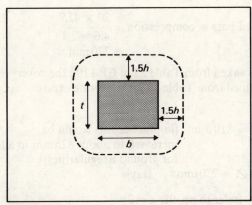

Fig. 5.16

V_B are the shear forces used in the local bond equation:

$$f_{bs} = \frac{V}{\sum u_s d}$$

f_{bs} must be less than the values found from Table 21 of CP 110.

Thus the complete design procedure for pad footings may be summarised as follows:

1. Choose the depth of the base on the basis of the anchorage length needed for the main steel of the column. To this is added the necessary cover to the steel, which is found in Table 19 of CP 110 under the moderate category.
2. Determine the base dimensions knowing column loads and moments and from the soil pressures allowed. A quick approximation may be made if the effective eccentricities of the column load are kept within the kern.
3. Calculate the actual values of earth pressure.
4. Calculate the bending moments and from them the areas of steel required. Then distribute the steel in accordance with Cl. 3.10.4.1.
5. Check the base for both types of shear and for local bond.

Example 5.3 A column 275 mm wide × 775 mm deep carries a 1500 kN axial load and a major axis moment of 400 kNm. When concrete of $f_{cu} = 25\,N/mm^2$ and steel of $f_y = 425\,N/mm^2$ is used, eight 20-mm $\varnothing$ bars are needed. The pad footing rests on clay which has a bearing capacity at the depth of the footing of $0.5\,N/mm^2$. Design the footing.

Stage I: Depth of the base.

The anchorage length $= \dfrac{\varnothing \cdot f_y}{4.6 f_{ba}}$

for 20 mm deformed bars in compression $= \dfrac{20 \times 425}{4.6 \times 2.4}$

$= 770\,mm$

(The bond stress is taken from Table 22 of CP 110); the cover to the steel is obtained from Table 19 for moderate exposure $= 40\,mm$

Overall depth $= h = 810\,mm$ (in practice this would be increased to about 825 mm to allow for ground irregularities)

$d = 770 - 0.5\,bar\;\varnothing = 750\,mm$ (say).

Stage II: Length of base   (refer to pp. 98; $\alpha = 3$).

It is advisable on clay to restrict f_2/f_1 to 3 or 4

Table 19 Nominal cover to reinforcement

Condition of exposure	Nominal cover Concrete grade				
	20	25	30	40	50 and over
	mm	mm	mm	mm	mm
Mild: e.g. completely protected against weather, or aggressive conditions, except for brief period of exposure to normal weather conditions during construction	25	20	15	15	15
Moderate: e.g. sheltered from severe rain and against freezing whilst saturated with water. Buried concrete and concrete continuously under water	—	(40)	30	25	20
Severe: e.g. exposed to driving rain, alternate wetting and drying and to freezing whilst wet. Subject to heavy condensation or corrosive fumes	—	50	40	30	25
Very severe: e.g. exposed to sea water or moorland water and with abrasion	—	—	—	60	50
Subject to salt used for de-icing	—	—	50*	40*	25

* Only applicable if the concrete has entrained air (*see* **6.3.6**).

Thus e must not exceed $L/12$

$$e = \frac{M}{N} = \frac{400 \times 10^6}{1500 \times 10^3} = 266.67 \text{ mm}$$

$L > = 3200$ mm.

Stage III: Width of base.

$$f_2 = \frac{N}{A} + \frac{M}{Z} < = 0.5 \text{ N/mm}^2$$

$$\frac{N}{BL} + \frac{6M}{BL^2} < = 0.5$$

$$B > = \frac{3000 \times 10^3}{3200} + \frac{12 \times 400 \times 10^6}{3200^2} = 1406.25 \text{ mm}$$

$$\boxed{\text{let } B = 1450 \text{ mm} \quad \text{and} \quad L = 3200 \text{ mm}}$$

$$f_1 = \frac{1.5 \times 10^6}{1450 \times 3200} - \frac{6 \times 4 \times 10^8}{1450 \times 3200^2} = 0.162 \text{ N/mm}^2$$

$$f_2 = 0.485 \text{ N/mm}^2$$

$$f_2/f_1 = 3.$$

Stage IV: Bending moments and Flexural Shear Forces.
Section ABCD (Fig. 5.13)   Distance from edge to column
face $= 1212.5 \text{ mm}$

$$f_3 = 0.162 + \frac{1987.7}{3200} \times (0.485 - 0.162)$$

$$= 0.362 \text{ N/mm}^2$$

$$V_L = 0.5(0.362 + 0.485) \times 1212.5 \times 1450 = 744.86 \text{ kN}$$

$$\text{Centroid distance} = \frac{0.362 + 2 \times 0.485}{3(0.162 + 0.485)} \times 1212.5 = 832.65 \text{ mm}$$

$$M_L = 620.2 \text{ kNm}$$

Therefore $A_s = 2319.28 \text{ mm}^2 \qquad x = 58 \text{ mm};$
$$f_{yd2} = 369.565 \text{ N/mm}^2$$

$$= 1600 \text{ mm}^2/\text{m}$$

Use 16-mm bars at 125 mm spacing $= 1610 \text{ mm}^2/\text{m}$

$100 A_s/bd = 0.215$

Section STUV (Fig. 5.13)
Average stress $= 0.5(0.162 + 0.485) = 0.3235 \text{ N/mm}^2$.

$$V_B = 0.3235 \times 3200 \times 0.5(1450 - 275) = 608.18 \text{ kN}$$

$$M_B = 608.18 \times 0.294 = 178.65 \text{ kNm}$$

Effective depth $= 750 - 16 = 734 \text{ mm}$

$$A_s = 661.65 \text{ mm}^2$$

Refer to Cl. 3.10.4.1 $\beta_1 = \dfrac{3200}{1450} = 2.207$

$$\frac{1}{1 + \beta_1} = 0.624$$

Hence 412.87 mm^2 of the steel to be placed in central strip 1450 mm wide

248.78 mm^2 of the steel to be placed in outer strips.

However, both these areas are less than the minimum area of steel required by Cl. 3.11.4.1 of CP 110.

$A_s = 0.0015 \times 3200 \times 750 = 3600$ mm^2 evenly distributed

Use 16-mm bars at 175 mm spacing

Section ABCD (shear) (Fig. 5.15)
Distance from ABCD to edge $= 1212.5 - 1.5 \times 750$.
$$= 87.5 \text{ mm}$$

$f_4 = 0.476$ N/mm^2

$V = 0.5(0.476 + 0.485) \times 87.5 \times 1450 = 60.97$ kN

$v = \dfrac{60\,974}{1450 \times 750} = 0.056$ N/mm^2 which is much less than the allowed v_c

Section STUV (Fig. 5.15) Distance from column to edge
$$= 587.5 \text{ mm}$$
This is less than $1.5d$, therefore shear is no problem in the transverse direction.

Stage V: Punching shear.
The effective perimeter lies outside the base for part of its length, therefore punching shear is no problem to this base.

Stage VI: Local bond stresses.

Section ABCD (Fig. 5.13)

Using V_L: $f_{bs} = \dfrac{744\,862}{12 \times \pi \times 16 \times 750}$

$$= 1.65 \text{ N/mm}^2$$

(There are twelve bars across the section)

From Table 21 of CP 110   allowed $f_{bs} = 2.5$ N/mm^2

Section STUV (Fig. 5.13)

$f_{bs} = \dfrac{608\,180}{19 \times \pi \times 16 \times 734} = 0.87$ N/mm^2

The base is safe in local bond

The design is now complete.

Finally in this section concerning pad footings, consider the volume of concrete used in the construction of the base in the last example – some 3.48 m³. Even if the column were 10 m high the volume of concrete in the column itself would only be 2.13 m³. Therefore, the base might justly be considered over-large. However, the dimensions B and L cannot be changed without increasing the soil pressures beyond their allowed limits. Figure 5.17 illustrates one way in which the depth of the slab may be reduced, thus saving a major proportion of the volume of concrete. The column base is stepped by constructing a plinth under the column. The consequence of this plinth is that the column steel may be securely anchored with effect from the surface of the plinth and not the surface of the base. Therefore the depth of the base is no longer directly influenced by the necessary anchorage.

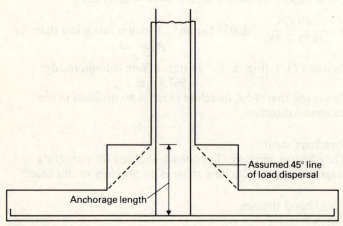

Assumed 45° line of load dispersal

Anchorage length

Fig. 5.17

The plinth should be so proportioned that the assumed 45° load dispersal line is within it. Therefore the height of the plinth must be > = to the outstand from the column faces. If the base in the last example had been constructed in this fashion, the situation shown in Fig. 5.18 would arise and the following would be the result:

(Letting height of plinth = 500 mm)

Overall pad depth = 310 mm $\qquad$ t = 1775 mm

$d = 250$ mm $\qquad$ b = 1275 mm

Section ABCD (Fig. 5.13) $\qquad$ $f_3 = 0.413$ N/mm²

$M_L = 169.68$ kNm $\qquad$ $A_s = 2023$ mm²

Percentage of steel = 0.558.

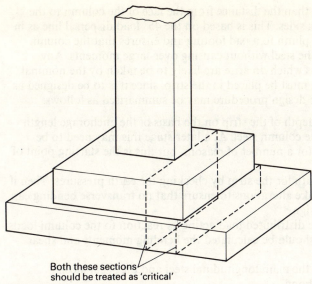

Both these sections
should be treated as 'critical'

Fig. 5.18

The area of steel at this section is, in fact, less than that derived for the section at the face of the column. This means that the critical section is at the column face; therefore, in general both sections must be checked. This principle applies to M_B and to the shear considerations as well.

The other sections are also less critical, therefore the design as created is satisfactory using the same amount of steel.

II. Strip footings

Whereas the main tendency of a pad footing is to act as a slab, a strip footing will have the characteristics of a beam, though a wide strip may also have some tendency to slab action. The consideration of moments and shear forces can therefore be more complicated and requires a further simplification of loading at this level. Loads on this type of foundation are in any case usually predominantly vertical and this is the only load case which is considered in this book.

Referring again to Fig. 5.6(b), the maximum moment induced, assuming that the strip is simply supported between the columns is given by:

$$M = uS^2/8 = PS/8 \qquad [5.19]$$

Maximum shear force $\qquad V = 0.5P \qquad [5.20]$

Bearing pressure $\qquad f = P/SB \qquad [5.21]$

These three equations are the basis of strip footing design when they carry only vertical loads. The only problem lies in ensuring that the width of the strip is not so large as to make transverse bending a problem. A simple approach to this is to arrange that the depth of the footing is no less than the distance from the face of the column to the edge of the strip's sides. This is based on the 45° load dispersal line as in the addition of a plinth to a pad footing and ensures that the column load is taken to the steel without causing over-large moments. Any bending moments which do arise are likely to be taken by the nominal shear steel which must be placed in the strip, since it is to be designed as a beam. Thus the design procedure may be summarised as follows:

1. Choose the depth of the strip on the basis of the anchorage length needed by the column steel. At a later stage this may need to be made larger for a number of reasons, but this is the starting point of the design.
2. Choose a width for the strip by checking the earth pressures. Then if necessary make alterations to ensure that no transverse bending can affect the design.
3. Calculate the distributed load acting as reaction to the column loads. From these should be calculated the bending moments and shear forces.
4. Design both the main longitudinal steel and the shear steel.
5. Check local bond.

Example 5.4 A series of short braced, axially loaded columns, spaced 4.5 m apart carry loads of 300 kN each. The foundation to be used is a strip. Design both the column and the strip footing. $f_{cu} = 30 \, \text{N/mm}^2$ $f_y = 460 \, \text{N/mm}^2$; allowed bearing pressure = 0.5 N/mm^2

Stage I: Design the column.
Try a section 75×75 mm

$$N = 0.4 f_{cu} A_c + 0.67 A_{sc} f_y$$

$$3 \times 10^5 = 0.4 \times 30 \times 75^2 + 0.67 \times A_{sc} \times 460$$

$$A_{sc'} = 754 \, \text{mm}^2 \text{ four bars, each} = 188.6 \, \text{mm}^2$$

Use four 16-mm $\emptyset$ bars.

Stage II: Depth of strip.

$$\text{Anchorage length required} = \frac{16 \times 460}{4.6 \times 2.7} = 592.6 \, \text{mm (say 600)}$$

$(f_{bs} = 2.7 \, \text{N/mm}^2$ from Table 22 of CP 110)

Cover = 30 mm from Table 19 $h = 630$ mm; $d = 590$ mm.

Stage III: Width of section.

$$f = 0.5 > = P/BS = \frac{300\,000}{B \times 4500}$$

$$B > = 133.33 \, \text{mm}$$

Let $B = 150$ mm.

Stage IV: Check that transverse bending is minimal.

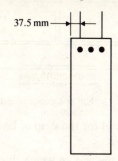

37.5 mm

Satisfactory.

Stage V: Bending and shear.

$M = PS/8 = 300 \times 4.5/8 = 168.75\,\text{kNm}$

$\dfrac{M}{bd^2} = 3.23$ using Chart 4 of CP 110 part 2

$100\,A_s/bd = 0.937$ $A_s = 829.3\,\text{mm}^2$

Use two 20 and one 16-mm $\varnothing$ bars $= 829.4\,\text{mm}^2$

$V = 150\,\text{kN}$ $v = \dfrac{150\,000}{150 \times 590} = 1.695\,\text{N/mm}^2$

$v_c = 0.681\,\text{N/mm}^2$

Thus shear steel is necessary

$\dfrac{A_{sv}}{S_v} > = \dfrac{b(v - v_c)}{0.87 f_{yv}}$ $f_{yv} = 425\,\text{max}$

$> = 0.41$

Note that the presence of links requires that there be at least two bars of compression steel to anchor the links in the compression zone. Therefore, these bars could be taken into account in the bending design if it is so desired.

Use 6 mm links and two 6-mm $\varnothing$ compression bars

$S_v < = \dfrac{18\pi}{0.41} = 138\,\text{mm}$

Hence use 6-mm links at 125-mm spacing.

Stage VI: Local bond.

$f_{bs} = \dfrac{150\,000}{590 \times 175.93} = 1.445\,\text{N/mm}^2$

Allowed $f_{bs} = 2.8\,\text{N/mm}^2$

Therefore the design is satisfactory.

120

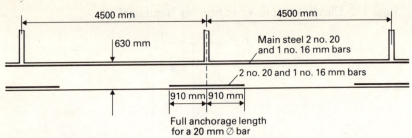

Fig. 5.19 Steel arrangement for the strip of Example 5.4

Figure 5.19 illustrates the way in which this steel is placed in the foundation. Note that underneath the columns some steel has been placed at the bottom of the section. This is done to prevent the occurrence of the large cracks mentioned on pp. 102. Usually the amount of steel placed in this position is simply the same as is used for the middle of each span, since this is conservative and involves the minimum trouble in its calculation. The steel should be continued on each side of the column for its full anchorage length.

Questions

1. Calculate the values of earth pressure at the corners of the pad footings shown in Fig. 5.20.
2. Calculate the values of distributed load under the columns 1–7 shown in Fig. 5.21 for the following loadings and spans. Determine the minimum strip width if the maximum bearing pressure allowed is $0.4 \, \text{N/mm}^2$.
   (a) all loads equal to 500 kN and span $L = 3 \, \text{m}$
   (b) $W_1 = W_3 = W_5 = W_7 = 500 \, \text{kN}; W_2 = W_4 = W_6 = 375 \, \text{kN}$
   (c) $W_1 = W_4 = W_7 = 500 \, \text{kN}; W_2 = W_3 = W_5 = W_6 = 375 \, \text{kN}$ for (b) and (c) $L = 2.5 \, \text{m}$.
3. A base $3200 \times 3200 \, \text{mm}$ is designed to support a column $350 \times 750 \, \text{mm}$ carrying an axial load of 1600 kN and a major axis moment of 425 kNm. If $f_{cu} = 30 \, \text{N/mm}^2$ and $f_y = 250 \, \text{N/mm}^2$, the maximum allowed bearing stress is $0.24 \, \text{N/mm}^2$ and the ratio of maximum to minimum stress is to be less than 3, design the steel needed. If any of the shear or bond stresses are greater than allowed without shear steel suggest how the addition of shear steel may be avoided. Assume that the column is placed centrally upon the base. The column steel consists of eight 16-mm $\varnothing$ bars.
4. Redesign the base of Question 3 as shown in Fig. 5.22.
5. A column containing six 32-mm $\varnothing$ bars $(f_y = 410 \, \text{N/mm}^2)$, of dimensions $400 \times 900 \, \text{mm}$ and carrying a load of 1800 kN and a major axis moment of 1100 kNm is supported centrally on a base 2 m wide by 3 m long. Calculate the depth of base required and the earth pressures. Thus design the base if $f_{cu} = 40 \, \text{N/mm}^2$ and $f_y = 250 \, \text{N/mm}^2$.

6. Design a strip foundation to support a row of short braced axially loaded columns 5 m apart and carrying 250 kN each. The columns are 400 mm square containing four 8-mm $\emptyset$ bars of $f_y = 410\,\text{N/mm}^2$. Use $f_{cu} = 40\,\text{N/mm}^2$ and $f_y = 410\,\text{N/mm}^2$, assuming the allowed bearing stress is $0.15\,\text{N/mm}^2$. (*Hint:* before choosing an effective depth read the last sentence of Cl. 3.11.4.3 in CP 110.)

7. Design a strip foundation to support a row of columns 6 m apart, carrying 475 kN each. The column section is 150 mm square,

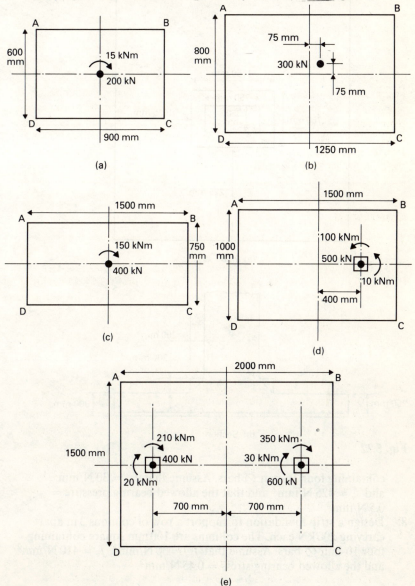

Fig. 5.20

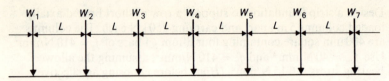

Fig. 5.21

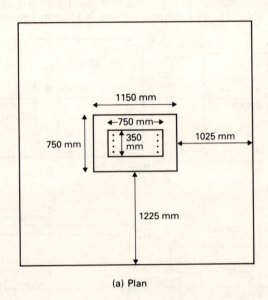

(a) Plan

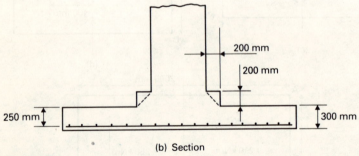

(b) Section

Fig. 5.22

containing four 16-mm $\emptyset$ bars. Assume that $f_{cu} = 30\,\text{N/mm}^2$, and $f_y = 425\,\text{N/mm}^2$ and that the allowed bearing pressure = $0.5\,\text{N/mm}^2$.

8. Design a strip foundation to support a row of columns 3 m apart carrying 200 kN each. The columns are 100 mm square containing four 10-mm $\emptyset$ bars. Assume that $f_{cu} = 30\,\text{N/mm}^2$, $f_y = 410\,\text{N/mm}^2$ and the allowed bearing stress = $0.45\,\text{N/mm}^2$.

Chapter 6

Cantilever retaining walls

In the last chapter, the weakness of soil in compression was demonstrated; its weakness in shear and tension means that a free standing, vertical face is at best dangerous and at worse impossible to construct. Unless such a face is supported, it will collapse to form a slope, generally at an angle of between 30° and 60°, depending on the type of soil and its moisture content. In many engineering situations, however, a vertical face is necessary, thus some means of support must be designed. When pipes are being laid, a trench must be excavated, the vertical sides of which are usually supported by means of temporary shoring – one face being used to support the other by the introduction of struts pressed between them. When a road is being excavated into a hillside, it is often more economical to have a vertical face than to allow the natural angle of slope, with the extra excavation which would be required. Sometimes, as in a railway cutting or in the case of some urban motorways, a carriageway must be sunk into the ground, leaving two vertical sides which must be supported. One of the many solutions to these latter situations is the cantilevered retaining wall as illustrated in Fig. 6.1. The vertical face of earth is supported by the reinforced concrete stem which experiences both bending and shear. This stem is then supported by the base which is designed in a fashion similar to that outlined in the previous chapter.

Consider first the main ways in which these walls may fail, as illustrated in Fig. 6.2.

1. *Bending or shear failure in the stem*
   The stem does not usually contain shear reinforcement, thus the shear stress, as with slabs, must not exceed v_c, and it is checked at

124

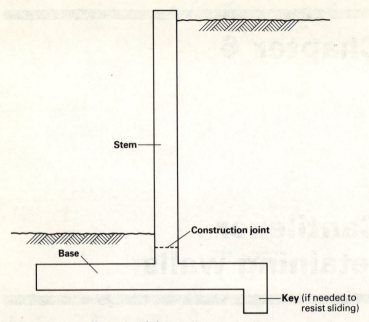

Fig. 6.1 Cantilever retaining wall

a section 1.5 times the effective depth of the stem up from the surface of the base. The stem is usually reinforced in tension only, as otherwise the depth of the base would have to be dictated by the anchorage requirements of the compression steel. Thus the thickness of the stem must be great enough to allow it to be singly reinforced.

2. *Bending or shear failure in the base*

   This may occur either in front or behind the stem. In front the bottom steel is in tension, while the top steel to the rear of the stem is the tension steel. Compression steel may be used without restriction, so long as anchorage requirements are met. As for the stem, the shear stress should in general be less than v_c, and it is checked at sections 1.5 times the effective depth of the base from the faces of the stem.

3. *Earth bearing and overturning failures*

   The base dimensions must be such as to limit the bearing stresses under the base to acceptable levels. However, this is not a sufficient requirement as there is the possibility of the wall being toppled if it is not heavy enough. Therefore both earth pressures *and* overturning failure must be prevented.

4. *Sliding failure*

   The coefficient of friction beneath a wall is usually between 0.35 and 0.6. If the pressure from the retained material is too great, the wall

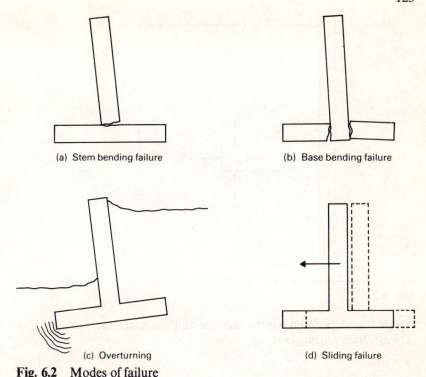

(a) Stem bending failure

(b) Base bending failure

(c) Overturning

(d) Sliding failure

Fig. 6.2 Modes of failure

will slide, thus either the wall must be heavy enough to provide adequate frictional resistance, or a key must be used.

Note: In Fig. 6.1 it is shown that the base of the wall is not usually visible and is buried beneath the earth in front of the stem. This earth is obviously helping to resist the forces behind the wall, but – **its effect must *always* be neglected**. Unfortunately, the soil in front of the wall is admirably suitable for the concealment of sundry services such as electricity cables, gas pipes, telephone ducting, water and even sewerage pipes. In order to place these, trenches must be dug and these destroy the resistance of this soil to the movement of the wall, thus it must always be neglected. Hence it is never economical to bury the base too deep – usually 250 – 500 mm will be sufficient.

Lateral earth pressure

Consider the small, cubic element a depth, h, below the free surface of a retained material (liquid or earth) as shown in Fig. 6.3. Consider first the two horizontal faces; there will be a pressure on these faces due to the weight of material above the element. If w is the material ultimate unit

126

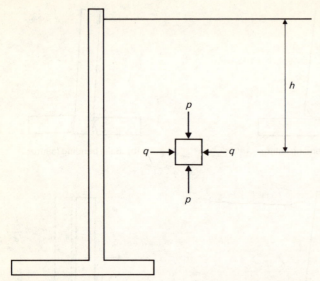

Fig. 6.3

weight* in kN/m^3 and p is the pressure on these horizontal faces in kN/m^2, then, h being in m,

$$p = wh \tag{6.1}$$

If the material is a liquid then the pressure on the vertical faces will also be equal to p. If, however, the material is soil, the pressure on the vertical faces is not necessarily equal to p and may be more than or less than p. Indeed, the pressure on two faces along one axis may be different from the pressure on the two faces along the axis at right angles to the first. However, the values of the horizontal pressures, q, must lie between two extremes, the lower known as the 'Active Pressure' and the higher known as the 'Passive Pressure'. If the horizontal pressure is less than the active value then the soil will collapse by 'flowing' outwards horizontally. If the horizontal pressure is at least equal to the active value then the soil

---

* Since CP 110 is a limit-state code, all loads applied to reinforced concrete members should be factored by the partial safety factors for loads. However, these partial safety factors are not specifically meant to be applied to earth pressures. The pressure of earth on a retaining wall is caused to vary by such effects as heavy rain, swelling and ground movement. It is therefore difficult to obtain the correct values for these factors.

The term 'Ultimate Unit Weight' used in this chapter is an attempt to alleviate this problem. It may for the moment be taken to mean the normal unit weight of the soil as measured in the laboratory multiplied by the partial load factors given in CP 110, with the understanding that this is not the ideal situation.

mass will be in equilibrium and thus it is this value which is used in the design of retaining walls. If the horizontal pressure is greater than the passive value then the vertical pressure p will not be enough to prevent the soil mass from heaving upwards. This is the value of pressure which is experienced when something is pushed against a soil face until the soil gives way. Thus this is the pressure which is used to resist frictional movement when a key is used, since the leading face of the key is being pushed into the soil. However, some designers (and the author is one of them) prefer to use half of the passive value since in order to fully develop it the wall would have to move forward quite some distance, whereas the development of active pressure needs only a very small movement. However, this will be considered in more detail later.

The value of active and passive pressures and their relation to the vertical pressure p is in reality not easily discovered. They depend on the soil type, its moisture content, its degree of consolidation and many other factors. Many theories have been put forward as a means to their prediction, and the one which is used in this book has to be used with care in certain situations. It is called the Rankine Theory of Earth Pressures and relates the horizontal extreme pressures to the vertical pressure and to an 'Internal Angle of Friction', the physical interpretation of which is beyond the scope of this book. Thus, if θ is the angle of internal friction:

Active pressure $= P_a = k_a p$ [6.2]

Passive pressure $= P_p = k_p p$ [6.3]

where $\quad k_a = \dfrac{1 - \sin \theta}{1 + \sin \theta}$ [6.4]

and $\quad k_p = \dfrac{1 + \sin \theta}{1 - \sin \theta}$ [6.5]

or $\quad k_p = 1/k_a$.

Forces and moments on retaining wall stems

Figure 6.4 illustrates the way in which the horizontal pressure varies linearly or triangularly with depth. The total force which is acting is F.

$F = 0.5 k_a w h^2$ kN/m length of wall [6.6]

If h is equal to the total height of material behind the stem then this is the force causing bending in the stem and the moment is given by:

$M = Fh/3 = k_a w h^3/6$ kNm/m length [6.7]

If the height from the section at which shear is considered is used instead then eqn [6.6] gives the value of shear to be used in design. Alternatively, if h is equal to the depth to the bottom of the base (including the

128

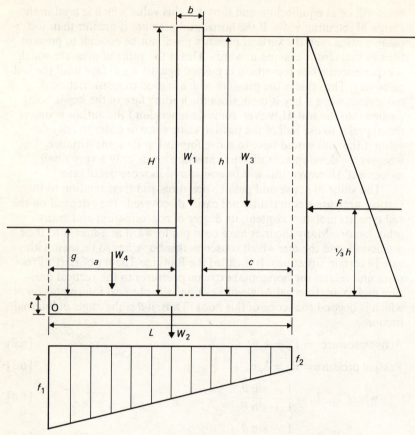

Fig. 6.4

depth of a key if it is used), then eqn [6.6] gives the value of the force to be resisted by the frictional resistance or by the key.

Earth pressures under the base

These are found in a manner similar to that of Chapter 5, thus those equations will be modified to suit. The weight of the wall must now be taken into account, but the equations obtained are not over-complicated. However, the student is well advised not to remember formulae but to remember how they were derived and design from first principles at all times unless he is writing a computer program.

Let the unit weight of reinforced concrete = w_c kN/m³

= 24 kN/m³ usually
referring to Fig. 6.4.

$W_1 = w_c bH$ kN/m length of wall $\qquad$ [6.8]

$W_2 = w_c tB$ kN/m length of wall $\qquad$ [6.9]

$W_3 = wch$ kN/m length of wall $\qquad$ [6.10]

$W_4 = wag$ kN/m length of wall

Note: W_4 is usually small enough to be neglected and will be so in this book. However, it may have been thought that since the contribution from the lateral pressure on that side was being neglected that W_4 would have been discounted automatically. This is not the case, since W_4 tends to increase the value of f_1 and thus its presence is detrimental to the stability of the wall. The lateral pressure's presence would tend to stabilise the wall, hence since it is possible that it is absent it is discounted.

Let $W = W_1 + W_2 + W_3$ kN/m $\qquad W_4$ neglected

Taking moments about the centre of the base

$$We = M - W_1 \times (a + 0.5b - 0.5L) - W_3 \times (a + b + 0.5c - 0.5L)$$
$$= M - 0.5W_1(a - c) - 0.5W_3(a + b) \qquad [6.11]$$

Thus the effective eccentricity can be found as before, if $e < = L/6$

$$\left. \begin{array}{l} f_1 = \dfrac{W}{L} + \dfrac{6We}{L^2} \text{ kN/m}^2 \\[3mm] f_2 = \dfrac{W}{L} - \dfrac{6We}{L^2} \text{ kN/m} \end{array} \right\} \qquad [6.12]$$

If $e > L/6$

Loaded length of base $= 3(0.5L - e)$

$$f_1 = \frac{2W}{3(0.5L - e)} \qquad [6.13]$$

Moments and forces on the base

Section to the front of the stem (Fig. 6.5(a))

$$f_3 = f_1 - a(f_1 - f_2)/L \qquad [6.14]$$

$$M_{BF} = a^2 (2f_1 + f_3)/6 \text{ kNm/m} \qquad [6.15]$$

$$f_4 = f_1 - \frac{a - 1.5d}{L}(f_1 - f_2) \qquad [6.16]$$

$$V_F = 0.5(f_1 + f_4)(a - 1.5d) \text{ kN/m} \qquad [6.17]$$

Section to the rear of the stem

$$f_5 = f_2 + c(f_1 - f_2)/L \qquad [6.18]$$

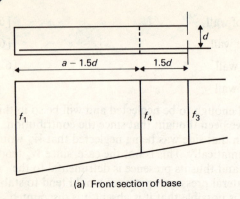

(a) Front section of base

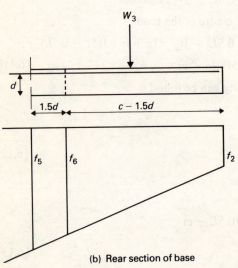

(b) Rear section of base

Fig. 6.5

$$M_{BR} = c^2(2f_2 + f_5)/6 \text{ kNm/m} \tag{6.19}$$

$$f_6 = f_2 + \frac{c - 1.5d}{L}(f_1 - f_2) \tag{6.20}$$

$$V_R = 0.5(f_2 + f_6)(c - 1.5d) \text{ kN/m} \tag{6.21}$$

Local bond stresses are also easily checked.

Overturning and sliding

Force causing sliding $= F_s = 0.5k_a w(h + t)^2$ kN/m $\tag{6.22}$

Frictional resistance $= \mu W$ $\tag{6.23}$

The Factor of Safety against sliding $= \dfrac{\text{frictional resistance}}{F_s}$

This should be greater than 1.5.
Taking moments about O in Fig. 6.4:

Overturning moment $= F_s(h + t)/3 = k_a w(h + t)^3/6$ [6.24]

Restoring moment $= W_1 \times (a + 0.5b) + 0.5W_2 L + W_3 \times (L - 0.5c)$

                                                        [6.25]

The factor of safety against overturning $= \dfrac{\text{restoring moment}}{\text{overturning moment}}$

This also should be greater than 1.5.

If the factor of safety against sliding is less than 1.5 a key should be designed to resist the extra sliding force needed to provide a factor of safety of 1.5.

If the factor of safety against overturning is less than 1.5 then the length of the base should be increased, or the stem moved forward on the base.

Example 6.1 Design a retaining wall to support a difference in level of 4.5 m, the base being buried 0.5 m below the lower level. Assume the ultimate unit weight of the retained material $= 19 \, \text{kN/m}^3$, and use concrete of $f_{cu} = 30 \, \text{N/mm}^2$ and steel of $f_y = 410 \, \text{N/mm}^2$. The allowed bearing pressure $= 0.1 \, \text{N/mm}^2$, the coefficient of friction $= 0.4$, and the angle of internal friction $= 30°$.

Stage I: Stem design.

$$k_a = \frac{1 - 0.5}{1 + 0.5} = 1/3$$

Total force on stem $= F = 0.5 k_a w h^2$ $w = 19 \, \text{kN/m}^3$

                                           $h = 4.5 + 0.5 = 5 \, \text{m}$

$$= \frac{19 \times 25}{6} = 79.167 \, \text{kN/m}$$

Moment in the stem $= M = Fh/3 = 79.167 \times 5/3$

                                      $= 131.94 \, \text{kNm/m}$

                                      $= 131.94 \times 10^3 \, \text{Nmm/mm}$

A rough value for the stem's effective depth may be obtained using eqn [1] of CP 110. This gives

$$d_{min} = \frac{M}{0.15 f_{cu}}$$ [6.26]

This formula assumes that the neutral axis depth, $x = 0.5d$ and uses the simplified concrete stress block. Note that the b found in the eqn in CP 110 is 1 m, and thus it does not appear in this eqn. In this example

$$d_{min} = \frac{131.94 \times 10^3}{0.15 \times 30} = 171.2 \text{ mm}$$

Therefore an effective depth of 200 mm would be sufficient, but this example shall use $d = 250$ mm. Question 1 at the end of this chapter involves redesigning this example using $d = 200$ mm.

$$M/bd^2 = 2.11 \qquad 100\,A_s/bd = 0.65 \qquad A_s = 1625 \text{ mm}^2/\text{m}$$

Use 25-mm ϕ bars at 300 mm spacing

Overall thickness of wall $= 250 + \dfrac{25}{2} + \text{cover} \approx 300$ mm

Local bond: $V = F = 79.167 \text{ kN/m};$ $\qquad \sum U_s = \dfrac{1000}{300} \times \pi 25$

$$= 261.8 \text{ mm}$$

$$f_{bs} = \frac{79\,167}{250 \times 261.8} = 1.21 \text{ N/mm}^2$$

Allowed $= 2.8 \text{ N/mm}^2$

Shear: checked at a section 375 mm up from the base

$$V = 0.5k_a w(h - 0.375)^2$$

$$= \frac{19 \times 4.625^2}{6} = 67.74 \text{ kN/m}$$

$$v = \frac{67\,740}{1000 \times 250} = 0.271 \text{ N/mm}^2$$

$$v_c = 0.595 \text{ N/mm}^2$$

Satisfactory.

Stage II: Base design.
This is largely a matter for trial and error since W_2 depends on the base depth and length, but these cannot be obtained until W_2 is known. Also the relative values of a and c are not known. Hence the design is a matter of experience to a great degree; however, a rough idea is obtained by placing the stem centrally and letting $W_2 = 0$. Then a suitable section may be chosen. The ultimate unit weight of concrete is taken as $1.4 \times 24 = 33.6 \text{ kN/m}^3$

Try $d = 250\,\text{mm}$; $\quad t = 300\,\text{mm}$; $\quad L = 3\,\text{m}$;
$a = 1\,\text{m}$; $\quad c = 1.75\,\text{m}$

$W_1 = 33.6 \times 0.3 \times 5 = \quad 50.40 \;\text{kN/m}$

$W_2 = 33.6 \times 0.3 \times 3 = \quad 30.24 \;\text{kN/m}$

$W_3 = \; 19 \times 1.75 \times 5 = \underline{166.25} \;\text{kN/m}$

$$W = 246.89 \;\text{kN/m}$$

$We = 131.94 - W_1 \times (0) - \dfrac{166.25}{2} \times (1.0 + 0.25)$

$\quad = 28.038\,\text{kNm/m}$

Thus $\quad e = 0.114\,\text{m} \quad$ and $e < L/6$

$$f_1 = \frac{246.89}{3} + \frac{28.038 \times 6}{9} = 100.99\,\text{kN/m}^2$$

$$f_2 = 63.60\,\text{kN/m}^2$$

(a) **Front section:** $f_3 = 101 - (101 - 63.60)/3 = 88.53\,\text{kN/m}^2$

Distance of centroid from $f_3 = \dfrac{88.53 + 2 \times 101}{3(88.53 + 101)} = 0.511\,\text{m}$

$M_{BF} = 0.511 \times 0.5(88.53 + 101) \times 1 = 48.4\,\text{kNm/m}$

$\dfrac{M_{BF}}{bd^2} = 0.775 \qquad 100 A_s / bd = 0.225 \qquad A_s = 563\,\text{mm}^2/\text{m}$

Use 10-mm $\varnothing$ bars at 125 mm spacing
or 12-mm $\varnothing$ bars at 200 mm spacing

$f_4 = 101 - \dfrac{0.625}{3}(101 - 63.60) = 93.21\,\text{kN/m}^2$

$V_F = 60.7\,\text{kN/m}$

$v = 0.243\,\text{N/mm}^2$

$v_c = 0.315\,\text{N/mm}^2$

Satisfactory

Local bond: $V = 0.5(f_1 + f_3)a = 94.765\,\text{kN/m}$

Using 10-mm bars: $\sum U_s = \dfrac{1000}{125} \times \pi 10 = 251.33\,\text{mm}$

Using 12-mm bars: $\sum U_s = 5\pi 12 = 188.5\,\text{mm}$

$f_{bs} = \dfrac{94\,765}{250 \times \Sigma U_s} = 1.51\,\text{N/mm}^2$ using 10-mm bars

$\qquad\qquad\qquad = 2.01\,\text{N/mm}^2$ using 12-mm bars

(b) **Rear section:** $f_5 = 63.6 + \dfrac{1.75}{3}(101 - 63.6)$

$$= 85.42\,\text{kN/m}^2$$

$M_{BR} = 108.52\,\text{kNm/m}$

$\dfrac{M_{BR}}{bd^2} = 1.736 \qquad 100\,A_s/bd = 0.523 \qquad A_s = 1308.6\,\text{mm}^2/\text{m}$

Use 16-mm ∅ bars at 150 mm spacing

$f_6 = 63.6 + \dfrac{1.375}{3} + (101 - 63.6) = 80.74\,\text{kN/m}^2$

$V_R = 99.23\,\text{kN/m}$

$v = 0.397\,\text{N/mm}^2$

$v_c = 0.55\,\text{N/mm}^2$

Satisfactory

Local bond: $V = 130.39\,\text{kN/m} \quad \sum U_s = \dfrac{1000}{150} \times \pi16 = 335.1\,\text{mm}$

$\qquad\qquad f_{bs} = 1.56\,\text{N/mm}^2$

Satisfactory.

Stage III: Check sliding and overturning.

Sliding force $= 19 \times 5.3^2/6 = 88.95\,\text{kN/m}$

Resisting force $= 0.4 \times 246.89 = 98.76\,\text{kN/m}$

Therefore a key must be provided

Overturning moment $= 1/6 \times 1/3 \times 19 \times 5.3^3$
$$= 157.148\,\text{kNm/m}$$

Restoring moment $= (50.4 \times 1.125) + (30.24 \times 1.5)$
$$+ (166.25 \times 2.125)$$
$$= 455.34\,\text{kNm/m}$$

Factor of safety $= 2.9$
Acceptable

Stage IV: Design of the key (refer to Fig. 6.6).
It will be noted that the key has been placed directly below the stem. There is no overriding reason for this to be the case, but sometimes the construction is made simpler by positioning the key here.

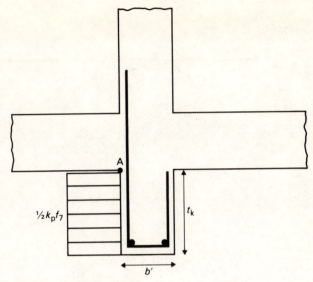

Fig. 6.6 Design of a key

Let the vertical pressure under the base at A be f_7

$$f_7 = f_1 - \frac{a}{L}\cdot(f_1 - f_2) = f_3 \quad \text{so long as the thickness of the key, b' is close to the stem thickness, } b$$

Thus the value of passive pressure used at A $= 0.5k_p f_7$. Theoretically, the passive pressure varies as the depth of the key increases, but it is conservative and simpler to assume that the passive pressure is constant over the depth of the key.

Thus, passive force effective in resisting sliding $= 0.5k_p f_7 \cdot t_k$
In this example, $f_7 = 88.53\,\text{kN/m}^2$, $\quad k_p = 3$
Force resisting sliding $= 132.8t_k + 98.76\,\text{kN/m}$
Actual sliding force required $= 1.5 \times 88.95 = 133.43\,\text{kN/m}$
Thus $132.8t_k = 34.67 \quad t_k = 0.26\,\text{m}$
Since the value of t_k is so close to the effective depth bending in the key need not be checked, so long as minimum reinforcement is used as shown in Fig. 6.6.

Section II

Structural steelwork

Chapter 7

Beams

Introduction

In contrast to the design of reinforced concrete in accordance with CP 110, the design of steelwork to BS 449: Part 2: 1969 is carried out at working loads. BS 449 was first issued as long ago as 1932, and although it has undergone a large number of changes and amendments since then, the basic philosophy of design has remained. In essence, the code is concerned, not with the collapsing structure (although the so-called 'plastic' design method is allowed but not covered), but with the working or in-service condition. Thus, the thrust of the method is slightly different; whereas when using CP 110, the aim is to ensure that the actual ultimate *capacity* is greater than the required ultimate *capacity*, with BS 449 the aim is to ensure that the section provided does not experience a stress larger than a specified allowable stress.

BS 449 is currently in the final stages of a major revision, which hopes to put steel on the same philosophical ground as that of concrete – a limit-state philosophy. However, until this is issued, and indeed for a number of years afterwards, the present design methods will be used for a large number of structures. This book therefore is confined entirely to the 'elastic' design process, although it is intended to revise it at some future date when the limit-state code has had time to gain sufficient adherents to make this worth while.

The universal beam

Unlike concrete, steel is equally strong in compression and tension, therefore all of the material incorporated in a beam is stressed, with the parts furthest from the neutral axis being more highly stressed than those in the centre. Thus most of the strength of a beam rests in these outer fibres, the central ones being more or less dead weight. For this reason, the familiar webbed sections illustrated in Fig. 7.1 have evolved to provide adequate economy with great strength. This chapter is primarily

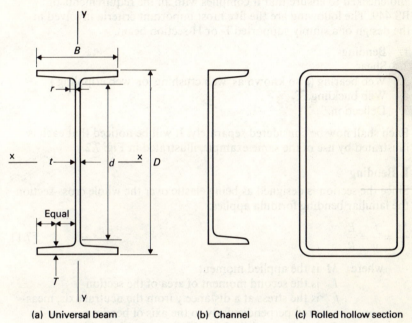

(a) Universal beam   (b) Channel   (c) Rolled hollow section

Fig. 7.1 Types of section

concerned with the section of Fig. 7.1(a) – the universal beam (UB), also referred to as the I-beam – as this is the most common form of steel beam in use today. However, a comparison will later be made with a rolled hollow section (RHS), which in certain circumstances has advantages over the I-beam. The other familiar I-shaped section is the H-column, and its design as a beam follows the same procedure as laid down for the UB.

As with reinforced concrete beams, the first step is to construct the shear force and bending moment diagrams, exactly the same procedure being followed. The definition of effective span is given in Cl. 24 and is roughly the same as given in CP 110 except that where a beam is supported by a column the span may be taken as between the points at which it is assumed that the reactions on the columns are applied. These points are usually assumed to be 100 mm from the faces of the column, thus reducing the span of the beam by perhaps as much as 200 mm over the apparent span.

Note: The loads used to construct the bending moment and shear force diagrams are the **unfactored** loads found in CP 3. This is as a consequence of the fact that BS 449 is not a limit-state code.

Once these diagrams have been drawn, a suitable section is chosen and checked to ensure that it complies with all the requirements of BS 449. The following are the five most important criteria involved in the design of a simply supported I- or H-section beam:

1. Bending.
2. Shear.
3. Web bearing (also known as 'web crushing' or 'web crippling').
4. Web buckling.
5. Deflection.

Each shall now be considered separately. It will be noticed that each is illustrated by use of the same example, illustrated in Fig. 7.2.

I. Bending

Since the section is designed as being elastic over the whole cross-section, the familiar bending formula applies:

$$\frac{M}{I} = \frac{f}{y} = \frac{E}{R} \qquad [7.1]$$

where M is the applied moment
I is the second moment of area of the section
f is the stress at a distance y from the neutral axis, measured perpendicularly to the axis of bending
E is the modulus of elasticity, which for steel is 210 kN/mm² (or 210 000 N/mm²)
R is the radius of curvature measured to the neutral axis

Therefore, knowing the maximum value of y, the maximum stress in the outermost fibre may be found:

$$f_{max} = \frac{M y_{max}}{I} \qquad [7.2(a)]$$

usually the quantity I/y_{max} is called the 'section modulus' and is given the symbol Z. (In Appendix B this is referred to as the 'Elastic modulus'.)

$$f_{max} = M/Z \qquad [7.2(b)]$$

Values for I and Z about both the major $x-x$ and minor $y-y$ axes may be found in various sources[1,2] and in the tables in Appendix B at the end of this book.

Example 7.1 Calculate the maximum stress which occurs in a 406 × 178 mm UB 74 kg/m at loaded as shown in Fig. 7.2.

$$M = \frac{0.726 \times 15^2}{8} + 29.75 \times 5 = 169.169 \text{ kNm}$$

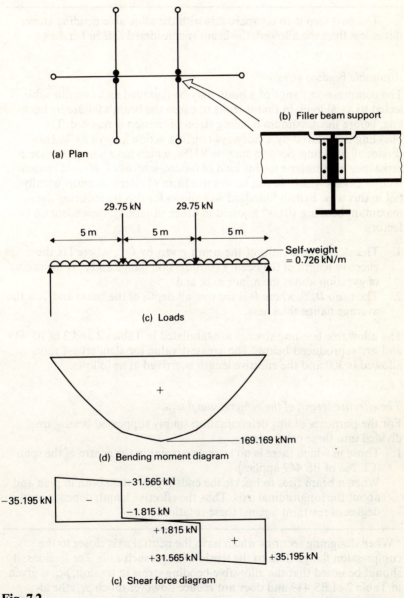

(a) Plan

(b) Filler beam support

29.75 kN   29.75 kN

5 m   5 m   5 m

Self-weight = 0.726 kN/m

(c) Loads

169.169 kNm

(d) Bending moment diagram

−35.195 kN   −31.565 kN   −1.815 kN   +1.815 kN   +31.565 kN   +35.195 kN

(c) Shear force diagram

Fig. 7.2

From tables $\quad Z = 1322\,\text{cm}^3 = 1.322 \times 10^6\,\text{mm}^3$

Thus $\quad f_{\text{max}} = \dfrac{169.169 \times 10^6}{1.322 \times 10^6}\,\text{N/mm}^2$

$$= 127.95\,\text{N/mm}^2.$$

The next step is to compare this with the allowable bending stress; if it is less than the allowed, the beam is considered safe in bending.

Allowable bending stresses

The compression flange of a beam may be regarded as a column subjected to axial load, in that it tends to cause the beam's failure by buckling, before the maximum bending stress in tension is reached. The buckling is achieved by a sideways twisting action known as 'lateral-torsional' buckling. Section such as RHSs, which have a large torsional resistance are immune to this kind of failure, whereas *I*, *H*- and channel-section beams which do not have some form of lateral support usually fail in this way. British Standard 449 allows for this by reducing the maximum bending stress* allowed in other situations, dependent on two factors:

1. The slenderness ratio of the beam given by l/r_y, where l is the effective length of the beam's compression flange and r_y is the radius of gyration about the minor axis; and
2. The ratio D/T, where D is the overall depth of the beam and T is the average flange thickness.

The allowable bending stresses are tabulated in Tables 2 and 3 of BS 449 and are reproduced below. The greatest value for slenderness ratio allowed is 300 and the effective length is arrived at as follows.

The effective length of the compression flange

For the purposes of this determination simply supported beams are divided into three categories:
1. Those in which there is no lateral restraint in the centre of the span (Cl. 26a of BS 449 applies).
    When a beam tries to buckle the ends try to rotate both in plan and about the longitudinal axis. Thus the effective length depends on the degree of restraint against these rotations.

---

*When designing sections which have the neutral axis closer to the compression flange than to the tension flange, such as in Tee sections, it should be noted that the allowable bending stress in tension, p_{bt}, is given in Table 2 of BS 449 and does not reduce however much p_{bc}, the allowable bending stress in compression, is reduced.

Table 2 Allowable stress p_{bc} or p_{bt} in bending
(*See also Clauses 19 and 20, and Tables 3a, 3b and 3c*)

Form	Grade	Thickness of material	p_{bc} or p_{bt}
		mm	N/mm²
Rolled I-beams and channels	43	All	165
Compound girders composed of rolled I-beams or channels plated, with thickness of plate	43	Up to and including 40	165
		Over 40	150
Plates, flats, rounds, squares, angles, tees and any sections other than above	43	Up to and including 40	165
		Over 40	150
Plate girders with single or multiple webs	43	Up to and including 40	155
		Over 40	140
Universal beams and columns	43	Up to and including 40	165
		Over 40	150
Plates, flats, rounds, squares and other similar sections, rolled I-beams, double channels forming a symmetrical I-section which acts as an integral unit, compound beams composed of rolled I-beams or channels plated, single channels, angles and tees	50	Up to and including 65	230
		Over 65	$Y_s/1.52$
	55	Up to and including 40	280
		Over 40	260
Plate girders with single or multiple webs	50	Up to and including 65	215
		Over 65	$Y_s/1.63$
	55	Up to and including 40	265
		Over 40	245
Hot rolled hollow sections	43	All	165
Hot rolled hollow sections	50	All	230
Hot rolled hollow sections	55	All	280
Slab bases		All steels	185

where Y_s = yield stress agreed with manufacturer, with a maximum value of 350 N/mm².

The effective length is given by:

$$l = k_1 k_2 L \qquad [7.3]$$

where l is the flange effective length
        k_1 is the plan rotation factor of the compression flanges
        k_2 is the longitudinal (torsional) rotation factor
        L is the effective span of the beam

Table 3a   Allowable stress p_{bc} in bending (N/mm^2) for beams of grade 43 steel

l/r_y	D/T							
	10	**15**	**20**	**25**	**30**	**35**	**40**	**50**
90	165	165	165	165	165	165	165	165
95	165	165	165	163	163	163	163	163
100	165	165	165	157	157	157	157	157
105	165	165	160	152	152	152	152	152
110	165	165	156	147	147	147	147	147
115	165	165	152	141	141	141	141	141
120	165	162	148	136	136	136	136	136
130	165	155	139	126	126	126	126	126
140	165	149	130	115	115	115	115	115
150	165	143	122	104	104	104	104	104
160	163	136	113	95	94	94	94	94
170	159	130	104	91	85	82	82	82
180	155	124	96	87	80	76	72	71
190	151	118	93	83	77	72	68	62
200	147	111	89	80	73	68	64	59
210	143	105	87	77	70	65	61	55
220	139	99	84	74	67	62	58	52
230	134	95	81	71	64	59	55	49
240	130	92	78	69	61	56	52	47
250	126	90	76	66	59	54	50	44
260	122	88	74	64	57	52	48	42
270	118	86	72	62	55	50	46	40
280	114	84	70	60	53	48	44	39
290	110	82	68	58	51	46	42	37
300	106	80	66	56	49	44	41	36

Intermediate values may be obtained by linear interpolation.
Note: For materials over 40 mm thick the stress shall not exceed 150 N/mm^2.

Table 3*b* Allowable stress p_{bc} in bending (N/mm^2) for beams of grade 50 steel

l/r_y	D/T							
	10	**15**	**20**	**25**	**30**	**35**	**40**	**50**
80	230	230	230	230	230	230	230	230
85	230	230	230	227	227	227	227	227
90	230	230	228	220	220	220	220	220
95	230	230	222	212	212	212	212	212
100	230	230	215	204	204	204	204	204
105	230	226	209	196	196	196	196	196
110	230	221	203	188	188	188	188	188
115	230	216	196	181	181	181	181	181
120	230	211	190	173	173	173	173	173
130	230	202	177	157	157	157	157	157
140	225	193	165	142	142	142	142	142
150	219	183	152	126	126	126	126	126
160	213	174	139	112	110	110	110	110
170	207	165	126	106	97	94	94	94
180	201	155	114	101	91	85	80	77
190	195	146	109	96	86	80	75	68
200	189	136	104	91	82	75	70	64
210	183	127	100	87	77	71	66	60
220	177	118	96	83	74	67	62	56
230	171	112	92	79	70	64	59	53
240	165	108	89	76	67	61	56	50
250	159	105	86	73	64	58	53	47
260	153	102	83	70	62	56	51	45
270	147	99	80	68	59	53	49	43
280	141	96	77	65	57	51	47	41
290	135	93	75	63	55	49	45	39
300	129	90	72	61	53	47	43	37

Intermediate values may be obtained by linear interpolation.
Note: For materials over 65 mm thick the stress shall not exceed
$\frac{y_s}{1.52}$ N/mm^2, where y_s = yield stress agreed with manufacturer, with a maximum value of 350 N/mm^2.

Table 3c Allowable stress p_{bc} in bending (N/mm^2) for beams of grade 55 steel

l/r_y	D/T							
	10	15	20	25	30	35	40	50
75	280	280	280	280	280	280	280	280
80	280	280	280	277	277	277	277	277
85	280	280	277	267	267	267	267	267
90	280	280	269	257	257	257	257	257
95	280	278	261	247	247	247	247	247
100	280	272	253	238	238	238	238	238
105	280	266	245	228	228	228	228	228
110	280	260	237	218	218	218	218	218
115	280	254	228	208	208	208	208	208
120	280	248	220	198	198	198	198	198
130	273	236	204	178	178	178	178	178
140	266	224	188	158	158	158	158	158
150	258	212	171	138	138	138	138	138
160	250	200	155	121	119	119	119	119
170	243	188	139	114	103	99	99	99
180	235	176	123	107	97	89	84	80
190	227	164	117	101	91	83	78	71
200	219	152	111	96	86	78	73	66
210	212	140	106	91	81	74	69	62
220	204	128	102	87	77	70	65	58
230	196	120	97	83	73	66	61	54
240	189	116	94	79	70	63	58	51
250	181	112	90	76	67	60	55	49
260	173	108	87	73	64	57	53	46
270	165	105	84	70	61	55	50	44
280	158	101	81	68	59	53	48	42
290	150	98	78	65	57	51	46	40
300	142	95	76	63	55	49	44	38

Intermediate values may be obtained by linear interpolation.
Note: For materials over 40 mm thick the stress shall not exceed 260 N/mm.

Degree of restraint	k_1	k_2
None	1.0	1.2
Partial	0.85	—
Full	0.7	1.0

Figure 7.3 illustrates two types of end connection which provide full torsional restraint but less than full plan rotation restraint. Full plan restraint can be provided by welding a plate on to the end of the beam and then bolting to the column or by using very large angle cleats where the thickness of the angle is equal to the flange thickness or greater. Alternately the beam could be completely welded to the column. *Finally, k_2 will be equal to 1.2 if the load is applied to the compression flange.*

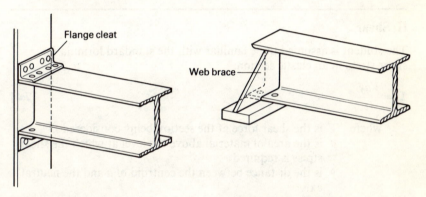

(a) Full torsion restraint   (b) Full torsion restraint
    partial plan restraint   no plan restraint

Fig. 7.3

2. Those in which effective lateral restraint is provided by tie beams at intervals along the span, in addition to full torsional end restraint.

    The effective length is taken as the actual length between the tie beams. Note that these tie beams must give restraint to the compression flange's tendency to twist sideways. Thus if the filler beams are attached to the web as illustrated in Fig. 7.2(a), and to other beams parallel to the main beam, lateral restraint at these points may be assumed.

3. Those in which the compression flange supports a floor or other slab, where frictional resistance between the slab and compression flange may be assumed.

    The maximum bending stress of Table 2 in BS 449 may be used so long as the frictional force can resist 2.5 per cent of the maximum force in the compression flange. Effective, full lateral restraint is thus provided.

Example 7.2 What is the allowable bending stress in the beam of Example 7.1? Assume that the beam is made from steel of grade 43. There is full restraint against torsion at the ends of the beam.

The effective length of the compression flange = 5 m.

From tables $\quad r_y = 3.91\,\text{cm} = 39.1\,\text{mm}$

and $\qquad\qquad D/T = 25.8$

Thus $\qquad\qquad l/r_y = 5000/39.1 = 127.9$

And from Table 3a of BS 449 $\quad p_{bc} = 128.1\,\text{N/mm}^2$
Thus the beam is safe in bending since the actual stress is less than this value.

II. Shear

The student is assumed to be familiar with the standard formula for shear stress in an elastic section:

$$f_q = \frac{Va\bar{y}}{It} \qquad\qquad [7.4]$$

where V is the shear force at the section being considered
 a is the area of material above the level at which the shear stress is required
 $\bar{y}$ is the distance between the centroid of a and the neutral axis
 I is the second moment of area of the whole section
 t is the thickness of the section at the level for which the shear stress is required
 f_q is the required shear stress

Figure 7.4 illustrates how these terms apply to an I-beam. The value of $\bar{y}$ may be obtained from the tables for structural tees which are the result

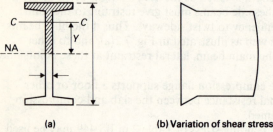

(a) (b) Variation of shear stress

Fig. 7.4

of cutting a UB in two along the centre of the web. As can be seen from Fig. 7.4(b), the value of shear stress in the web is almost constant. From this it can be seen that almost all of the shear force is taken by the web, and the contribution of the flanges is minimal. For this reason BS 449 allows the designer to check a beam for shear in one of two ways. Firstly, the value of maximum shear may be checked, or secondly a nominal 'Average Shear' stress may be calculated and checked against a different set of allowed stresses. Both these procedures will be illustrated for the last example.

Example 7.3 For the beam of Example 7.1 check the values of maximum and average shear stress.

Maximum shear stress, (occurs at the neutral axis of the UB)
From the UB tables of BS 449 $I = 272.79 \times 10^6 \, \text{mm}^4$ (gross)

$$t = 9.7 \, \text{mm} \qquad D = 412.8 \, \text{mm}$$

From the tables on structural tees:

$$c_x = 48.1 \, \text{mm}$$

$$\bar{y} = 0.5D - c_x$$

$$= 158.3 \, \text{mm}$$

$$a = 4740 \, \text{mm}^2$$

From Fig. 7.2 $\qquad V = 35.195 \, \text{kN}$

Thus $\quad f_q = \dfrac{35\,195 \times 4740 \times 158.3}{9.7 \times 272\,790\,000} = 9.98 \, \text{N/mm}^2$

The allowable maximum shear stress is obtained from Table 10 of BS 449, and this is reproduced along with Table 11 needed for average shear considerations.

$p_q = 115 \, \text{N/mm}^2$

Average shear stress
Area of web $= t \times D = 9.7 \times 412.8 = 4004.16 \, \text{mm}^2$

Average shear stress $= f'_q = \dfrac{V}{tD} = \dfrac{35\,195}{4004.16} = 8.79 \, \text{N/mm}^2$

From Table 11 of BS 449 $p'_q = 100 \, \text{N/mm}^2$

Note: The fact that $f'_q < f_q$ is reflected in the fact that $p'_q < p_q$
Thus the beam is very safe in shear.

Table 10 Allowable maximum shear stress p_q

Form	Grade	Thickness or diameter mm	p_q N/mm^2
Plates, sections and bars	43	Up to and including 40	115
		Over 40	105
Plates, sections and bars	50	Up to and including 65	160
		Over 65	$Y_s/2.2$
Plates, sections and bars	55	Up to and including 40	195
		Over 40	180
Hot rolled hollow sections	43	All	115
Hot rolled hollow sections	50	All	160
Hot rolled hollow sections	55	All	195

where Y_s = yield stress agreed with manufacturer, with a maximum value of 350 N/mm^2.

Table 11 Allowable average shear stress p'_q in unstiffened webs

Grade	Thickness mm	p'_q N/mm^2
43	Up to and including 40	100
	Over 40	90
50	Up to and including 65	140
55	Up to and including 40	170
	Over 40	160

III. Web bearing

In addition to the shearing and bending stresses in the web of a beam, there are compressive stresses in the vertical direction, since the loads

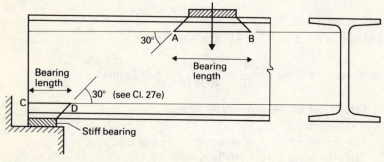

Fig. 7.5

bear directly on the flanges of the beam. In the last example this only applies at the ends of the beam since the tie beams transfer their load directly to the web by means of the web cleats. The critical section for bearing will obviously be where the web is thinnest. This occurs where the web joins the flange at the top of the root radius. As shown in Fig. 7.5 BS 449 allows the compressive load to be dispersed at an angle of 30° through the flange to this section, and the bearing stress is assumed to be constant along AB and CD. The value calculated in this way must not exceed the allowed bearing stress laid down in Table 9 of BS 449.

Example 7.4 Calculate the value of stiff bearing needed for the supports of the beam in Example 7.1.

Allowed bearing stress = 190 N/mm²

Bearing area = t × bearing length

Required bearing area = 35 195/190 = 185.24 mm²

Thus required bearing length = 185.24/9.7 = 19.1 mm

Vertical distance through flange
to top of root radius = $0.5 (D - d)$

$$= \frac{412.8 - 357.4}{2}$$

$$= 27.7 \text{ mm}$$

Therefore minimum bearing length available = $0.5(D - d)/\tan 30°$

$$= 48 \text{ mm}$$

Therefore only a nominal bearing is needed.

Table 9 Allowable bearing stress p_b

Form	Grade	p_b
		N/mm²
Plates, sections and bars	43	190
Plates, sections and bars	50	260
Plates, sections and bars	55	320
Hot rolled hollow sections*	43	190
Hot rolled hollow sections*	50	260
Hot rolled hollow sections*	55	320

* See Subclause 28c for provisions for stiffening tubes subject to bearing pressure.

152

IV. Web buckling

As well as the possibility of the web failing in bearing, the vertical compression may also cause a localised buckling at mid depth. This is typified by the web bulging outwards and the flange bowing downwards underneath the compressive load. Figure 7.6 illustrates the way in which

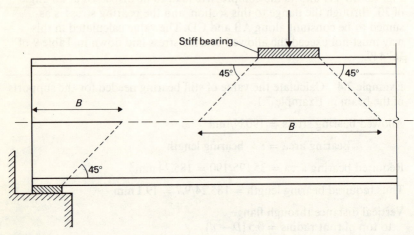

Fig. 7.6

BS 449 copes with this situation. The compressive load is allowed to disperse at an angle of 45° to the centre of the beam where a loaded length of B is found. This multiplied by the web thickness gives the area of web taken to resist web buckling. The allowed stress, which is assumed constant over the full length of B, is found from Table 17 of BS 449, which is reproduced on pp. 189 to 191. Table 17 is used for the design of columns and depends on the value of the columns' slenderness ratio, defined as the effective length divided by the radius of gyration about the minor axis. This term is simulated for web buckling by having a slenderness ratio given by:

$$\frac{d_3}{t}\sqrt{3}$$

where t = the web thickness
  d_3 = the clear depth between root radii, this is the same as d in the tables

However this value of slenderness ratio may only be used if:

1. The flange through which the load (or reaction) is applied is effectively restrained against lateral movement relative to the other flange; and
2. Rotation of the loaded flange relative to the web is prevented.

153

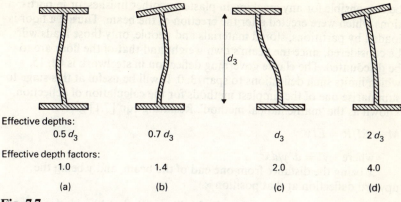

Effective depths:
0.5 d_3 0.7 d_3 d_3 2 d_3

Effective depth factors:
1.0   1.4   2.0   4.0

(a)   (b)   (c)   (d)

Fig. 7.7

Figure 7.7 illustrates how the slenderness ratio above should be increased if 1. and 2. are not met (Fig. 7.7d), if 1. is not met but 2. is met (Fig. 7.7c), and if 2. is met but not 1. (Fig. 7.7b).

Example 7.5 Calculate the value of the length of stiff bearing needed in the beam of Example 7.1 to resist web buckling.

$$\text{Slenderness ratio} = \frac{d}{t} \times 1.732 = \frac{357.4}{9.7} \times 1.732$$

$$= 63.8$$

This does not need to be increased since there is full torsional restraint at the end of the beam where the web buckling is likely to take place.

From Table 17a of BS 449

$p_c = 122 \, \text{N/mm}^2$

$$\text{Thus area of web needed to resist buckling} = \frac{35\,195}{122}$$

$$= 288.5 \, \text{mm}^2$$

$$\text{Thus } B \text{ required} = \frac{288.5}{9.7} = 29.74 \, \text{mm}$$

Minimum B available $= 0.5D/\tan 45° = 0.5D = 206.4 \, \text{mm}$

Thus the beam is safe against web buckling.

V. Deflection

When calculating deflections in accordance with BS 449, only the live loads are taken into account since it is argued that any deflection which takes place due to self-weight will: 1. not be so easily noticed, and 2. not be responsible for any cracking in plaster or other finishes, or in partitions which were erected after the erection of the beam. Thus, if a floor is loaded by partitions, stored materials and people, only those loads will be considered, since the beam's own weight and that of the floor are to be discounted. The clause governing deflection in steelwork is Cl. 15, which limits such deflections to span/360. It will be useful at this stage to summarise one of the simplest methods for the calculation of deflection, known as the 'mathematical method'. Restating eqn [7.1] gives:

$$M = EI/R = EIy''$$

where $y'' = d^2y/dx^2$
x being the distance from one end of the beam, and y being the upward deflection at that position x.

Note: The student is assumed to be familiar with the abbreviations y'' to stand for d^2y/dx^2, and y' to stand for dy/dx.

The method is best illustrated by an example.

Example 7.6 Calculate the deflection in the beam shown in Fig. 7.8. Taking the origin at A:

Between A and B: $M = P(L - a)x/L$ [7.5]

Between B and C: $M = P(L - a)x/L - P(x - a)$ [7.6]

These two equations are then combined in a special type of equation as shown in eqn [7.7]
Across the whole of the beam

$$M = P(L - a)x/L - P[x - a]$$ [7.7]

The significance of the square brackets is as follows:

for $x > a$: the term in the square brackets is positive
for $x < a$: the term in the square brackets is **zero**

Note: The square brackets term is *never* allowed to be negative.
Thus for $x < a$ eqn [7.5] results, while for $x > a$ eqn [7.6] results.

The deflections along the beam are found by integrating eqn [7.7] twice – once to give EIy' and a second time to give EIy. However, the integration of the square-bracketed term must be done in an unusual fashion.

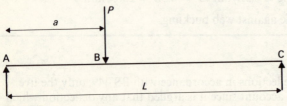

Fig. 7.8

It might be expected that each term inside it should be integrated separately, but this is not the case. Instead, the **whole** term is treated as if it was x. In other words, the integral of $[x - a]$ is $0.5[x - a]^2$ and not $(0.5x^2 - ax)$. This point must be carefully watched and adhered to.

Hence
$$EIy' = \frac{P(L - a)x^2}{2L} - 0.5\,P[x - a]^2 + c_1 \qquad [7.8]$$

and
$$EIy = \frac{P(L - a)x^3}{6L} - \frac{P}{6}[x - a]^3 + c_1 x + c_2 \qquad [7.9]$$

where c_1 and c_2 are constants of integration

when $x = 0$, $y = 0$

Thus $c_2 = 0$ from eqn [7.9]

Similarly

when $x = L$ $y = 0$

Thus $c_1 = \dfrac{P}{6L}(L - a)^3 - \dfrac{PL}{6}(L - a)$

Substituting this into eqn [7.9] gives:
$$EIy = \frac{P(L - a)x^3}{6L} - \frac{P}{6}[x - a]^3 + \frac{Px}{6L}(L - a)^3 - \frac{PLx}{6}(L - a) \qquad [7.10]$$

Thus the mid span deflection is given by:
$$d = \frac{PL^3}{48EI}\left\{3\,\frac{a}{L} - 4(a/L)^3\right\} \qquad [7.11]$$

It can be shown that this lies within 2.5 per cent of the true maximum, and thus this equation is accurate enough for the purposes of the calculation of deflection.

Exactly the same method may be applied to many situations, but can sometimes become tedious. Other methods of calculating deflections can be found in *The Steel Designers' Manual*.[3]

It can be shown that the maximum deflection occurs at
$$x = L - \sqrt{(L^2 - a^2)/3} \qquad [7.12]$$
which is never more than $0.077\,35L$ from the centre.

For example,

Let $P = 120\,\text{kN}$; $a = 1\,\text{m}$; $L = 3\,\text{m}$

$I = 4427\,\text{cm}^4$; $E = 2.1 \times 10^5\,\text{N/mm}^2$

Mid span deflection = 6.185 mm

Maximum deflection occurs at $x = 1.376\,\text{m}$ from A

Maximum deflection = 6.245 mm

Note: Equation [7.10] gives this as negative since it is a downwards deflection.

Table 7.1 Deflection formulae for beams and cantilevers

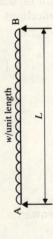

$$d_{\max} = \frac{5wL^4}{384EI}$$

$$M_{\max} = \frac{wL^2}{8}$$

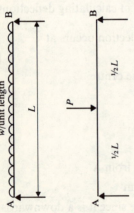

$$d_{\max} = \frac{PL^3}{48EI}$$

$$M_{\max} = \tfrac{1}{4}PL$$

$$d_{\max} = \frac{PL^3}{6EI}\left[\frac{3a}{4L} - \left(\frac{a}{L}\right)^3\right]$$

$$M_{\max} = Pa$$

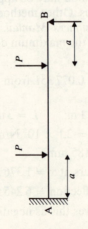

$$M_{\max} = \tfrac{1}{2}wa^2, \quad d_c = \frac{wa^4}{8EI}, \quad d_{\max} = \frac{wa^4}{8EI}\left(1 + \frac{4b}{3a}\right)$$

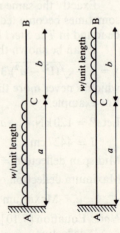

$$M_{\max} = wb(a + \tfrac{1}{2}b), \quad d_{\max} = \frac{wb(8a^3 + 18a^2b + 12ab^2 + 3b^3)}{24EI}$$

157

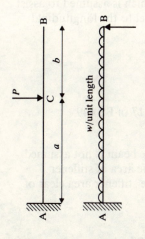

$$M_{max} = Pa, \quad d_c = \frac{Pa^3}{3EI}, \quad d_{max} = \frac{Pa^3}{3EI}\left(1 + \frac{3b}{2a}\right)$$

$$M_A = -\frac{wL^2}{8}, \quad M_{max} = \frac{9wL^2}{128}, \quad d_{max} = \frac{wL^4}{185EI}$$

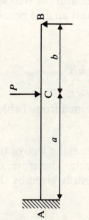

$$M_A = -\frac{Pb(L^2 - b^2)}{2L^2}, \quad M_c = \tfrac{1}{2}Pb\left(2 - \frac{3b}{L} + \frac{b^3}{L^3}\right), \quad d_{max} = \frac{wL^4}{384EI}$$

$$M_A = M_B = -\frac{wL^2}{12}, \quad M_c = \frac{wL^2}{24}, \quad d_c = \frac{Pa^3b^2}{12EIL^3}(4L - a)$$

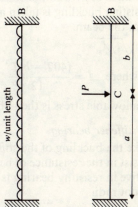

$$M_A = -\frac{Pab^2}{L^2}, \quad M_B = \frac{Pba^2}{L^2}, \quad M_c = \frac{2Pa^2b^2}{L^3}$$

$$d_{max} = \frac{2Pa^2b^3}{3EI(3L - 2a)^2}$$

when $x = L^2/(3L - 2a)$ from A

Web stiffeners

When the web of a beam is not strong enough to resist failure in bearing or buckling one of two things may be done to remedy the situation. First, the length of bearing on which the beam rests or on which the load bears may be increased, or second, the web may be stiffened by the insertion of pieces of steel at right angles to the plane of the web and fitting tightly between the flanges and webs. Nowadays, these stiffeners are welded to the flanges and webs but sometimes they are bolted and then it is most important to have the ends of them bearing on the surfaces of the flanges as tightly as possible. The clause relating to their design is 28. They must be checked for resistance to buckling and bearing just as the web itself is checked, each being considered separately below.

Web stiffener buckling

The load which it is required to resist in buckling is assumed to impose a uniform stress across an area consisting of:

   The cross-sectional area of the stiffeners
      plus
   The area of web to either side of the stiffeners' centre line which is of length twenty times the web thickness, where this length is available to either side. Thus if the stiffeners are taken to the full width of the beam and are of breadth b, the area acting to resist buckling is given by:

$$A_{st} = (B - t) \times b + 40t^2$$

   where B is the overall width of the beam
           t is the thickness of the web

The effective length of the stiffener is taken as 0.7 times the actual length of the stiffener:

$$l_{st} = 0.7(D - 2T)$$

The radius of gyration of stiffener and the web which is assumed to assist in resisting buckling is taken about an axis parallel to the longitudinal axis of the beam.

$$r_{sy} = \sqrt{I/A_{st}}$$

   where $I = \dfrac{(40T - b)T^3}{12} + \dfrac{bB^3}{12}$

The allowable stress is then obtained from Table 17 of BS 449.

Web stiffener bearing

Unlike the buckling of the stiffener, the web of the beam is not assumed to assist in the resistance of bearing. Therefore, the area of stiffener effective in resisting bearing is simply given by the stiffener area clear of the root radii.

$$A_{sb} = (B - t - 2r)b$$

   assuming that the stiffeners are the full width of the beam.

The allowable bearing stress is obtained as before.

If the web length available for buckling is less than twenty times the web thickness either side of the stiffener centre line, then the above formulae will need to be adjusted to take account of the reduced length.

Example 7.7 The end of a 686 × 254 mm at 140 kg/m UB is required to resist a load of 650 kN on a bearing length of 20 mm. If the beam is made from grade 50 steel, design the stiffeners required.

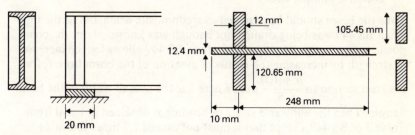

Fig. 7.9

The situation is illustrated in Fig. 7.9.

Allowed bearing stress = $260 \, \text{N/mm}^2$

Length of stiffener assumed effective in bearing = $B - t - 2r$

$$= 210.9 \, \text{mm}$$

Thickness of stiffener needed in bearing $= \dfrac{650\,000}{210.9 \times 260}$

$$= 11.85 \, \text{mm}$$

Therefore use width = 12 mm
Check buckling:

$$I = \frac{(258 - 12) \times 12.4^3}{12} + \frac{12 \times 253.7^3}{12}$$

$$= 16.368 \times 10^6 \, \text{mm}^4$$

$$A = 12.4 \times 258 + 241.3 \times 12$$

$$= 6094.8 \, \text{mm}^2$$

$$r = 51.82$$

From Table 17b of BS 449: $p_c = 181.35 \, \text{N/mm}^2$

Thus the allowed load before buckling = 1105.3 kN.

Cased beams (Cl. 21 of BS 449)

In all domestic buildings where steel is used as part of the structural frame, it must be protected from fire. Among the many ways in which this may be accomplished, perhaps the one which seems to offer the greatest economy, is that in which the beam is entirely encased in concrete. Thus the concrete serves two purposes:

1. It protects the beam against the heat of the fire; and
2. It contributes some strength, by virtue of the bond between the concrete and the steel.

Thus the beam should be designed as a composite beam, but at the time when BS 449 was being drafted not enough was known about the composite action in this situation. Therefore, BS 449 allows for the increase in strength by increasing the radius of gyration of the beam from r_y for the steel section to $\dfrac{b + 100}{5}$ mm, where b is the overall width of the beam flanges. Then the allowable stress in bending is obtained as usual from Table 3 of BS 449, except that it must not exceed 1.5 times the value obtained for the uncased section.

Having made this allowance, the steel beam is assumed to act alone in shear and deflection, without any allowance for the contribution of the concrete. As for web buckling and bearing, no specific guidance is given but it is the author's opinion that web buckling should be neglected since the concrete surrounding the web will effectively prevent any tendency of the web to buckle. As for bearing, it will be sufficient to assume that the load spreads through the concrete to the surface of the steel at an angle of 45°, unless the load or reaction bears directly on the steel surface. Then the design proceeds as normal.

British Standard 449 places the following further restrictions on the use of structural casing to beams:

Only beams and girders with equal flanges may be designed as cased beams when the following conditions are fulfilled:

1. The section is of single web and I-form or of double open channel form with the webs not less than 40 mm apart.
2. The beam is unpainted and is solidly encased in ordinary dense concrete, with 10 mm aggregate (unless solidity can be obtained with a larger aggregate), and a cube strength of not less than 21 N/mm² at twenty-eight days when tested in accordance with BS 1881, 'Methods of Testing Concrete'.
3. The minimum width of solid casing is equal to $b + 100$ mm.
4. The surface and edges of the flanges of the beam have a concrete cover of not less than 50 mm.
5. The casing is effectively reinforced with wire complying with BS 4482 'Hard drawn mild steel wire for the reinforcement of concrete'. The wire shall be of at least 5 mm diameter and the reinforcement shall

be in the form of stirrups or binding at not more than 200 mm pitch, and so arranged as to pass through the centre of the covering to the edges and soffit of the lower flanges. Alternatively, the casing may be reinforced with fabric complying with BS 4483 'Steel fabric for the reinforcement of concrete' or with bars complying with BS 4449 'Hot rolled steel bars for the reinforcement of concrete', provided in either case that the same requirements of diameter spacing and positioning are met.

Example 7.8 A 686×254 mm at 170 kg/m UB has a simply supported span of 7 m and carries a UDL inclusive of its own weight and that of any casing. Assuming that the beam is of grade 50 steel, and that concrete has a unit weight of 23 kN/m³, compare the value of UDL which may be carried if the beam is uncased and if it is cased. Assume no lateral restraint or end torsional restraint and that the load is applied to the compression flange.

Uncased

Effective length $= 1.0 \times 1.2 \times 7.0 = 8.4$ m

$r_y = 53.6$ mm; $\quad l/r_y = 156.72;$ $\quad D/T = 29.2$

From Table 3b of BS 449: $p_{bc} = 115.46$ N/mm²; $\quad Z = 4.902 \times 10^6$ mm³;

$$M = Zp_{bc} = 565.98 \text{ kNm} = \frac{WL}{8}$$

Maximum UDL $= 646.84$ kN

Weight of beam $= 7 \times 170 \times 9.81$ N

$\qquad\qquad\qquad\quad = 11.67$ kN

Thus the maximum imposed UDL which can be supported by the uncased section $= 646.84 - 11.67 = 635.17$ kN.

Cased

$r_y = (b + 100)/5 = 355.8/5 = 71.16$ mm

$l/r_y = 8400/71.16 = 118.0;$ $\quad p_{bc} = 176.2$ N/mm²

This is greater than 1.5 times the value of the allowable bending stress for the uncased section, therefore 1.5 times the uncased value must be used.

$p_{bc} = 1.5 \times 115.46 = 173.19$ N/mm²

$M = 848.98$ kNm

UDL $= 970.26$ kN

Dimensions of concrete casing $= 355.8 \times 792.9$ minimum

$\qquad\qquad\qquad\qquad\qquad\qquad\quad = 360 \times 800$ actual

162

Area of section = 360 × 800 − area of steel beam
$$= 288\,000 - 21\,630 = 266\,370\,\text{mm}^2$$

Weight of beam = weight of concrete + weight of steel beam
$$= 0.266\,37 \times 7 \times 23 + 11.67$$
$$= 54.56\,\text{kN}$$

The maximum imposed UDL which can be supported by the case section = 970.26 − 54.56 = 915.70 kN.

Shear
Allowed average shear stress = 140 N/mm²

Allowed shear force = 140 × 14.5 × 692.9 N
$$= 1406.59\,\text{kN}$$

This is more than the shear for either cased or uncased section.

Deflection
Allowed deflection = 7000/360 = 19.444 mm

$$\text{Deflection} = \frac{5WL^3}{384EI}$$

Hence $W_{\text{max}} = 19.44 \times 384EI/5L^3$
$$= 1552.85\,\text{kN}$$

Thus both sections are safe in deflection.

Web buckling does not apply

Web bearing
Uncased section   without a stiff bearing

Length of web in bearing = $0.5(D - d)/\tan 30°$
$$= 71.274\,\text{mm}$$

Web bearing capacity = 71.274 × 14.5 × 260
$$= 268.7\,\text{kN}$$

Stiff bearing required = 15 mm (rounded up)

Cased section   without stiff bearing

Length of web in bearing = 71.274 + cover to steel
$$= 71.274 + 53.55 = 124.824\,\text{mm}$$

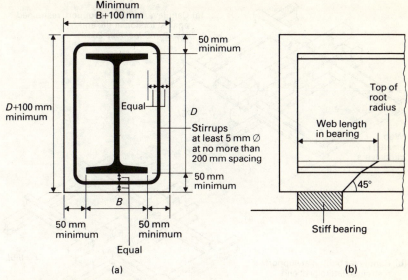

Fig. 7.10 Cased sections

$$\text{Web bearing capacity} = 124.824 \times 14.5 \times 260$$
$$= 470.59\,\text{kN}$$

Stiff bearing required = 4 mm (rounded up)

Thus the cased section is able to carry 44 per cent more live load than the uncased section. However, due to the conservative nature of the design process, the actual difference is likely to be much greater and it is to be hoped that a less conservative design method will appear in the near future. Figure 7.10 summarises some of the more salient points touched upon in this consideration of cased beams.

Design of cantilevers

True cantilevers – projections from truly fixed supports – are rare as a result of the very high moments which must be resisted by the supporting wall. More often they occur as continuations of simply supported beams beyond their supports, in which case the supporting wall need only resist the vertical shear while the beam at the wall support resists the moment. However, this situation results in less easily determined deflections since the beam may rotate at the support. However, by the application of the theory advanced in a previous section the deflections may be calculated without too much trouble. Apart from this, the only difference between the design of simply supported beams and the design

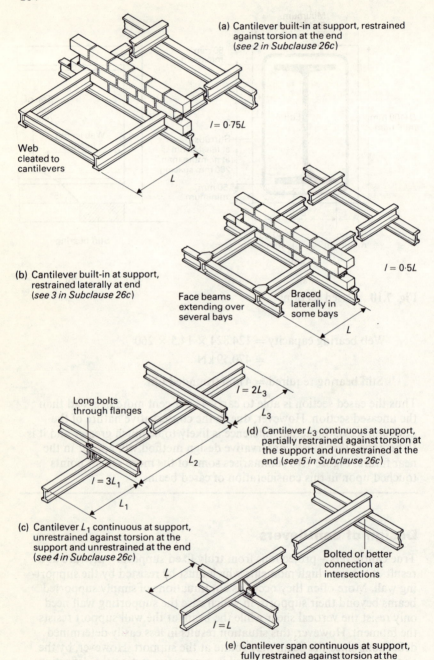

(a) Cantilever built-in at support, restrained against torsion at the end (*see 2 in Subclause 26c*)

$l = 0.75L$

Web cleated to cantilevers

L

(b) Cantilever built-in at support, restrained laterally at end (*see 3 in Subclause 26c*)

$l = 0.5L$

Face beams extending over several bays

Braced laterally in some bays

L

Long bolts through flanges

$l = 2L_3$

L_3

(d) Cantilever L_3 continuous at support, partially restrained against torsion at the support and unrestrained at the end (*see 5 in Subclause 26c*)

L_2

$l = 3L_1$

L_1

(c) Cantilever L_1 continuous at support, unrestrained against torsion at the support and unrestrained at the end (*see 4 in Subclause 26c*)

L

Bolted or better connection at intersections

$l = L$

(e) Cantilever span continuous at support, fully restrained against torsion at the support and unrestrained at free end (*see 6 in Subclause 26c*)

Fig. 7.11

of cantilevers is the determination of the effective length of the compression flange, which is, of course, the bottom flange for a cantilever. This aspect of their design is covered in Cl. 26c of BS 449, and states that the effective length shall be determined from a consideration of the support and free end restraint conditions as follows:

At the cantilevered support of such a beam, the web of the beam just above the bottom root radius is subjected simultaneously to bending, bearing and shear stresses. Therefore it should, in theory at least, be checked in accordance with Cl. 14d of BS 449. However, the author has on a number of occasions, performed this calculation and compared it with the more simple approach of checking each stress separately. Therefore it is recommended that the simpler approach be used since the author has not found any case where the more 'exact' approach was necessary.

If the actual projecting length is L, then the effective length l shall be taken as dictated by the following:

1. Built in at the support, free at the end $\qquad l = 0.85L$
2. Built in at the support, restrained against torsion at the end by contiguous construction (see Fig. 7.11a) $\qquad l = 0.75L$
3. Built in at support, restrained against lateral deflection and torsion at the end (see Fig. 7.11b) $\qquad l = 0.5L$
4. Continuous at the support, unrestrained against torsion at the support and free at the end (see Fig. 7.11c) $\qquad l = 3L$
5. Continuous at the support with partial restraint against torsion of the support and free at the end (see Fig. 7.11d) $\qquad l = 2L$
6. Continuous at the support, restrained against torsion at the support and free at the end (see Fig. 7.11e) $\qquad l = L$

If in cases 4, 5 and 6 above the free end is restrained in a fashion similar to that of Figs. 7.11a or 7.11b, the value of effective length obtained from 4, 5 or 6 can be multiplied by $\dfrac{0.75}{0.85}$ or $\dfrac{0.5}{0.85}$ respectively.

Having thus determined the effective length of the cantilever, the design proceeds in a similar fashion to that of the simply supported beams already dealt with.

Example 7.9 Design beams in grade 50 steel for the following situation – beams span simply supported for a distance of 10 m and project beyond one support to form a balcony 4 m wide. Their free ends are unrestrained while the supports are built into a concrete wall. The loads on the balcony consist of 8 kN/m dead load and 50 kN/m live load. The loads on the simple span consist of 20 kN/m dead load and 20 kN/m live load. The simple span supports a slab providing full lateral restraint. Try a section 533 × 210 mm at 92 kg/m.

Stage I: Permissible bending stresses.

Cantilever – $L = 4$ m   case 1. applies

$$l = 0.85 \times 4 = 3.4 \text{ m}$$

$r_y = 43.4$ mm; $l/r_y = 78.34$; $p_{bc} = 230$ N/mm^2

Simple span – $L = 10$ m
Full lateral restraint
Hence $p_{bc} = 230$ N/mm^2.

Stage II: Bending moment and shear force diagrams.
See Fig. 7.12.
There are three load cases depending on whether the live load is applied or not. The dead load is always present. The load cases are:

1. Live load on balcony only
2. Live load on simple span only
3. Live load throughout the beam

The student should verify the diagrams shown in Fig. 7.12 before continuing further.
Beam load = 92×9.81 N/m = 0.903 kN/m

Cantilever: $M_{max} = 471.224$ kNm; $V_{max} = 235.612$ kN

Simple span: $M_{max} = 476.296$ kNm; $V_{max} = 251.637$ kN

Thus in bending and shear, the simple span dictates the design, since the values are larger.

Stage III: Check bending and shear stresses.
Checking the simple span also satisfies the cantilever in this case. Usually, the check would be carried out for both.

$Z = 2.072 \times 10^6$ mm^3;

$$f_{bc} = \frac{476.296 \times 10^6}{2.072 \times 10^6}$$

$$= 229.87 \text{ N/mm}^2$$

$D = 533.1$ mm; $t = 10.2$ mm;

$$f'_q = \frac{V}{tD} = \frac{251\,637}{10.2 \times 533.1}$$

$$= 46.28 \text{ N/mm}^2$$

Thus the beam is safe in shear and bending.

167

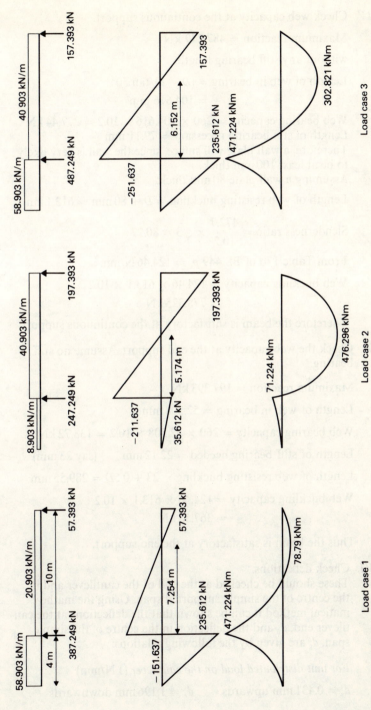

Fig. 7.12

Stage IV: Check web capacity at the continuous support.

Maximum reaction = 487.249 kN

without any stiff bearing length;

Length of web in bearing $= (D - d)/\tan 30°$
$$= 104.616 \text{ mm}$$

Web bearing capacity $= 260 \times 104.616 \times 10.2 = 277.44 \text{ kN}$
Length of stiff bearing necessary $= 79.11 \text{ mm}$
Therefore, a wall plate will suffice, since the wall is very likely to be at least 100 mm thick.
Assuming a wall plate 80 mm thick:

Length of web resisting buckling $= D + 80 \text{ mm} = 613.1 \text{ mm}$

Slenderness ratio $= \dfrac{472.7}{10.2} \times \sqrt{3} = 80.27$

From Table 17b of BS 449 $p_c = 124.46 \text{ N/mm}^2$

Web buckling capacity $= 124.46 \times 613.1 \times 10.2$
$$= 778.325 \text{ kN}$$

Therefore the beam is satisfactory at the continuous support.

Stage V: Check the web capacity at the end support. Assume no stiff bearing.

Maximum reaction = 197.393 kN

Length of web in bearing = 52.308 mm

Web bearing capacity $= 260 \times 52.308 \times 10.2 = 138.72 \text{ kN}$

Length of stiff bearing needed $= 22.12 \text{ mm}$ (say 23 mm)

Length of web resisting buckling $= 23 + 0.5D = 289.55 \text{ mm}$

Web buckling capacity $= 124.46 \times 613.1 \times 10.2$
$$= 367.58 \text{ kN}$$

Thus the beam is satisfactory at the end support.

Stage VI: Check deflections.
These should be checked at the end of the cantilever and in the centre of the simply supported span. Using the mathematical method it can be shown that the deflections at the cantilever end, d_c and the deflection in the centre of the simple span, d_s are given by the following relations:

For unit distributed load on the cantilever (1 N/mm)

$d_s = 0.431 \text{ mm upwards}$; $d_c = 1.196 \text{ mm downwards}$

For unit distributed load on the simple span (1 N/mm)

$d_s = 1.123$ mm downwards; $\qquad d_c = 1.437$ mm upwards

Case 1 loads: $\qquad d_c = 59.8$ mm $\qquad\qquad d_s = 21.55$ mm up

Case 2 loads: $\qquad d_c = 28.74$ mm up $\qquad d_s = 22.46$ mm

Case 3 loads: $\qquad d_c = 31.06$ mm $\qquad\qquad d_s = 0.91$ mm

Allowed deflections $\qquad d_c = 11.11$ mm

$\qquad\qquad\qquad\qquad\qquad\ d_s = 27.78$ mm.

Therefore although the simple span is within the limits, the cantilever is not. In fact the minimum value of I that is needed for the cantilever is 2.973×10^9 mm^4. The lightest UB which has this value is 914×305 mm at 201 kg/m, but to use this beam for the whole of the length of the beam would be very uneconomic. This illustrates that the main difficulty with cantilever construction is often deflection. The simplest way around this problem is to add additional plates to the flanges of the smaller 'span' beam by welding. This forms what is known as a 'compound' beam, the design of which is considered in Volume 2 of this book.

Comparison of universal beams with rolled hollow sections and H beams

When a beam is constructed from a rectangular RHS there is a possibility of a number of advantages over both a UB or an H-section beam designed to perform the same task. The allowable bending stress for a RHS does not depend on the effective length of the compression flange since the high torsional rigidity of the section prevents any tendency towards lateral torsional buckling. Thus the full Table 2 stresses may be used, without any reduction regardless of span. However, this advantage is only fully realised on longer spans where the reduction in allowed stress for the I- and H-sections is of the order of 3. This is due to the way in which the flange thickness of RHS is the same as the web thickness. The flange thickness of I-section beams is greater than the web thickness and this gives greater bending strength than if the two thicknesses were the same.

For example, if the beam of Example 7.1 were designed as being RHS (rectangular) the required section would be 400×200 mm at 90.7 kg/m. This is markedly heavier than the 406×178 mm at 74 UB used.

Example 7.10 Consider the beam illustrated in Fig. 7.13. Design beams of I-section, H-section and RHS to perform the task adequately. Assume that the steel is of grade 43. Assume no lateral support with the load

170

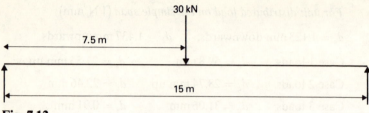

Fig. 7.13

applied to the compression flange. There is partial restraint at the ends of the beam against lateral bending.

$$k_1 = 0.85 \qquad k_2 = 1.2$$

Effective length $= 0.85 \times 1.2 \times 15 = 15.3\,\text{m}$

I. Universal beams

Try $610 \times 305\,\text{mm}$ at $149\,\text{kg/m}$

$$Z = 4.079 \times 10^6\,\text{mm}^3 \qquad D/T = 30.9 \qquad r_y = 66.8\,\text{mm} \qquad l/r_y = 229$$

$$M = 153.61\,\text{kNm} \qquad f_{bc} = \frac{153.61 \times 10^6}{4.079 \times 10^6}$$

$$= 37.7\,\text{N/mm}^2$$

From Table 3a of BS 449 $\qquad p_{bc} = 64\,\text{N/mm}^2$

Similarly, the following two sections will suffice in bending

686×254 at $140\,\text{kg/m}$ and 762×267 at $147\,\text{kg/m}$.

All three sections are found to be satisfactory in shear, bearing and web buckling, and in deflection.

II. Universal column sections as beams (H-sections)
Only one section works reasonably well:

$356 \times 368\,\text{mm}$ at $129\,\text{kg/m}$.

III. Rolled hollow sections

$$Z_{req} = \frac{140 \times 10^6}{165} = 850\,\text{cm}^3 \text{ (approx.)}$$

Try $300 \times 200\,\text{mm}$ at $92.6\,\text{kg/m}$ $\qquad M = 112.5 + \dfrac{0.926 \times 9.81 \times 15^2}{8}$

$$= 138.05\,\text{kNm}$$

$$Z = 964\,\text{cm}^3 \qquad f_{bc} = 143.2\,\text{N/mm}^2$$

Material saving = 0.711 tonnes over the lightest UB

= 0.546 tonnes over the universal column section.

However, the advantages of the RHS do not end there. Consider the painted area:

For an I- or H-section the painted area may be estimated as

$$A_p = (2D + 4B - 2t)L$$

where L is the actual span

Section	Painted area
RHS	14.6 m²
610 × 305 at 149	36.2 m²
686 × 254 at 140	35.4 m²
762 × 267 at 147	38.1 m²
356 × 368 at 129	32.4 m²

Hence the second advantage of the RHS lies in the reduction of paint used to protect the beam. At current prices this could mean a saving of £400 or £500 for this beam and since painting must be repeated every few years the saving will continue to be realised. Two more minor advantages lies in the simpler calculations and in the better looks afforded by the use of RHS.

Finally, it should be noted that the single webbed section is not completely shadowed by RHS for the longer spans, since by the use of plate girders, beams can be fabricated which will more closely match the weight of the RHS beam. However, the RHS beam will still need less paint and in any case the plate girder will require greater fabrication in the form of welding. In short, RHS beams are not used often enough at present, but it is to be hoped that they will be so in the near future.

Questions

1. Design the beam required to span 2 m simply supported carrying a point load of 150 kN, 0.3 m from one end, using grade 43 steel. There is no lateral restraint, but the load is applied to the compression flange. Calculate the length of stiff bearing needed to prevent web buckling and bearing.
2. A simply supported beam spanning 18 m carries a slab, the load from which is 50 kN/m, of which half is the imposed load. Design three beams in grades 43, 50 and 55 steel to suit this purpose assuming full lateral restraint.
3. A 457 × 152 mm at 82 kg/m UB of grade 50 steel has a simply supported span of 8 m and carries a distributed load of 260 kN in ad-

addition to the load of 90 kN due to the mass of the structural floor, beam, etc. Is the beam correctly designed if the length of stiff bearing is 75 mm? Assume full lateral support.

4. A 610 × 229 mm at 101 kg/m UB of grade 43 steel has a simply supported span of 4 m and carries a point load of 719 kN (live) at 0.45 m from one end, on the compression flange. If there is full torsional end restraint and the compression flange is fully restrained against lateral bending at either end, determine whether or not the beam conforms to the requirements of BS 449. The length of stiff bearing is 75 mm.

5. Design the web stiffeners needed in Question 4.

6. The critical span of a beam is defined as that effective span below which the maximum value of p_{bc} may be used, even if the beam is not supported laterally. For the following sections calculate L_c, the critical span for the grades of steel specified.

   1. 914 × 419 mm at 388 kg/m in grade 50 steel.
   2. 457 × 191 mm at 67 kg/m in grade 55 steel.
   3. 203 × 133 mm at 25 kg/m in grade 43 steel.
   4. 406 × 178 mm at 67 kg/m in grade 50 steel.
   5. 356 × 171 mm at 67 kg/m in grade 50 steel.

7. A 533 × 210 mm at 109 kg/m UB of grade 50 steel has a simply supported span of 10 m and carries a UDL, two thirds of which is live load. The other third is made up of its own weight and the weight of the slab which it supports. By considering bending, shear and deflection, determine the maximum value of this load. Calculate the length of stiff bearing required and if it is greater than 15 mm, design the web stiffeners needed.

8. A 305 × 127 mm at 42 kg/m UB in grade 43 steel has a simply supported span of 6.5 m, cased in concrete of dimensions 225 × 410 mm and is to support a column at mid span. The column load consists of 15 kN dead load and 30 kN live load. Assume lateral restraint at the column and end torsional restraint. Check that the beam is correctly designed and calculate the length of stiff bearing needed. The unit weight of concrete = 23 kN/m³.

9. A 406 × 178 mm at 74 UB has a simply supported span of 8 m. It supports two point loads at 1 m and 5 m from one end as well as its own weight. Calculate the mid span deflection if both loads are of P kN live load.

10. Find the maximum value of P which can be supported by the beam in Question 9 if a cased construction is used and the beam is of grade 43 steel, the concrete dimensions being 520 × 280 mm. Assume no lateral restraint or end restraint, nor any stiff bearing length.

11. A beam spanning 6 m across the ceiling of a warehouse projects a further 1 m beyond one wall, thus acting as a hoist. A movable crab hangs from the lower flange and can be positioned anywhere along the beam. If the largest load to be carried is 75 kN when the crab is

over the warehouse floor and 50 kN when on the cantilever, design a beam in grade 43 steel to suit the purpose. Assume that the cantilever is continuous at the support with partial restraint against torsion of the support. The beam is built into the wall at the other end.

References

1. *Handbook on Structural Steelwork*, published by The British Constructional Steelwork Association and Constrado.
2. *The Steel Designers' Manual*, published by Crosby Lockwood Staples for Constrado.
3. Ibid.

Chapter 8

Stanchions

The design of compression members

In the design of any compression member, two questions must be asked:
1. when shall a column be deemed to have failed?, and 2. what factors
affect that failure? In the section on the design of reinforced concrete
columns, since only short columns were considered, the failure was
deemed to have occurred when the ultimate concrete and steel stresses
were reached. The design of the equivalent steel column being somewhat
easier from a mathematical point of view, this section will probe into the
subject of column design in slightly greater detail, making the answer to
the first question more difficult. First consider the factors affecting the
failure of columns, especially steel columns.

1. The effective length and the radius of gyration
As has been illustrated on p. 84, although the Euler buckling formula
may not be used in the design of any real column, it does serve to illustrate the effects of length and cross-sectional geometry on the behaviour
of a column. It may be restated in the form:

$$f_e = \frac{\pi^2 E}{(kL/r_y)^2}$$

where f_e is the critical or 'Euler' stress
   k is the effective length factor
   L is the actual length
   r_y is the radius of gyration (usually taken about the weaker
     or 'minor' axis)

Therefore the two factors effective length and radius of gyration may be combined in one factor called the slenderness ratio.

Thus effective length $= kL = l$

and slenderness ratio $= l/r_y = kL/r_y$

Note: The values of 'k' found by applying the Euler theory for various end restraint conditions have already been illustrated in Fig. 4.1, but as a result of the effect of some of the following factors, BS 449 modifies them as shown in Fig. 8.1.

2. The initial straightness

Both local and overall misalignments will have very great effect on the load carrying capacity of any column. If the central portion is out of line

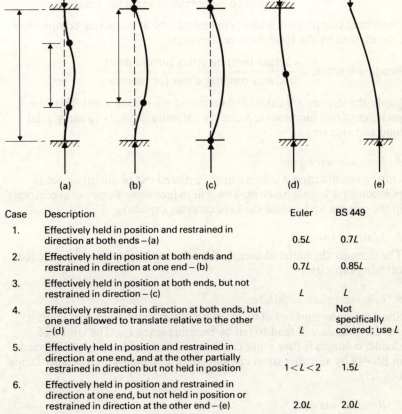

Case	Description	Euler	BS 449
1.	Effectively held in position and restrained in direction at both ends – (a)	0.5L	0.7L
2.	Effectively held in position at both ends and restrained in direction at one end – (b)	0.7L	0.85L
3.	Effectively held in position at both ends, but not restrained in direction – (c)	L	L
4.	Effectively restrained in direction at both ends, but one end allowed to translate relative to the other – (d)	L	Not specifically covered; use L
5.	Effectively held in position and restrained in direction at one end, and at the other partially restrained in direction but not held in position	1 < L < 2	1.5L
6.	Effectively held in position and restrained in direction at one end, but not held in position or restrained in direction at the other end – (e)	2.0L	2.0L

Note: In case 5 the value of the effective length truly depends on the degree of partial restraint, but it is most commonly around the figure of 1.5L

Fig. 8.1

with the ends the eccentricity caused will induce extra moments in the column and failure will occur at lower loads than an initially straight column of similar dimensions.

3. End moments or Eccentricity of Reaction

If bending stresses are imposed on a section which is also carrying thrust, then the interaction formula given in BS 449 must be used. Thus

$$\frac{f_c}{p_c} + \frac{f_{bc}}{p_{bc}} \text{ must not exceed unity}$$

In other words, the action of the column experiencing both axial stress and bending stress may be considered to be made up of two fractions – the compressive fraction and the bending fraction

$$\text{Compressive fraction} = \frac{\text{actual compressive stress (or load)}}{\text{allowed compressive stress (or load)}}$$

(the actual compressive stress being calculated as the actual compressive load divided by the total cross-section area)

$$\text{Bending fraction} = \frac{\text{actual bending stress (or moment)}}{\text{allowed bending stress (or moment)}}$$

(again the stresses are calculated on the whole section area). Thus the presence of end moments reduces the column's capacity to carry axial load and vice versa.

4. Errors in setting out

During construction a column may be placed out of plumb or out of position by a few millimetres. This will induce some degree of eccentricity in the column and reduce the load carrying capacity.

5. Material strength

The stronger the material used in the column the greater will be the load carrying capacity.

6. Behaviour close to failure

Short columns tend to fail by crushing of the column material while a slender column will tend to fail by buckling sideways. The failure of slender columns is thus more catastrophic and sudden. This is reflected in BS 449 by allowing short columns to be designed with a smaller factor of safety.

7. Residual stresses

As a thick section cools from an elevated temperature the outer fibres cool more quickly and reach the solid state more quickly. As the centre fibres cool they force the outer fibres to contract and this induces compression in them, while a balancing tension is experienced in the centre. Clearly these stresses will affect the carrying capacity of the column.

How then does BS 449 approach the design of a compression member? The relevant clauses are 30 to 40 inclusive, Cl. 31 dividing compression members into two classes: 1. struts, and 2. stanchions with which this chapter is concerned. The definition of these two categories is not helped by the statement at the beginning of the code, where a strut is defined as 'A pillar, **stanchion**, column or other compression member' (author's emphasis). However, by reference to Appendix D of BS 449 it may be seen that a stanchion is understood to be the type of column used in a framed structure. Basically this consists of a column and beam arrangement whereby the beams are connected to the columns at right angles and only at floor levels. In short, it is a rectangular, three-dimensional arrangement. Struts may be used to describe all other types of compression member, such as the compression chord in a truss.

The design depends on the calculation of an allowable axial stress for the column dependent on the maximum slenderness ratio of the column. Once the slenderness ratio has been obtained, the allowable axial stress p_c, may be found from Table 17 of BS 449 for the grade of steel being used. This table, reproduced in this chapter (pp. 189–191), is based on the Perry-Robertson formula for column buckling. This formula allows for the lack of initial straightness and assumes that failure will occur when the stress in any fibre reaches the yield stress. Thus this is the failure criterion adopted by BS 449 and is the answer to the first question posed on p. 174. The equation is given on p. 96 in Appendix B of BS 449, and may be restated more simply as:

$$(C_o - k_2 p_c)(Y_s - k_2 p_c) = \eta C_o Y_s$$

where p_c = the permissible average stress
         k_2 = load factor or coefficient, taken as 1.7 in BS 449
         Y_s = the minimum yield stress
         C_o = the Euler critical stress f_e
         η = $0.3(l/r_y)^2$

It is interesting to observe this equation in graphical form alongside the Euler formula, and this is done in Fig. 8.2.

Clearly the design process must be one of trial and error. This is necessary due to the need to calculate the permissible stress dependent on the actual slenderness ratio. This cannot be accomplished until a section is chosen, but a section cannot be chosen until the permissible stress is known. The designer must therefore use his experience of designing other columns in order to pick a trial section. Once this trial section is chosen, the first task is to calculate the effective length and the

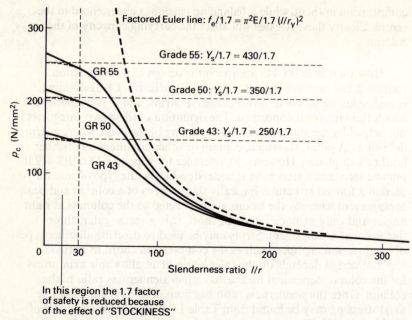

Fig. 8.2

In this region the 1.7 factor
of safety is reduced because
of the effect of "STOCKINESS"

slenderness ratio. This done, the axial strength is compared with the
strength which is needed and modifications are made if necessary.

The effective length of stanchions

At the outset of this section it must be realised that the calculation of
an effective length for any stanchion is now recognised as being a little
contrived. The behaviour of buildings is much more complex than would
be suggested by this method. If an accurate effective length were to be
calculated for stanchions in many buildings, it would be found that the
effective length was at variance with that which will be used for design to
BS 449. Indeed, the effective length will vary according to the load and
its distribution on the rest of the structure. However, akin to many
techniques of modern structural engineering, this method is used simply
because it works and has done for many years. Nevertheless, the inherent
inaccuracies must result in the use of oversized columns and it is to be
hoped that the new code will present a radically different approach to
that which is about to be set forth in this section.

Once the dimensions and overall layout of a building have been
decided, the designer can calculate each column's effective length by
referring to Cl. 31b and Appendix D of BS 449. Appendix D consists of a
number of diagrams showing typical stanchions as they would be in a

frame. To these diagrams BS 449 attaches the relevant effective length factor which should be used in the design of a column such as the one shown. It is then usually a simple matter for any designer to match the specific column with which he is dealing to one of the types given in this appendix. Samples of these diagrams are reproduced in Figs. 8.3 to 8.10 and all of the examples and questions given in this section are of types similar to one of these columns.

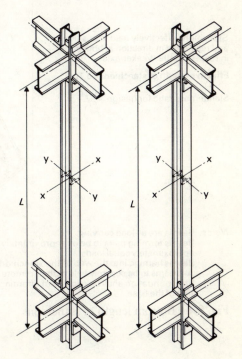

Stanchions effectively held in position and restrained in direction at both ends in respect of the weaker axis

Effective length of stanchions = 0.7L

Slenderness ratio for design $= \dfrac{0.7L}{r_y}$

Notes: Beams are all load carrying.
Beams forming pairs to be of approximately equal size and span and carrying approximately equal loads.
Beams framing into the web of the stanchions to have moment connections.
All beams to be securely held at their remote ends.

Fig. 8.3 Continuous intermediate lengths or top lengths of stanchions

So far it has been assumed that to each column there belongs one and only one effective length, but this is not necessarily the case as is illustrated in Fig. 8.10. This column is supported at top and bottom, but has additional restraints along its length. However, these restraints are only effective in reducing the tendency of the column to buckle about one of its axes – in this case the minor axis. The column may still buckle about the major x–x axis without the side purlins affording any restraint in that direction. Thus the effective length of the column about the minor

180

Stanchions effectively held in position
and restrained in direction at both ends
in respect of the weaker axis

Effective length of stanchions=0.7L

Slenderness ratio for design$=\dfrac{0.7L}{r_y}$

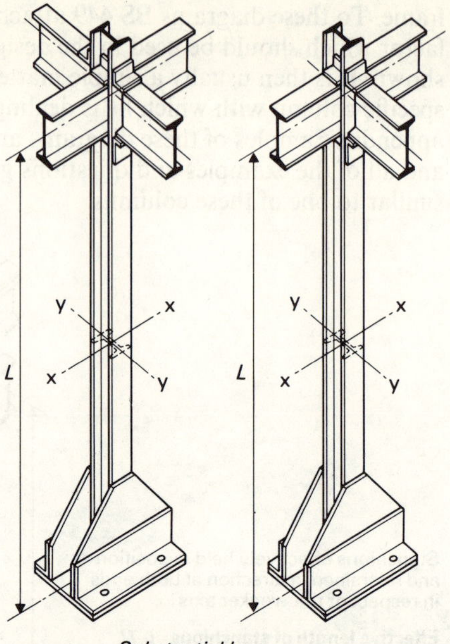

Substantial base

Notes: Beams are all load carrying.
Beams forming pairs to be of approximately equal size and span and carrying approximately equal loads.
Beams framing into the web of the stanchion to have moment connections.
All beams to be securely held at their remote ends.
The foundation shall be capable of affording restraint commensurate with that of the base.

Fig. 8.4   Bottom length of stanchions

axis is less than its effective length about the major axis. Two values of slenderness ratio are therefore calculated:

$$l_x/r_x \qquad \text{and} \qquad l_y/r_y$$

where l_x and l_y are respectively the effective lengths about the major and minor axes, and r_x and r_y are the radii of gyration about the major and minor axes respectively

The greater of these two values will be used to determine the permissible axial stress to be used in designing this column.

Example 8.1   A column similar to that shown in Fig. 8.3 carries an axial load of 500 kN. Design a section in grade 43 steel to carry this load. The actual length of the column is 5 m.

Effective length = 0.7 × 5 = 3.5 m

This end is not effectively restrained in direction about the y–y axis

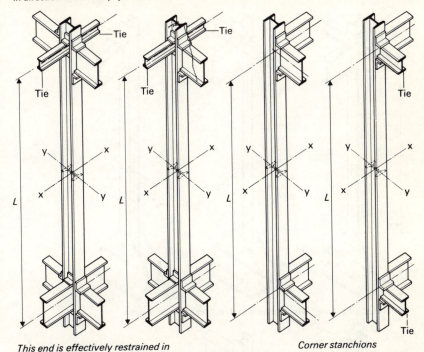

This end is effectively restrained in direction about both axes

Corner stanchions

Stanchions effectively held in position at both ends and restrained in direction at one end

Effective length of stanchions = 0.85L

Slenderness ration for design $= \dfrac{0.85L}{r_y}$

Notes: Beams are load-carrying except tie beams.
Beams forming pairs to be of approximately equal size and span and carrying approximately equal loads.
All beams shall be securely held at their remote ends.

Fig. 8.5 Continuous intermediate lengths or top lengths of stanchions

Try a section: 152×152 mm at 37 kg/m universal column

$$r_y = 38.7 \text{ mm}$$

$$l/r_y = 90.4$$

From Table 17a of BS 449 $p_c = 90.5$ N/mm^2

Section area $= 4740$ mm^2

Hence allowed compressive force $= 4740 \times 90.5/1000$ kN
$$= 428.97 \text{ kN (too small)}$$

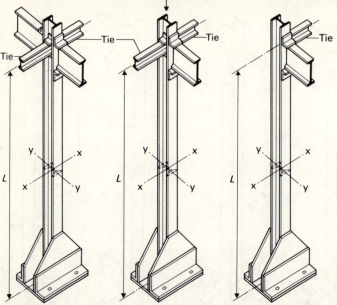

This end is not effectively restrained
in direction about the y–y axis

Tie

Tie

Tie

Tie

L L L

Substantial base giving restraint in direction about both axes

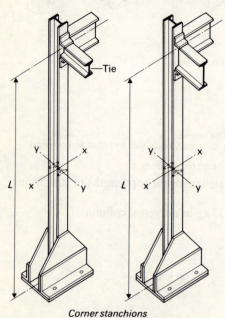

Tie

L L

Stanchions effectively held in position
at both ends and restrained in
direction at one end

Effective length of stanchions = 0.85L

Slenderness ratio for design $= \dfrac{0.85L}{r_y}$

Notes: Beams are load-carrying
except tie beams.
Beams forming pairs to be of
approximately equal size
and span and carrying
approximately equal loads.
All beams shall be securely
held at their remote ends.
The foundation shall be
capable of affording restraint
commensurate with that of
the base.

Corner stanchions

Fig. 8.6   Bottom lengths of stanchions

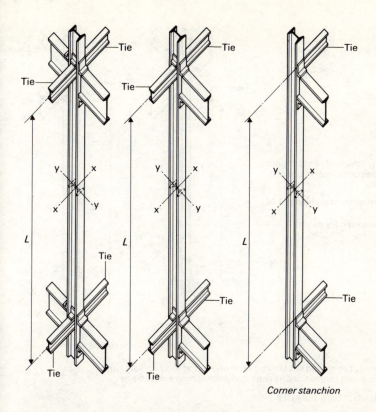

Corner stanchion

Stanchions effectively held at both ends in position but not restrained in direction (about the weaker axis)

Effective length of stanchions = 1.0L

Slenderness ratio for design $= \dfrac{1.0L}{r_y}$

Notes: All the beams shall be securely held at their remote ends.
Tie beams may have no moment connections.

Fig. 8.7 Continuous intermediate length

Try a section: 203 × 203 mm at 46 kg/m UC

$$r_y = 51.1 \text{ mm}$$

$$l/r_y = 68.5$$

From Table 17a in BS 449 $p_c = 117.5 \text{ N/mm}^2$

Section area $= 5880 \text{ mm}^2$

Hence allowed axial compressive force $= 5880 × 117.5/1000 \text{ kN}$
$$= 690.0 \text{ kN}$$

This is the lightest universal column section which will suffice

184

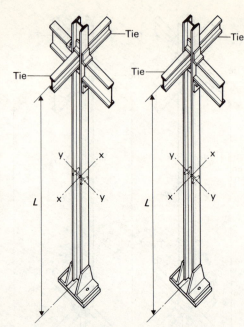

Stanchions effectively held at both ends in position but not restrained in direction

Effective length of stanchions = 1.0L

Slenderness ratio for design = $\dfrac{1.0L}{r_y}$

Small base (or slab base)

Notes: All the beams and ties shall be securely held at their remote ends.
Tie beam connections have no appreciable moment restraint.

Fig. 8.8   Bottom lengths

Alternatively
Try a section: 150×150 mm at 35.4 kg/m RHS

$$r_y = 57.8 \text{ mm}$$
$$l/r_y = 60.5$$
$$p_c = 125.5 \text{ N/mm}^2$$

Section area = 4510 mm²

Allowed axial compressive force = 566 kN

Try a section: 180×180 mm at 34.2 kg/m RHS

$$r_y = 70.8$$
$$l/r_y = 49.4$$
$$p_c = 133.5 \text{ N/mm}^2$$

Section area = 4360 mm²

Allowed axial compressive force = 582 kN

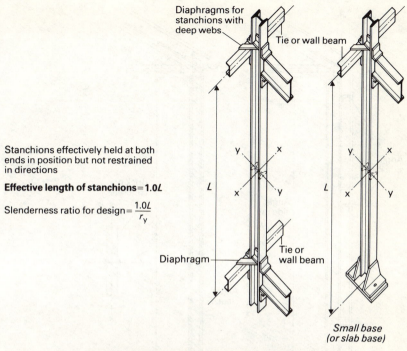

Diaphragms for stanchions with deep webs

Tie or wall beam

y · x

x · y

L

Stanchions effectively held at both ends in position but not restrained in directions

Effective length of stanchions=1.0*L*

Slenderness ratio for design=$\dfrac{1.0L}{r_y}$

Diaphragm

Tie or wall beam

y · x

x · y

L

Small base (or slab base)

Notes: All the beams and ties shall be securely held at their remote ends.
Tie beam connections have no appreciable moment restraint.

Fig. 8.9 Stanchions with tie beams attached to one flange

As can easily be seen both these columns satisfy the loading conditions and are lighter than the universal column section which must be used. They are square hollow sections, and it will be found that hollow sections often give a marked saving in material over the H-section alternative. The main drawback in using a hollow section as a column lies in the difficulty of connecting beams to it. These must either be site-welded directly onto the face of the RHS or brackets must be welded to the face whilst still in the fabrication shop. However, they can also be filled with concrete which increases the saving made and makes them an attractive choice. Then again a hollow section column may be filled with water or other fire-fighting fluid saving on the amount of services needed in the building and at the same time increasing the fire resistance of the hollow section itself.

Example 8.2   A column with a base similar to that shown in Fig. 8.8 but with the upper end supported as in Fig. 8.3 is loaded by an axial compressive force of 1500 kN. Design a suitable section in grade 50 steel if the actual length of column is 5 m.

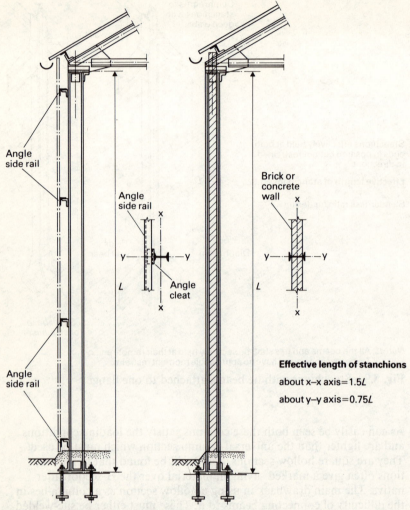

Effective length of stanchions

about x–x axis=1.5L

about y–y axis=0.75L

Fig. 8.10 Side stanchions in single-storey building

Universal column sections

Effective length: Top end held in position and in direction bottom end held in position but not in direction

$l = 0.85 \times L = 0.85 \times 5 = 4.25$ m

Try a section: 254 × 254 mm at 89 kg/m universal column

$$r_y = 65.2 \text{ mm}$$
$$l/r_y = 65.2$$
$$p_c = 159.5 \text{ N/mm}^2$$

Section area = 11 400 mm^2

Allowed axial compressive force = 1818 kN.

It will be found that the lighter 254 × 254 mm at 73 kg/m universal column can support a load of 1477 kN thus the section above is the lightest section which can carry the load specified.

Rolled hollow sections

Try a section: 200 × 200 mm at 73 kg/m square hollow section

$$r_y = 76.3 \text{ mm}$$

$$l/r_y = 55.7$$

$$p_c = 176.3 \text{ N/mm}^2$$

Section area = 9300 mm^2

Allowed axial compressive force = 1639.6 kN

Try a section: 250 × 250 mm at 75 kg/m square hollow section

$$r_y = 97.8 \text{ mm}$$

$$l/r_y = 43.46$$

$$p_c = 190.5 \text{ N/mm}^2$$

Section area = 9550 mm^2

Allowed axial compressive force = 1819 kN.

As before, both these columns are lighter than the equivalent universal column section.

Example 8.3 A column is supported in a manner similar to that shown in Fig. 8.10. It is to carry a load of 1500 kN; design a suitable section in grade 43 steel. The actual length of the column is 5 m.

Effective lengths

About the x–x axis $l_x = 1.5 \times 5 = 7.5 \text{ m}$

About the y–y axis $l_y = 0.75 \times 5 = 3.75 \text{ m}$

Universal column sections

Try a section: 254 × 254 mm at 107 kg/m UC

$$r_x = 113 \text{ mm} \qquad r_y = 65.7 \text{ mm}$$

$$l_x/r_x = \frac{7500}{113} = 66.4 \qquad l_y/r_y = \frac{3750}{65.7} = 57.1$$

$$\overline{p_c = 119.6 \text{ N/mm}^2}$$

Section area = 13 660 mm²

Allowed axial compression force = 1633.7 kN.

Try a section: 305 × 305 mm at 97 kg/m UC

$$r_x = 134 \text{ mm} \qquad r_y = 76.8 \text{ mm}$$

$$\underline{l_x/r_x = 56} \qquad l_y/r_y = 48.8$$

$$p_c = 129 \text{ N/mm}^2$$

Section area = 12 330 mm²

Allowed axial compressive force = 1590.6 kN.

Rolled hollow sections
Try a section: 250 × 250 mm at 117 kg/m

$$r_x = r_y = 95.3 \text{ mm}$$

$$\underline{l_x/r_x = 78.7} \qquad l_y/r_y = 39.35$$

$$p_c = 105.3 \text{ N/mm}^2$$

Section area = 14 900 mm²

Allowed axial compressive force = 1567 kN.

Try a section: 400 × 200 mm at 90.7 kg/m

$$r_x = 145 \text{ mm} \qquad r_y = 83.9 \text{ mm}$$

$$\underline{l_x/r_x = 51.7} \qquad l_y/r_y = 44.7$$

$$p_c = 132.3 \text{ N/mm}^2$$

Section area = 11 600 mm²

Allowed axial compressive force = 1533.5 kN.

Local buckling in rolled hollow sections

Rolled hollow sections which have large width/thickness ratios may fail by local buckling of the walls before the yield point is reached. Greater loads may be carried only if some form of internal stiffening is provided. Whilst filling the tube with concrete will perform this stiffening admirably, since it also increases the load carrying capacity of the column so formed beyond that which can be supported by the steel alone, it is usual to arrange that no column fails by local buckling without the addition of stiffeners. The problem is not so great as might at first be imagined since only eleven of the many sections produced by British Steel (Tubes Division) are affected. They are listed in Table 8.1 along with the limiting loads above which local buckling will cause failure.

Table 17a Allowable stress p_c on gross section for axial compression

l/r	p_c (N/mm^2) for grade 43 steel									
	0	**1**	**2**	**3**	**4**	**5**	**6**	**7**	**8**	**9**
0	155	155	154	154	153	153	153	152	152	151
10	151	151	150	150	149	149	148	148	148	147
20	147	146	146	146	145	145	144	144	144	143
30	143	142	142	142	141	141	141	140	140	139
40	139	138	138	137	137	136	136	136	135	134
50	133	133	132	131	130	130	129	128	127	126
60	126	125	124	123	122	121	120	119	118	117
70	115	114	113	112	111	110	108	107	106	105
80	104	102	101	100	99	97	96	95	94	92
90	91	90	89	87	86	85	84	83	81	80
100	79	78	77	76	75	74	73	72	71	70
110	69	68	67	66	65	64	63	62	61	61
120	60	59	58	57	56	56	55	54	53	53
130	52	51	51	50	49	49	48	48	47	46
140	46	45	45	44	43	43	42	42	41	41
150	40	40	39	39	38	38	38	37	37	36
160	36	35	35	35	34	34	33	33	33	32
170	32	32	31	31	31	30	30	30	29	29
180	29	28	28	28	28	27	27	27	26	26
190	26	26	25	25	25	25	24	24	24	24
200	24	23	23	23	23	22	22	22	22	22
210	21	21	21	21	21	20	20	20	20	20
220	20	19	19	19	19	19	19	18	18	18
230	18	18	18	18	17	17	17	17	17	17
240	17	16	16	16	16	16	16	16	16	15
250	15									
300	11									
350	8									

Intermediate values may be obtained by linear interpolation.
Note: For material over 40 mm thick, other than rolled I-beams or channels, and for Universal columns of thickness exceeding 40 mm, the limiting stress is 140 N/mm^2.

190

Table 17b Allowable stress p_c on gross section for axial compression

l/r	p_c (N/mm²) for grade 50 steel									
	0	**1**	**2**	**3**	**4**	**5**	**6**	**7**	**8**	**9**
0	215	214	214	213	213	212	212	211	211	210
10	210	209	209	208	208	207	207	206	206	205
20	205	204	204	203	203	202	202	201	201	200
30	200	199	199	198	197	197	196	196	195	194
40	193	193	192	191	190	189	188	187	186	185
50	184	183	181	180	179	177	176	174	173	171
60	169	168	166	164	162	160	158	156	154	152
70	150	148	146	144	142	140	138	135	133	131
80	129	127	125	123	121	119	117	115	113	111
90	109	107	106	104	102	100	99	97	95	94
100	92	91	89	88	86	85	84	82	81	80
110	78	77	76	75	74	72	71	70	69	68
120	67	66	65	64	63	62	61	60	60	59
130	58	57	56	55	55	54	53	52	52	51
140	50	50	49	48	48	47	47	46	45	45
150	44	44	43	43	42	42	41	41	40	40
160	39	39	38	38	37	37	36	36	36	35
170	35	34	34	34	33	33	33	32	32	31
180	31	31	30	30	30	30	29	29	29	28
190	28	28	27	27	27	27	26	26	26	26
200	25	25	25	25	24	24	24	24	23	23
210	23	23	23	22	22	22	22	22	21	21
220	21	21	21	20	20	20	20	20	20	19
230	19	19	19	19	19	18	18	18	18	18
240	18	18	17	17	17	17	17	17	17	16
250	16									
300	11									
350	8									

Intermediate values may be obtained by linear interpolation.
Note: For materials over 65 mm thick, the allowable stress p_c on gross section for axial compression shall be calculated in accordance with the procedure in Appendix B taking Y_s equal to the value of the yield stress agreed with the manufacturer, with a maximum value of 350 N/mm².

Table 17c Allowable stress p_c on gross section for axial compression

l/r	0	1	2	3	4	5	6	7	8	9
	p_c (N/mm^2) for grade 55 steel									
0	265	264	264	263	262	262	261	260	260	259
10	258	258	257	256	256	255	254	254	253	252
20	252	251	250	250	249	248	248	247	246	246
30	245	244	244	243	242	241	240	239	239	238
40	236	235	234	233	232	230	229	227	226	224
50	222	220	219	217	214	212	210	208	205	203
60	200	197	195	192	189	186	183	180	178	175
70	172	169	166	163	160	157	154	151	148	146
80	143	140	138	135	133	130	128	125	123	121
90	118	116	114	112	110	108	106	104	102	100
100	99	97	95	93	92	90	89	87	86	84
110	83	82	80	79	78	76	75	74	73	72
120	71	69	68	67	66	65	64	63	62	62
130	61	60	59	58	57	56	56	55	54	53
140	53	52	51	50	50	49	49	48	47	47
150	46	45	45	44	44	43	43	42	42	41
160	41	40	40	39	39	38	38	37	37	37
170	36	36	35	35	34	34	34	33	33	33
180	32	32	32	31	31	31	30	30	30	29
190	29	29	28	28	28	28	28	27	27	27
200	26	26	26	25	25	25	25	25	24	24
210	24	24	23	23	23	23	23	22	22	22
220	22	22	21	21	21	21	21	20	20	20
230	20	20	20	19	19	19	19	19	19	18
240	18	18	18	18	18	18	17	17	17	17
250	17									
300	12									
350	9									

Intermediate values may be obtained by linear interpolation.
Note: For materials over 40 mm thick, other than rolled I-beams or channels, and for Universal columns of thickness exceeding 40 mm, the limiting stress is 245 N/mm^2.

192

Table 8.1 Rolled hollow sections susceptible to local buckling

Steel grade	Serial size mm × mm	Wall thickness mm	Mass per metre kg/mm	Local buckling load kN
43C	300 × 200	6.3	48.1	881
	450 × 250	10.0	106	2081
50C	200 × 100	5.0	22.7	598
	250 × 150	6.3	38.2	996
	300 × 200	6.3	48.1	1034
	300 × 200	8.0	60.5	1627
	400 × 200	10.0	90.7	2401
	450 × 250	10.0	106	2516
	250 × 250	6.3	48.1	1151
	350 × 350	10.0	106	2842
	400 × 400	10.0	122	2917

The maximum length of stanchions (Cl. 33 of BS 449)

This clause states that neither of the following two ratios should exceed 180 if the axial loads arise from dead and imposed loads or 250 if the axial loads arise from wind load only.

Neither kL/r nor L/r should exceed 180 (or 250)

where k is as before, the effective length factor
L is the length between connections to the beams
r is the appropriate radius of gyration

Therefore, in the three previous examples the ratio which is involved in this consideration is that of actual length divided by r_y. None of the columns contravenes this regulation.

The effect of end moments on the carrying capacity of stanchions

Bending moments at the ends of columns arise in one of two ways:

1. In a frame designed to have rigid connections between columns and beams, the fixity moments at the ends of the beams must be carried by the columns also.
2. Even in a non-rigidly designed frame the reactions at the ends of the beam are not purely shear. In effect, its as if the reactions were applied at a small distance from the face of the column. This is referred to as the eccentricity of the beam reactions. BS 449 in Cl. 34a, imposes a standard value of 100 mm on this eccentricity, thus all

stanchions will be subject to axial loads and to bending moments at their ends. However, in two special cases, this eccentricity may be reduced or even eliminated.

(a) In the case of cap connections, viz. those connections made to the top of a column, where the column connection is in the form of a steel plate welded to the top of the column, the load shall be assumed to be applied at the face of the stanchion section, or at the edge of any packing, if used, towards the span of the beam.

(b) In the case of a roof truss connection made to a column cap, no eccentricity need be taken for simple bearings without moment capability.

The next question must be to determine the proportion of these moments which are taken by the column projecting above the beams and that which is taken by the downward projection of the column. In case 1 above where the moment arises from the inherent rigidity of construction the method of analysis will automatically yield the answer to this question, but for case 2 BS 449 gives the following rules:

1. Divide the moment due to eccentricity at any one floor or beam level equally between the upper and lower projections of the column unless the ratio of upper to lower stiffnesses exceeds 1.5. In cases exceeding this ratio, the moment is to be divided in proportion to these stiffnesses. The stiffness of any column is defined as the moments of inertia of the column sections divided by the actual lengths.

2. The moments developed at one level do not affect the levels immediately above or below. In other words they need only be considered as acting at the level where they are developed.

Example 8.4 Determine the moments acting at all three levels of the column in Fig. 8.11. All the beams shown are producing moment about the x–x axis.
Working from the roof downwards:

Level 3
Actual depth of the column section $= D = 203.2$ mm
Since it is a column cap connection, the eccentricity is taken as being at the face of the column
Eccentricity $= 0.5D = 101.6$ mm
Taking moments about the centre of the column:

$M = 150 \times 101.6/1000 - 200 \times 101.6/1000$ kNm clockwise

$= 5.08$ kNm clockwise

Level 2
The eccentricity will be taken from the edge of the larger section
Actual depth of the column section $= 289.1$ mm

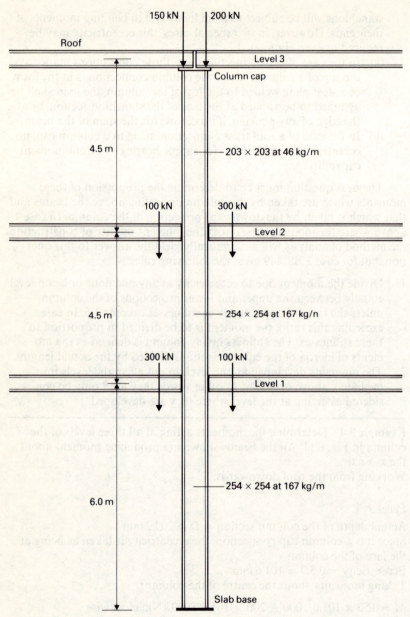

Fig. 8.11

Eccentricity $= 100 + 0.5D$ mm $= 244.55$ mm

$M = (300 - 100) \times 0.24455$ kNm clockwise

$\quad = 48.91$ kNm clockwise divided between upper and lower sections

Stiffness of upper column $= 45.64 \times 10^6/4500 = 10\,142.2$ mm$^3 = s_1$

Stiffness of lower column $= 299.14 \times 10^6/4500 = 66\,475.6$ mm$^3 = s_2$

Moment in upper column $= \dfrac{s_1}{(s_1 + s_2)} \times M = 6.47$ kNm clockwise

Moment in lower column $= \dfrac{s_2}{(s_1 + s_2)} \times M = 42.43$ kNm clockwise

Level 1

Eccentricity $= 244.55$ mm

$M = 48.91$ kNm anticlockwise

The ratio of stiffnesses will be found to be less than 1.5 hence the moment is divided equally between the upper and lower column sections

Finally, since the base of the column is a 'Slab Base' the moment there will be zero.

Figure 8.12 shows how this distribution of moments gives rise to a bending moment diagram and results in deflection. For the upper and lower columns this deflection is characterised by having a point of

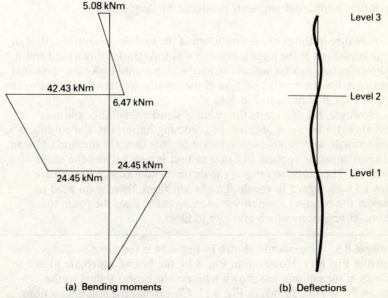

(a) Bending moments (b) Deflections

Fig. 8.12 Column bending

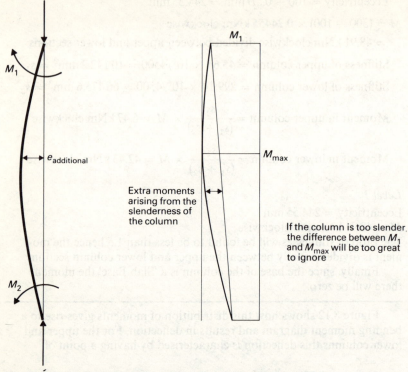

Fig. 8.13 Additional moments produced by slenderness

contraflexure in the column length, while the middle column length is in single curvature. If the middle column was too slender this would result in the deflection producing eccentricity which would make the moments experienced in the central portion of the column greater than those at the ends. This is illustrated in Fig. 8.13.

However, BS 449 limits the value of slenderness ratio, and this prevents extra moment of this type becoming important. If a column is designed with appreciable secondary moments then the method of 'Beam Columns' must be applied, but that subject is well beyond the scope of this book. Therefore, so long as the recommendations of BS 449 are followed with regard to the design of stanchions, there is no need to consider the bending moments at sections other than the beam to column connections which give rise to them.

Example 8.5 The column shown in Fig. 8.14 is the same one as is shown in Fig. 8.11. However, in Fig. 8.14, the beams which are joined to the web of the column are shown whereas the beams joined to the column flanges are shown in Fig. 8.11. Check that the column sections chosen are sufficient, if the steel used is of grade 43.

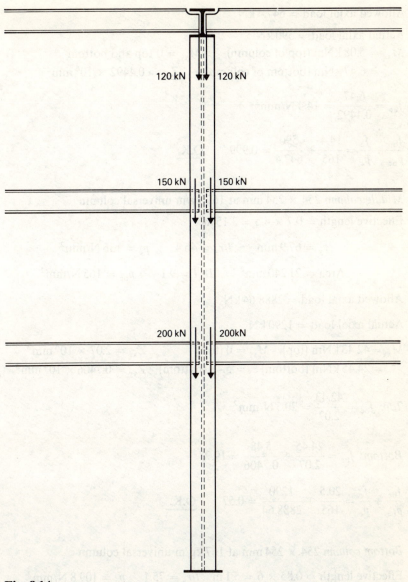

Fig. 8.14

Roof column 203 × 203 mm at 46 kg/m universal column
Effective length = 0.85 × 4.5 = 3.825 m

$$r_y = 51.1 \text{ mm} \qquad l/r_y = 74.9 \qquad p_c = 110.1 \text{ N/mm}^2$$
$$\text{Area} = 5880 \text{ mm}^2 \qquad D/T = 18.5 \qquad p_{bc} = 165 \text{ N/mm}^2$$

Allowed axial load = 647.4 kN

Actual axial load = 590 kN

M_{xx} = 5.08 kNm (top of column) M_{yy} = 0 top and bottom

     = 6.47 kNm (bottom of column) Z_{xx} = 0.4492 × 10⁶ mm³

$$f_{bc} = \frac{6.47}{0.4492} = 14.4 \text{ N/mm}^2$$

$$\frac{f_{bc}}{p_{bc}} + \frac{f_c}{p_c} = \frac{14.4}{165} + \frac{590}{647.4} = 0.999 \quad \underline{\text{O.K.}}$$

Middle column 254 × 254 mm at 167 kg/m universal column

Effective length = 0.7 × 4.5 = 3.15 m

$$r_y = 67.9 \text{ mm} \qquad l/r_y = 46.4 \qquad p_c = 136 \text{ N/mm}^2$$

$$\text{Area} = 21\,240 \text{ mm}^2 \qquad D/T = 9.1 \qquad p_{bc} = 165 \text{ N/mm}^2$$

Allowed axial load = 2888.64 kN

Actual axial load = 1290 kN

M_{xx} = 42.43 kNm (top) M_{yy} = 0 (top) Z_{xx} = 2.07 × 10⁶ mm³

     = 24.45 kNm (bottom) = 5.48 (bottom) Z_{yy} = 0.7406 × 10⁶ mm³

Top: $f_{bc} = \dfrac{42.43}{2.07} = 20.5 \text{ N/mm}^2$

Bottom: $f_{bc} = \dfrac{24.45}{2.07} + \dfrac{5.48}{0.7406} = 19.21$

$$\frac{f_{bc}}{p_{bc}} + \frac{f_c}{p_c} = \frac{20.5}{165} + \frac{1290}{2888.64} = 0.57 \quad \underline{\text{O.K.}}$$

Bottom column 254 × 254 mm at 167 kg/m universal column

Effective length = 0.85 × 6 = 5.1 m $l/r_y = 75.1$ $p_c = 109.8 \text{ N/mm}^2$

            Area = 21 240 mm² $D/T = 9.1$ $p_{bc} = 165 \text{ N/mm}^2$

Allowed axial load = 2332.15 kN

Actual axial load = 1990 kN

M_{xx} = 24.45 kNm (top) $M_{yy} = (200 - 100) \times 0.1096/2 = 5.48$ kNm (top)

     = 0 (bottom)   = 0 (bottom)

$$Z_{xx} = 2.07 \times 10^6 \, \text{mm}^3 \qquad Z_{yy} = 0.7406 \times 10^6 \, \text{mm}^3$$

$$f_{bc} = \frac{24.45}{2.07} + \frac{5.48}{0.7406} = 19.21 \, \text{N/mm}^2$$

$$\frac{f_{bc}}{p_{bc}} + \frac{f_c}{p_c} = \frac{19.21}{165} + \frac{1990}{2332.15} = 0.97 \qquad \underline{\text{O.K.}}$$

Note: **Self-weight should have been included, but is omitted for the sake of clarity. If included the column is still satisfactory.**

It may seem that the middle column section could be reduced, but for practical reasons it is often the case that column sections are changed at every second storey. The cost of erection is thus reduced making the final structure more economic.

Cased stanchions (Cl. 30b)

In order to protect a steel column from fire it is necessary to shield it from the direct heat of a flame. One method of accomplishing this is to encase the section in dense concrete, giving a cover of some 50 mm. However, as with an encased beam, this will also act to give added strength to the column in two ways:

1. The concrete will take some of the axial load.
2. The concrete will cause the column to be less slender.

The first of these involves the composite action of the concrete in conjunction with the steel, and results from the bond between the two materials. Thus the stress in the concrete is related to the strain and thus to the stress in the steel. So long as the concrete has a certain minimum strength, the actual concrete stress will be less than the ultimate stress that is possible. If the maximum possible concrete stress is f_c, then the actual concrete stress will be a fraction of this given by:

$$f_c \times \frac{p_c}{p_{bc}}$$

where the fraction p_c/p_{bc} represents the fraction of maximum steel stress that is experienced in the fully loaded column, and

p_c is obtained from Table 17 of BS 449 and depends on the slenderness ratio

p_{bc} is obtained from Table 2 of BS 449 and does not depend on the slenderness ratio

f_c is made equal to one quarter of the minimum permissible cube strength, which is specified by BS 449 as being 21 N/mm^2 at twenty-eight days. Therefore $f_c = 5.25 \, \text{N/mm}^2$

BS 449 expresses the above formula slightly differently as:

$$\text{Concrete stress} = \frac{p_c}{0.19 \times p_{bc}}$$

but $\dfrac{1}{0.19} \doteqdot 5.25$ as in the above formula

As to the area over which this stress acts, BS 449 stipulates that it be calculated without regard to the area occupied by the steel section and that a rectangle of sides 150 mm greater than the dimensions of the steel section be used, if there is concrete to that dimension available. Thus the maximum area of concrete, A_c is:

$$A_c = (B + 150)(D + 150)\,\text{mm}^2$$

and the load taken by the concrete

$$P_{cc} = (D + 150)(B + 150)p_c/0.19p_{bc}$$

As for the reduction in slenderness, this is allowed for in a manner exactly similar to that for the encased beam. British Standard 449 stipulates that the minimum width of casing should be $B + 100$ mm. The radius of gyration of the column about the minor axis may then be taken as $0.2(B + 100)$ mm. However, the radius of gyration about the major axis is assumed not to be affected by the casing and is equal to the radius of gyration of the section alone. Finally, BS 449 places the following restrictions on the use of the method:

1. In no case shall the axial load on the cased stanchion exceed twice that which would be permitted on the uncased section.
2. The slenderness ratio, or more correctly the length ratio, measured as the full length, centre to centre of connections divided by the steel section's radius of gyration, r_y shall not exceed 250.
3. The steel shall remain unpainted to allow adequate bond between steel and concrete to develop.
4. The concrete used shall be mixed from an aggregate of 10 mm unless full compaction and solidity can be obtained from a concrete of larger aggregate. The minimum cube strength at twenty-eight days shall be 21 N/mm².
5. The surface and edges of the steel section shall have a minimum cover of 50 mm. This also means that the minimum width of the casing will be $B + 100$ mm.
6. The casing shall be effectively reinforced with wire complying with BS 4482, 'Hard drawn mild steel wire for the reinforcement of concrete.' The wire shall be of at least 5 mm diameter and the reinforcement shall be in the form of stirrups or binding at not more than 200 mm pitch, and so arranged as to pass through the centre of the

covering to the edges and outer faces of the flanges, and to be supported by and attached to longitudinal spacing bars not fewer than four in number.

Alternatively, the casing may be reinforced with fabric complying with BS 4483, 'Steel fabric for the reinforcement of concrete,' or with bars complying with BS 4449, 'Hot rolled bars for the reinforcement of concrete,' provided in either case that the same requirements of diameter, spacing and positioning are met.

In summary of these requirements, the same diagram as used for the casing of beams, Fig. 7.10a may be used.

Example 8.6 Design the column in Fig. 8.15 first as an uncased and secondly as a cased section. The total vertical load is 1500 kN; the steel of

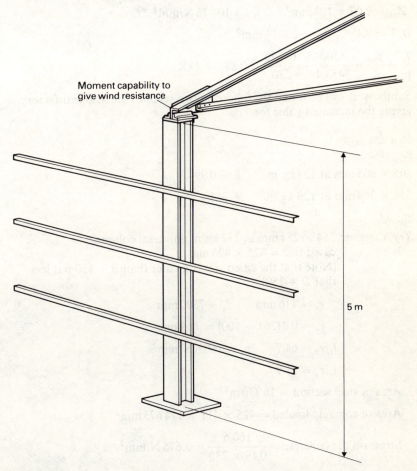

Moment capability to give wind resistance

5 m

Fig. 8.15

the column section is to be of grade 50. In order to give stability under wind loading, the top connection is capable of carrying moment.

Uncased

Try a section: 254 × 254 mm at 167 kg/m universal column

$$l_x = 7.5 \, \text{m} \quad l_y = 3.75 \, \text{m} \quad r_x = 119 \, \text{mm} \quad r_y = 67.9 \, \text{mm}$$

$$l_x/r_x = 63 \quad l_y/r_y = 55.2 \quad p_c = 164 \, \text{N/mm}^2$$

Area = 21 240 mm^2

Allowed axial load = $164 \times 21\,240 \, \text{N} = 3483.4 \, \text{kN}$

Actual axial load = 1500 kN

Eccentricity = $D/2 = 144.5 \, \text{mm} \qquad M_{xx} = 216.83 \, \text{kNm}$

$Z_{xx} = 2.07 \times 10^6 \, \text{mm}^3 \qquad f_{bc} = 104.75 \, \text{N/mm}^2$

$D/T = 9.1 \qquad p_{bc} = 230 \, \text{N/mm}^2$

$$\frac{f_c}{p_c} + \frac{f_{bc}}{p_{bc}} = \frac{1500}{3483.4} + \frac{104.75}{230} = 0.89 \qquad \underline{\text{O.K.}}$$

Similarly, it will be found the following two sections are also satisfactory, giving the indicated value for

$$\frac{f_c}{p_c} + \frac{f_{bc}}{p_{bc}} = F$$

305 × 305 mm at 137 kg/m $F = 0.994$

356 × 368 mm at 129 kg/m $F = 1$

Cased

Try a section: 254 × 254 mm at 132 kg/m universal column
    cased size = 425 × 425 mm
    (Note that the cased size is greater than $B + 150$ but less that $D + 150$)

$$r_x = 116 \, \text{mm} \qquad l_x = 7500 \, \text{mm}$$

$$r_y = 0.2 \, (261 + 100) = 72.2 \, \text{mm}$$

$$l_x/r_x = 64.7 \qquad p_c = 160.6 \, \text{N/mm}^2$$

$$l_y/r_y = 51.9$$

Area of steel section = 16 770 m^2

Area of concrete loaded = $425 \times 411 = 174\,675 \, \text{mm}^2$

Stress on the concrete = $\dfrac{160.6}{0.19 \times 230} = 3.675 \, \text{N/mm}^2$

Allowed axial load = $160.6 \times 16\,770 + 3.675 \times 174\,675 = 3335.2 \, \text{kN}$

Actual axial load $= 1500\,\text{kN}$

$M_{xx} = 1500 \times 0.1382 = 207.3\,\text{kNm}$ $\qquad Z_{xx} = 1.622 \times 10^6\,\text{mm}^3$

(Note that only the steel takes bending)

$f_{bc} = 127.81\,\text{N/mm}^2$ $\qquad p_{bc} = 230\,\text{N/mm}^2$

$$\frac{f_c}{p_c} + \frac{f_{bc}}{p_{bc}} = \frac{1500}{3335.2} + \frac{127.81}{230} = 1 \qquad \underline{\text{O.K.}}$$

Try a section: 305 × 305 mm at 118 kg/m universal column cased to
475 × 475 mm

It will be found that $\dfrac{f_c}{p_c} + \dfrac{f_{bc}}{p_{bc}} = 1.012$.

It is arguable whether this should be allowed or not. Personally, the author would not be inclined to do so. The load which this cased section could withstand is 1482 kN.

Try a section: 356 × 368 mm at 129 kg/m universal column cased to
500 × 520 mm.
It will be found that this section can withstand a load of
1720 kN.

Slab bases (Cl. 38b)

Considering that the design of concrete slab bases has been dealt with in an earlier chapter the term 'slab base' may be confusing when applied to a column in structural steel. In this context it implies a thin plate of steel welded or bracketed to the base of the column section. This then rests on the concrete foundation. Typical examples are shown in Figs. 8.16a and 8.16b, with the principal dimensions being given in Fig. 8.16c. This chapter is confined to the study of bases the load on which is concentric. In other words, no moments are assumed to be acting on the foot of the column. It is further assumed that the column is placed centrally on the slab base.

Analysis

The load from the column, W, is assumed to produce a uniformly distributed stress below the slab base which is given by:

$$w = W/LH \qquad\qquad [8.1]$$

If the plate of the base is treated as being a cantilever, the moments on sections t–t and s–s are:

$$M_{ss} = wAH \times \frac{A}{2} = 0.5aA^2H \qquad M_{tt} = wBL \times \frac{B}{2} = 0.5wB^2L$$

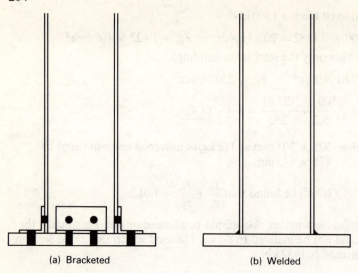

(a) Bracketed   (b) Welded

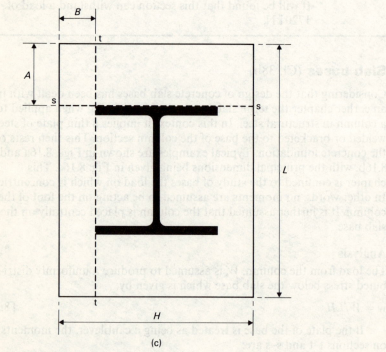

(c)

Fig. 8.16 Slab bases

The resulting bending stresses are:

$$f_{ss} = \frac{3wA^2}{t^2} \qquad f_{tt} = \frac{3wB^2}{t^2}$$

It is usual to assume that A is larger than B, thus the minimum value of t is:

$$t = \sqrt{\frac{3wA^2}{p_{bct}}}$$

p_{bct} being the allowed stress in slab bases $= 185\,\text{N/mm}^2$

Therefore, if the value of A is set equal to the maximum projection of the plate beyond the edges of the column, this formula will give a minimum value for the required plate thickness to withstand the axial load. However, this formula overestimates the thickness and greater economy can be obtained by allowing for the way in which the plate bends in two directions. One of the simplest ways of accomplishing this is by taking account of Poisson's ratio. The effect is illustrated in Fig. 8.17 where a cube of material is subjected to a bending moment about one of the axes along one face. As is to be expected, the two opposing faces subject to this bending moment are deformed, but the two faces at right angles are also 'squashed' in sympathy. However, the amount of squashing is less than the magnitude of the primary strains caused by the bending moment. Poisson's ratio is given by

$$n = \frac{\text{secondary strains}}{\text{primary strains}}$$

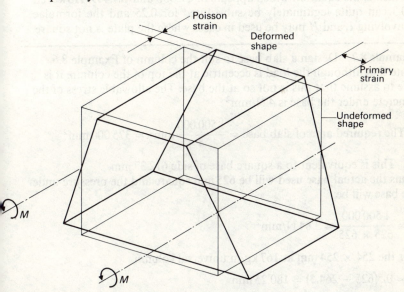

Since the material is assumed to act elastically, the term 'strain' can be substituted by 'stress'. Whence, by imposing a bending moment about the t–t axis (Fig. 8.16), a bending moment of the opposite sign is also imposed about the s–s axis equal to M_{tt} times the Poisson's ratio, n. Therefore the true, or effective moments are:

$$M'_{tt} = M_{tt} - n \cdot M_{ss} \qquad\qquad M'_{ss} = M_{ss} - n \cdot M_{tt}$$
$$= 0.5wB^2L - 0.5nwA^2H \qquad\qquad = 0.5wA^2H - 0.5nwB^2L$$
$$= 0.5w(B^2L - nA^2H) \qquad\qquad = 0.5w(A^2H - nB^2L)$$

$$f'_{tt} = 3w\left(B^2 - nA^2 \cdot \frac{H}{L}\right)\bigg/ t^2 \qquad f'_{ss} = 3w\left(A^2 - nB^2 \cdot \frac{L}{H}\right)\bigg/ t^2$$

equating the largest of these two stresses to p_{bct} will give an expression for the minimum thickness of plate.

When BS 449 was first issued, the value of Poisson's ratio was taken as being 0.25, but it is now recognised that 0.3 is a more correct figure. However, using 0.25 and assuming that a square plate is being designed, $L = H$, and letting A be the greatest projection:

$$f'_{ss} = \frac{3w(A^2 - 0.25B^2)}{t^2} = p_{bct}$$

$$\text{Thus} \quad t = \sqrt{\frac{3w}{p_{bct}}(A^2 - 0.25B^2)} \qquad\qquad\qquad [8.2]$$

which is the equation which appears in Cl. 38b and BS 449. However, 0.3 can quite legitimately be substituted for 0.25 and the formulae involving L and H may be used in cases where the plate is not square.

Example 8.7 Design a slab base to suit the column of Example 8.6. Note that although the load is eccentric at the top of the column, it is safe to assume that this is not so at the base. The allowable stress of the concrete under the base is $4\,N/mm^2$.

$$\text{The required area of slab base} = \frac{1\,500\,000}{4}\,mm^2 = 375\,000\,mm^2$$

This is equivalent to a square base of side 612.37 mm.
Thus the actual base used will be 625 mm square and the pressure under the base will be:

$$w = \frac{1\,500\,000}{625 \times 625} = 3.84\,N/mm^2$$

For the 254×254 mm at 167 kg/m universal column

$$A = 0.5(625 - 264.5) = 180.25\,mm$$
$$B = 0.5(625 - 289.1) = 167.95\,mm$$

From eqn [8.2]

$$t = \sqrt{\frac{3 \times 3.84}{185}(180.25^2 - 0.25 \times 167.95^2)}$$

$$= 39.8\,\text{mm} \quad (\text{say } 40\,\text{mm})$$

Or using $n = 0.3$ $\quad t = 38.68\,\text{mm} \quad (\text{again say } 40\,\text{mm}).$

Section	Plate thickness	
	$n = 0.25$	$n = 0.3$
305 × 305 at 137	32.47	31.24
356 × 368 at 129	29.55	28.67
254 × 254 at 132	37.10	36.39
305 × 305 at 118	34.66	33.56

Finally, when the slab base is welded directly onto the end of the column, the two surfaces must be machined to provide even bearing across the interface. This prevents uneven stress patterns being set up which might cause premature failure of the base.

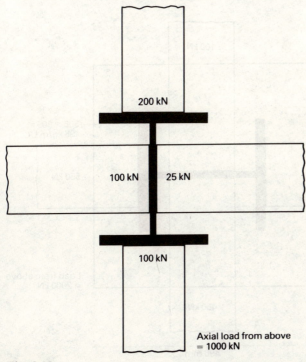

200 kN

100 kN   25 kN

100 kN

Axial load from above = 1000 kN

Fig. 8.18

208

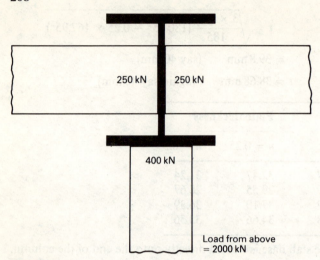

Fig. 8.19

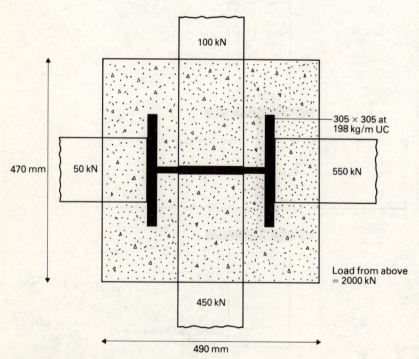

Fig. 8.20

Questions

1. A column similar to that shown in Fig. 8.4 of actual length 7.1 m, carries an axial load of 2000 kN. Design suitable stanchions in grades 43, 50 and 55 steel.

2. Repeat Question 1 for cased columns.

3. A column section carries four beams as shown in Fig. 8.18. The load from above is 1000 kN. If the effective length is 5 m, and the grade of steel used is 43, design a section remembering the minimum eccentricity rules of BS 449. Assume that the upper column is of the same section as the one you are designing.

4. A column carries three beams as shown in Fig. 8.19. The loads are as shown, the grade of steel used is 43, the actual length is 5 m, the column being supported as shown in Fig. 8.9. Design a suitable section if it continues to the storey above.

5. Check the column section shown in Fig. 8.20 if it is constructed from grade 50 steel and continues above and below the level of the beams shown. The effective length is 5 m.

6. Suggest two ways in which the section of Question 5 may be altered to reduce weight and cost.

7. Check the section shown in Fig. 8.21, if the steel is of grade 50 and the effective length = 4.8 m.

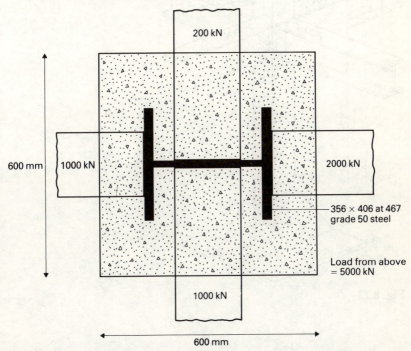

Fig. 8.21

8. Suggest alterations which may be made to the stanchion of Question 7 in order to reduce cost.

9. Check column ABC shown in Fig. 8.22 if the steel is of grade 50.

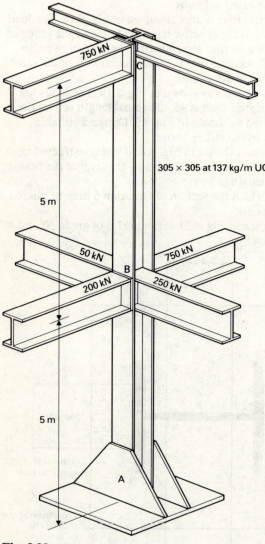

305 × 305 at 137 kg/m UC

Fig. 8.22

Chapter 9

Connections

Introduction

It is usual in structural design to calculate the forces, moments and shears in a framework, based on the assumption that the rigid joints can be achieved practically and with ease. Yet in terms of the percentage cost, it is these joints which can easily cost proportionately more than the rest of the frame. Although the joint is a place of very high stress concentrations it is the most empirically designed component for the whole frame. Very often the 'rigid' joints are not fully rigid and this is bound to affect the performance of the structure and result in wide differences between the predictions of the analysis and the truth of the situation.

In view of these and other shortcomings, how is it possible to design a joint to satisfy all the necessary criteria? One is tempted to reply facetiously – 'with difficulty'; yet this is not very far from the mark. Thus it will be understood that this chapter cannot be more than a brief introduction to the subject, yet without it the book would not in the author's view make any sense. Firstly, it will be necessary to classify the types of connection and joints and then consider a few of the major modes of failure adopted by connections.

Joints and connections

Joints may be first classified into two categories:
1. Those employing bolts or rivets – mechanical fasteners.
2. Those employing welding.
   Each calls for a different treatment and will be dealt with separately.

Bolted and riveted connections

Basically there are three classes of connectors: those which act in
1. shear, 2. tension or 3. friction induced by initial tension. Rivets are
confined to the first two, since it isn't really possible to calculate the
tension induced by the thermal contraction of the hot rivet. When two
or more plates are bolted together and clamped with an initial tension
in the bolt, the plates will develop a frictional shear resistance around
the bolt, equal to the coefficient of friction times the clamping force.
Provided that this frictional resistance is not exceeded, the stresses in the
bolt will not vary significantly from the initial tension, since there can be
no bearing of the bolt on the sides of the hole. This is the connection
which most closely approaches the fully 'rigid' assumption and must be
made with high-strength bolts of special construction and manufacture.

In many of the types of joint used nowadays, it is not possible to
utilise rivets or bolts acting in pure shear or in pure tension, a combination of stresses being the result. In fact, rivets are not often used now
because of their high cost. This is due to the need to have four men to
install them. The types of bolted and riveted joint which shall be considered are:

1. Moment joints causing shear and/or tension in the connectors. These
   are mostly found in portal frame and rigid frame construction, e.g.
   the moment connection of a beam to a column; the fixed base of a
   ground floor column, effected by using bolts placed in tension
   and embedded in the concrete of the foundation.
2. Force joints such as are to be found in pin-jointed trusses. These are
   usually constructed from connectors in shear and friction.
3. Eccentric joints which transmit both force and moment.

Although bolted joints are in many modern applications being
superseded by welding, there are still some major areas where the use of
bolts is desirable, e.g. if site welding is either dangerous or too slow;
where welding would cause deformation, where the steelwork is joined to
foundations. Therefore their design of great importance.

Modes of failure

A. Shear connections   A riveted (or non-friction bolted) joint may fail in
one of several ways as illustrated in Fig. 9.1:

1. By shearing of the rivet or bolt, either in single or double shear.
2. By crushing of the rivet or bolt, or of the material on which it bears.
3. By tension failure of the reduced section area either side of the
   connectors.
4. By shear-tearing of the edge.
5. By transverse tension of the front edge.

Failure by 4. and 5. may most easily be prevented by providing
sufficient end distance equal to twice the effective bolt or rivet diameter,
measured from the centre of the bolt to the edge of the plate. There is

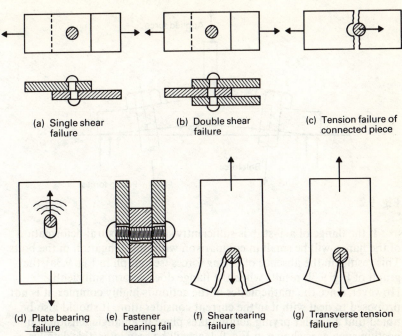

(a) Single shear failure

(b) Double shear failure

(c) Tension failure of connected piece

(d) Plate bearing failure

(e) Fastener bearing fail

(f) Shear tearing failure

(g) Transverse tension failure

Fig. 9.1 Modes of failure

also a minimum allowed edge distance given in Table 21 of BS 449, which is of the order of 1.5 to 1.75 times the effective diameter. Therefore the design process is concerned with 1., 2. and 3., calculated by the 'nominal stress' method.

B. Tension connections The failure of these connections is largely governed by the geometry of the connected pieces. It is not possible to consider all the various types of failure as there are as many types as there are types of joints, but a few important ones will be considered. The most obvious failure, that of directly induced tension failure in the bolt, rarely happens but should always be checked.

The most important mode of failure under this heading is the 'prying action' failure. One of the simplest connections with bolt groups in tension is the symmetric T-stub hanger with a simple line of fasteners parallel to, and on each side of, the web. The fasteners are assumed to be stressed equally because of symmetry of the connection. An external tensile load on the connection will reduce the contact pressure between the T-stub flange and the base. However, depending on the flexibility of the T-stub flange, additional forces may be developed near the flange tips. This is the phenomenon referred to as prying action and is illustrated in Fig. 9.2. Prying action increases the fastener force and may be detrimental to the strength and performance of the fasteners.

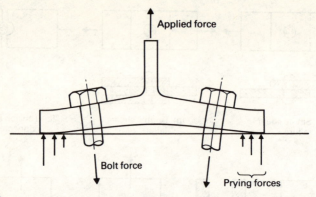

Fig. 9.2

If the flange of a T-stub is sufficiently stiff, the flexural deformation of the flange will be small in comparison with the elongation of the bolts. This results in the absence of prying forces (see graph in Fig. 9.3a), the graph of Fig. 9.3b results when the flange does deform sufficiently. However, since the mathematics of the action is highly complex, it is not proposed to deal with it in the current consideration. It should also be noted that a similar prying action takes place in an angle tension connection such as is shown in Fig. 9.4; as will be seen later, this detail is often used in the connections of beams to columns.

C. Friction connections The ultimate mode of failure of these connections will be similar to that for ordinary bolted or riveted connections, but the path to that failure will differ markedly. The main differences can be seen by examining the load-displacement curve for a single friction bolted connection in shear, as is shown in Fig. 9.5. Initially, from O to A, the connection is much more rigid than that made with a

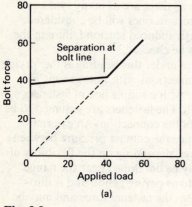

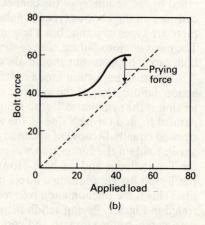

Fig. 9.3

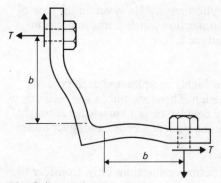

Fig. 9.4

non-friction type bolt, i.e. one in which there is only bearing. At
A the frictional resistance is just defeated and a slip occurs. This slip
allows the bolt to come into contact with the edges of the holes; the less
the hole clearance the less will this slip be. Increased load is now taken
by a combination of bearing and shear, and frictional resistance. **Note**
that if the test is being carried out on a tensometer where the elongation
is controlled, the transmitted force will actually drop after slip has occurred to a lower level indicated by B. On the other hand, in a real structure B' will be the load after slip, since the external load does not reduce.

From B' to C the deformation is almost elastic at a slope intermediate between OA and that of a bearing only connection. At C failure
by one of the modes described for the non-friction connection begins and
the joint becomes plastic from D to failure at E.

For bolts arranged in groups the failure is more complex but each
bolt will react as described above, the group having a number of slip
lines dependent upon load and individual bolt tension. The stage of the
curve from O to A is the one which is used for design, but this means

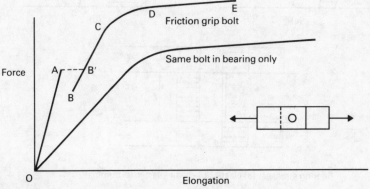

Fig. 9.5

that a greatly over-designed connection results. However, in the case of dynamic loading it is this type of connection which seems to perform best, since there is less chance of fatigue failure.

Design considerations

The design of bolted connections is highly complex and each type of joint has its own peculiarities of design. Therefore, only a few methods of analysis will be considered; if more detailed design knowledge is sought, then any of the references will give the student a better and fuller understanding.

The nominal stress method Non-friction connections only. Consider the bearing failure of the bolted connection shown in Fig. 9.6. If the designer

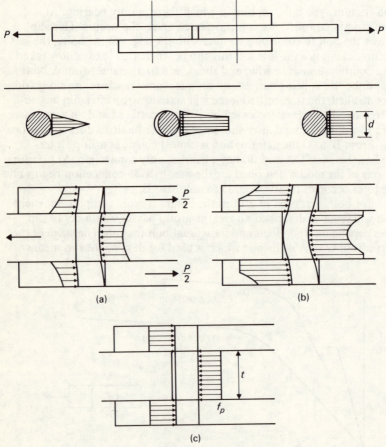

Bearing stresses: (a) elastic, (b) elastic-plastic, (c) nominal

Fig. 9.6

were called upon to calculate the actual stresses induced by the bolt bearing on the side of the hole, the task would be well-nigh impossible. The same applies to the induced shear and tension stresses in the bolt and the connected plates. It is for this reason that the designer simplifies design by assuming 'nominal stresses', based upon the adoption of uniformly distributed stresses across the section in question. In the case of a bolt bearing the stress is given by:

$$\text{Bearing Stress} = \frac{\text{Bolt Shearing Force}}{dt}$$

where d is the bolt's nominal diameter
t is the plate thickness

This gives the bearing stress on both the bolt (or rivet) and the metal through which it passes.

The nominal shearing stress on a bolt or rivet is given by:

$$\text{Bolt shear stress} = \frac{\text{bolt shear force}}{\text{cross-section area of bolt}}$$

Note: 'In calculating shear and bearing stresses, the effective diameter of a rivet shall be taken as the hole diameter, and that of a bolt as its nominal diameter' – quoted from Cl. 50a of BS 449.

Clauses 50–52 of BS 449 cover the allowable stresses, pitches and end distances for bolts and rivets, based on the nominal stress method.

In Fig. 9.6, the bolts are acting in what is known as double shear and enclosed bearing since there are two shear surfaces. If either the top or the bottom plate were removed the bolt would act in single shear and simple bearing. Obviously the bolt is twice as strong in double shear as it is in single shear; the allowable nominal bearing stress in single shear used to be reduced by 20 per cent from that used in enclosed bearing. However, since 1975, the approach has been altered to that described below. Firstly, however, it is necessary to discover the various grades of bolt and rivet which are in common use today.

There are two common grades of 'black bolt' as they are known, due to the fact that they are black in appearance. These are the 4.6 and 8.8 strength designated bolts. The significance of these numbers is as follows:

First figure

$$= \frac{1}{10} \times \text{minimum tensile strength expressed in kgf/mm}^2$$

Second figure

$$= \frac{1}{10} \times \frac{\text{minimum yield stress (or stress at permanent set)}}{\text{minimum tensile strength}}$$

expressed as a percentage.

e.g.

Grade 8.8 bolts: Minimum tensile strength = 80 kgf/mm^2

Minimum yield stress = not given in the code

Stress at permanent set limit, 0.2 = 64 kgf/mm^2

First figure $= \frac{1}{10} \times 80 = 8$

Second figure $= \frac{1}{10} \times \frac{64}{80} \times 100 = 8$

Note: (First figure) × (second figure) = minimum yield stress *or* stress at permanent set limit.

Only one type of rivet is in common use, and it is made from mild steel. British Standard 449 divides its consideration of allowed stresses into two parts: first it gives the allowed stresses on the connectors and then it gives the allowed bearing stresses on the plates through which the connectors pass. In the case of the stresses experienced by the connectors, the stresses from Table 20 in BS 449 are stated for mild steel rivets and grade 4.6 bolts. If other than these are used the stresses must be calculated as a ratio times the values from Table 20 of BS 449. In other words, the allowed stresses on the bolts or rivets are given by:

Allowed stress = K × (Table 20 stress)

For grade 4.6 bolts and mild steel rivets, $K = 1$
For grade 8.8 bolts

$$K = \frac{\text{yield stress of grade 8.8 bolt in N/mm}^2}{235}$$

or

$$K = \frac{0.7 \times \text{tension strength of 8.8 bolt in N/mm}^2}{235}$$

whichever is the lesser

For an 8.8 bolt $K = 2.338(0.7 \times \text{tension strength} = 785 \times 0.7)$.
Table 20 of BS 449 is reproduced below:

Table 20 Allowable stresses in rivets and bolts (N/mm^2)

Description of fasteners	Axial tension	Shear	Bearing
Power-driven rivets	100	100	300
Hand-driven rivets	80	80	250
Close tolerance and turned bolts	120	100	300
Bolts in clearance holes	120	80	250

In the case of the bearing stresses on the connected parts, this is set out in Table 20A of BS 449 and is related to the grade of the steel from which the plates are constructed.

Table 20A   Allowable bearing stresses on connected parts (N/mm^2)

Description of fasteners	Material of connected part		
	Grade 43	Grade 50	Grade 55
Power-driven rivets			
Close tolerance and turned bolts	300	420	480
Hand-driven rivets			
Bolts in clearance holes	250	350	400

However, these stresses may only be used where the end distance from the bolt's centre to the edge of the plate is greater than or equal to twice the effective diameter of the bolt or rivet. If the end distance is less than this factor of twice, the allowed stress must be reduced as:

$$\text{Allowed stress} = \frac{\text{actual end distance}}{2 \times \text{effective diameter}} \times (\text{Table 20A of BS 449 stress})$$

The friction grip method   The term 'high-strength friction-grip (HSFG) bolts' relates to bolts of high-tensile steel, used in conjunction with high-tensile steel nuts and hardened steel washers, which are tightened to a predetermined shank tension in order that the clamping force thus provided will transfer loads in the connected members by friction between the parts and not by shear in, or bearing on, the bolts or plies of the connected parts. There are two British Standards relating to these bolts. British Standard 4395 gives the dimensional requirements while BS 4604 gives the design recommendations. Note that BS 449 does not concern itself with these bolts save in the sense that it imposes the requirements of BS 4604 on the design of connections using these bolts. There are three types of HSFG bolts:

Class 1: general grade   – BS 4604:part 1
Class 2: higher grade with parallel shanks   – BS 4604:part 2
Class 3: higher grade with waisted shanks   – BS 4604:part 3

Of these three classes, Class 1 is the most important and is the only class to be considered in this book. High-strength friction grip bolts may be used in connections which impose: 1. pure shear; 2. pure tension; or 3. a combination of shear and tension on the bolts.

1. Connections imposing pure shear   In these connections the bolts are inserted into drilled holes in which they have a certain amount of play (2 mm to be precise). The nuts are then tightened in order that the tension in the shank induces friction between the plies, allowing loads to be carried by nothing more than friction. The number of bolts and their disposition in a joint shall be such that the resulting load at any bolt position does not exceed the value given by the following formula:

$$P_s = s_f \times n_i \times k \times p/F$$

where s_f is the slip factor (taken as 0.45 when the surfaces in contact are untreated in any way, and are grease-, oil-, and dirt-free)

n_i is the number of effective interfaces

k is 1.0 for Class 1 HSFG bolts

p is the proof load of one bolt

F is the load factor, which may be obtained from Table 9.1

Table 9.1 Load factors

Load type:	Static loads (self weight; stored materials; etc.)	Dynamic loads (machinery etc.)	Static + dynamic + wind loads
Class of bolt			
1	1.4	1.4	1.2
2	1.4	1.7	1.2–1.4
3	1.4	1.7	1.2–1.4

Table 9.2 Proof loads and shank tensions for bolts

Nominal size and thread diameter	Class 1 Proof load minimum shank tension	Class 2 Minimum proof load	Shank tension minimum 0.85 × p.l.	maximum 1.15 × p.l.	Class 3 Proof load minimum shank tension
	kN	kN	kN	kN	kN
M 12	49.4				
M 16	92.1	122.2	103.9	140.5	95.4
M 20	144.0	190.4	161.8	219.0	150.5
M 22	177.0	235.5	200.2	270.8	188.6
M 24	207.0	274.6	233.4	316.0	216.5
M 27	234.0	356.0	303.0	409.0	286.3
M 30	286.0	435.0	370.0	500.0	347.6
M 33	—	540.0	459.0	621.0	436.1
M 36	418.0				

Note: These design recommendations are consequent upon the tightening of **every** bolt to at least its minimum shank tension (equal to the proof load, *p* for Class 1 bolts) as set out in Table 9.2

2. Connections imposing pure tension   The maximum external tension which may be applied to any one bolt must not exceed 0.6 times the value of the proof load. Where fatigue conditions exist in the use of Class 1 bolts this figure must be reduced to a value of 0.5. By contrast, the Austrian regulations allow axial tension up to 0.8 of the shank tension. The external tension value must take account of prying action.

3. Connections imposing both tension and shear   Any external tension applied to a bolt serves to reduce its clamping action, thus simultaneously reducing its frictional resistance. However, the amount by which it does so is by no means certain. Experiment suggests that it varies in a fashion similar to the provisions in Clause 14c of BS 449, i.e. in a 'Von-Mises' ellipse. But BS 4604 is much more conservative, in its assumption of a straight-line relationship. The new value of shear capacity is given by:

$$P_s = s_f \times n_i \times k \times (p - 1.7 \times T)/F$$

where   the variables are as before except that T is the external tension

The factor of 1.7 arises from the fact that although the shank is stressed to the proof load, the maximum external tension which is allowed is 0.6 times the proof load. Therefore the external tension must be multiplied by $1/0.6 = 1.666$ (1.7 approx.) Under this rule the clamping action of a bolt is considered to cease when the externally applied tension reaches 0.6 times the proof load of the bolt.

Stresses induced by torque during tightening
High strength friction grip bolts are tightened by applying a torque to the nut, or occasionally, the bolt head. The torque and shank tension are related by the following formula:

$$M_a = K \cdot D \cdot P_v$$

where M_a = the applied torque
K = a non-dimensional coefficient
D = the nominal diameter of the bolt
P_v = the shank tension (or axial force)

The coefficient K depends upon a number of factors, among which are the type of thread and the coefficient of friction between the nut and the bolt, and between the nut and the washer or other bearing surface. For bolts as delivered in a lightly oiled condition, K varies between about

Table 9.3 Isometric precision hexagon bolts and screws

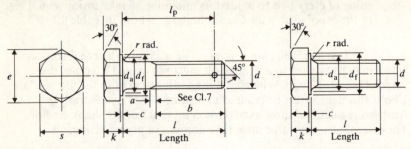

Hexagon head bolt, washer faced   Hexagon head screw, washer faced

All dimensions in millimetres.

1	2	3	4	5	6	7	8	9	10	11
Nominal size and thread dia. (d)	Pitch of thread (coarse pitch series)	Thread runout a (max.)	Diameter of unthreaded shank d		Width across flats s		Width across corners e		Diameter of washer face d_f	
			(max.)	(min.)	(max.)	(min.)	(max.)	(min.)	(max.)	(min.)
M 1.6	0.35	0.8	1.6	1.46	3.2	3.08	3.7	3.48	—	—
M 2	0.40	1.0	2.0	1.86	4.0	3.88	4.6	4.38	—	—
M 2.5	0.45	1.0	2.5	2.36	5.0	4.88	5.8	5.51	—	—
M 3	0.50	1.2	3.0	2.86	5.5	5.38	6.4	6.08	5.08	4.83
M 4	0.70	1.6	4.0	3.82	7.0	6.85	8.1	7.74	6.55	6.30
M 5	0.80	2.0	5.0	4.82	8.0	7.85	9.2	8.87	7.55	7.30
M 6	1.00	2.5	6.0	5.82	10.0	9.78	11.5	11.05	9.48	9.23
M 8	1.25	3.0	8.0	7.78	13.0	12.73	15.0	14.38	12.43	12.18
M 10	1.50	3.5	10.0	9.78	17.0	16.73	19.6	18.90	16.43	16.18
M 12	1.75	4.0	12.0	11.73	19.0	18.67	21.9	21.10	18.37	18.12
(M 14)	2.00	5.0	14.0	13.73	22.0	21.67	25.4	24.49	21.37	21.12
M 16	2.00	5.0	16.0	15.73	24.0	23.67	27.7	26.75	23.27	23.02
(M 18)	2.50	6.0	18.0	17.73	27.0	26.67	31.2	30.14	26.27	26.02
M 20	2.50	6.0	20.0	19.67	30.0	29·67	34·6	33·53	29.27	28.80
(M 22)	2.50	6.0	22.0	21.67	32.0	31.61	36.9	35.72	31.21	30.74
M 24	3.00	7.0	24.0	23.67	36.0	35.38	41.6	39.98	34.98	34.51
(M 27)	3.00	7.0	27.0	26.67	41.0	40.38	47.3	45.63	39.98	39.36
M 30	3.50	8.0	30.0	29.67	46.0	45.38	53.1	51.28	44.98	44.36
(M 33)	3.50	8.0	33.0	32.61	50.0	49.38	57.7	55.80	48.98	48.36
M 36	4.00	10.0	36.0	35.61	55.0	54.26	63.5	61.31	53.86	53.24
(M 39)	4.00	10.0	39.0	38.61	60.0	59.26	69.3	66.96	58.86	58.24
M 42	4.50	11.0	42.0	41.61	65.0	64.26	75.1	72.61	63.76	63.04
(M 45)	4.50	11.0	45.0	44.61	70.0	69.26	80.8	78.26	68.76	68.04
M 48	5.00	12.0	48.0	47.61	75.0	74.26	86.6	83.91	73.76	73.04
M 52	5.00	12.0	52.0	51.54	80.0	79.26	92.4	89.56	—	—
M 56	5.50	19.0	56.0	55.54	85.0	84.13	98.1	95.07	—	—
(M 60)	5.50	19.0	60.0	59.54	90.0	89.13	103.9	100.72	—	—
M 64	6.00	21.0	64.0	63.54	95.0	94.13	109.7	106.37	—	—
(M 68)	6·00	21·0	68·0	67.54	100.0	99.13	115.5	112.02	—	—

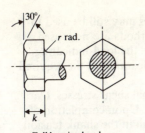

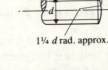

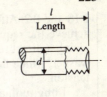

	30°					
r rad.				l Length		l Length

d

1¼ d rad. approx.

k

Full bearing head

Alternative type of head permissible on bolts and screws

Rounded end

Rolled thread end

Alternative types of end permissible on bolts and screws

12	13	14	15	16	17	18	19
Depth of washer face (c)	Transition diameter* d_a (max.)	Radius under head* r (max.)	(min.)	Height of head k (max.)	(min.)	Eccentricity of head (max.)	Eccentricity of shank and split pin hole to the thread (max.)
—	2.0	0.20	0.10	1.225	0.975	0.18	0.14
—	2.6	0.30	0.10	1.525	1.275	0.18	0.14
—	3.1	0.30	0.10	1.825	1.575	0.18	0.14
0.1	3.6	0.30	0.10	2.125	1.875	0.18	0.14
0.1	4.7	0.35	0.20	2.925	2.675	0.22	0.18
0.2	5.7	0.35	0.20	3.650	3.350	0.22	0.18
0.3	6.8	0.40	0.25	4.150	3.850	0.22	0.18
0.4	9.2	0.60	0.40	5.650	5.350	0.27	0.22
0.4	11.2	0.60	0.40	7.180	6.820	0.27	0.22
0.4	14.2	1.10	0.60	8.180	7.820	0.33	0.27
0.4	16.2	1.10	0.60	9.180	8.820	0.33	0.27
0.4	18.2	1.10	0.60	10.180	9.820	0.33	0.27
0.4	20.2	1.10	0.60	12.215	11.785	0.33	0.27
0.4	22.4	1.20	0.80	13.215	12.785	0.33	0.33
0.4	24.4	1.20	0.80	14.215	13.785	0.39	0.33
0.5	26.4	1.20	0.80	15.215	14.785	0.39	0.33
0.5	30.4	1.70	1.00	17.215	16.785	0.39	0.33
0.5	33.4	1.70	1.00	19.260	18.740	0.39	0.33
0.5	36.4	1.70	1.00	21.260	20.740	0.39	0.39
0.5	39.4	1.70	1.00	23.260	22.740	0.46	0.39
0.6	42.4	1.70	1.00	25.260	24.740	0.46	0.39
0.6	45.6	1.80	1.20	26.260	25.740	0.46	0.39
0.6	48.6	1.80	1.20	28.260	27.740	0.46	0.39
0.6	52.6	2.30	1.60	30.260	29.740	0.46	0.39
—	56.6	2.30	1.60	33.310	32.690	0.46	0.46
—	63.0	3.50	2.00	35.310	34.690	0.54	0.46
—	67.0	3.50	2.00	38.310	37.690	0.54	0.46
—	71.0	3.50	2.00	40.310	39.690	0.54	0.46
—	75.0	3.50	2.00	43.310	42.690	0.54	0.46

Note: Sizes shown in brackets are non-preferred.
*A true radius is not essential providing that the curve is smooth and lies wholly within the maximum radius, determined from the maximum transitional diameter, and the minimum radius specified.

0.18 and 0.2. Although torque-recording devices may still be used to achieve the required tension, more reliable methods are now used, giving a direct reading of the shank tension, e.g. the load-indicating washer, which flattens to a given feeler gauge clearance at the required tension.

Longitudinal tensile stresses, f, and torsional shear stresses, t, are induced in the shank of a bolt as it is tightened. Upon completion of tightening there is usually a very small reduction in the shank tension but there is often quite an appreciable drop in the torsion. This tends to

Table 9.4 Strength grade designations of steel bolts and screws

Strength grade designation	4.6	4.8	5.6	5.8	6.6	6.8	8.8	10.9	12.9	14.9
Tensile strength R_m min. kgf/mm^2	40	40	50	50	60	60	80	100	120	140
Yield stress R_e min. kgf/mm^2	24	32	30	40	36	48	—	—	—	—
Stress at permanent set limit $R_{0.2}$ min. kgf/mm^2							6.4	90	108	126

Table 9.5 Bolt tension areas and minimum pitches

Nominal size	Tensile stress area (mm^2)	Minimum pitch (mm)
M 6	20.1	15.0
M 8	36.6	20.0
M 10	58.0	25.0
M 12	84.3	30.0
M 16	157.0	40,0
M 20	245.0	50.0
(M 22)	303.0	55.0
M 24	353.0	60.0
(M 27)	459.0	67.5
M 30	561.0	75.0
(M 33)	694.0	82.5
M 36	817.0	90.0
(M 39)	976.0	97.5
M 42	1120.0	105.0
(M 45)	1300.0	112.5
M 48	1470.0	120.0
(M 52)	1760.0	130.0
M 56	2030.0	140.0

confirm the view that if a bolt doesn't fail during tightening, it will not fail subsequently. It is helpful and reassuring to realise that even under the most adverse conditions, at least 1.5 turns of the nut are needed to produce failure, but a figure of between 2 and 3 is more normal.

The equivalent stress f_e on the surface of the shank may be found by employing the formula in Cl. 14c of BS 449:

$$f_e = \sqrt{(f^2 + 3t^2)} = C \cdot f$$

where C = the resistance factor

The value of C is approximately 1.15, which may explain the maximum proof load requirements given for Class 2 bolts. Tests carried out in the USA showed that the clamping force achieved in bolts of various diameters, tightened with an impact wrench, was about 85 per cent of that obtained when similar bolts were loaded in direct tension. This may explain the minimum shank tension requirements for Class 2 bolts.

Design examples

The following three examples consist not of design but of checking. However, since much about the design of bolted and riveted connections involves a form of checking, and since the ability to check a bolted connection is equally important, no apology is needed for their inclusion.

Example 9.1   Check the maximum tensile force which may be taken by the butt-splice joint shown in Fig. 9.7, **if**: the connectors used are 18-mm ⌀ mild steel rivets in 20 mm clearance holes; the plates which are connected are of grade 50 steel.

Stage I: Check edge and end distances, and pitches between bolts.
1. End distance
   2 × hole diameter = 40 mm
   Actual end distance is greater than this value
   Thus the stresses in Table 20A of BS 449 may be used.
2. Edge distance
   From Table 21 of BS 449   minimum = 30
   Satisfactory.
3. Pitch
   From Cl. 51b of BS 449   minimum = 2.5 × 20 = 50
   maximum = 32 × 5 = 160 mm

Actual pitch is satisfactory since it equals 50 mm (measured from bolt to bolt, not in the direction of load).

Stage II: Check plate tension strength.
Gross section = 120 × 10
Net section = 80 × 10 = 800 mm^2
Allowed tension stress = 215 N/mm^2 (Table 19 of BS 449)

$P = 215 × 800$ N $= 172$ kN.

Table 9.6 Black washers to BS 4320

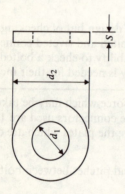

Nom. bolt dia.	Inside dia. d_1			Outside dia. d_2			Thickness S			Mass kg per 1000
	(Nom.)	(Max.)	(Min.)	(Nom.)	(Max.)	(Min.)	(Nom.)	(Max.)	(Min.)	
M 6	6.6	7.0	6.6	12.5	12.5	11.7	1.6	1.9	1.3	1.1
M 8	9.0	9.4	9.0	17.0	17.0	16.2	1.6	1.9	1.3	2.1
M 10	11.0	11.5	11.0	21.0	21.0	20.2	2.0	2.3	1.7	4.0
M 12	14.0	14.5	14.0	24.0	24.0	23.2	2.5	2.8	2.2	5.9
M 16	18.0	18.5	18.0	30.0	30.0	29.2	3.0	3.6	2.4	11.0
M 20	22.0	22.6	22.0	37.0	37.0	35.8	3.0	3.6	2.4	17.0
(M 22)	24.0	24.6	24.0	39.0	39.0	37.8	3.0	3.6	2.4	18.0
M 24	26.0	26.6	26.0	44.0	44.0	42.8	4.0	4.6	3.4	32.0
M 27	30.0	30.6	30.0	50.0	50.0	48.8	4.0	4.6	3.4	40.0
M 30	33.0	33.8	33.0	56.0	56.0	54.5	4.0	4.6	3.4	50.0
(M 33)	36.0	36.8	36.0	60.0	60.0	58.5	5.0	6.0	4.0	64.0
M 36	39.0	39.8	39.0	66.0	66.0	64.5	5.0	6.0	4.0	70.0

Normal dia. (Form E)

(Large dia. Form F)

Designation										Mass*
M 6										
M 8	9	9.4	9	21	21	20.2	1.6	1.9	1.3	3.5
M 10	11	11.5	11	24	24	23.2	2.0	2.3	1.7	5.6
M 12	14	14.5	14	28	28	27.2	2.5	2.8	2.2	9.1
M 16	18	18.5	18	34	34	32.8	3.0	3.6	2.4	16.0
M 20	22	22.6	22	39	39	37.8	3.0	3.6	2.4	20.0
(M 22)	24	24.6	24	44	44	42.8	3.0	3.6	2.4	26.0
M 24	26	26.6	25	50	50	48.8	4.0	4.6	3.4	45.0
(M 27)	30	30.6	30	56	56	54.5	4.0	4.6	3.4	55.0
M 30	33	33.8	33	60	60	58.5	4.0	4.6	3.4	60.0
(M 33)	36	36.8	36	66	66	64.5	5.0	6.0	4.0	87.0
M 36	39	39.8	39	72	72	70.5	5.0	6.0	4.0	112.0

* Due to thickness tolerance, mass can vary by as much as 30%.

All dimensions in mm

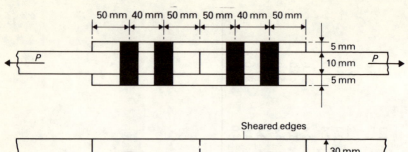

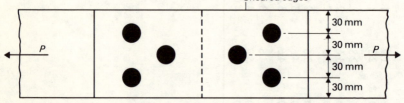

Sheared edges

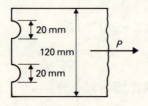

Fig. 9.7

Stage III: Check rivet shear strength.
Allowed stress = $100 \, \text{N/mm}^2$ (double shear)
Area = $2\pi \times 20^2/4 = 628.32 \, \text{mm}^2$

$P = 3 \times 100 \times 628.32 \, \text{N} = 188.5 \, \text{kN}$.

Stage IV: Check bearing.
Bearing area = $10 \times 20 = 200 \, \text{mm}^2$
Allowed bearing stress = $300 \, \text{N/mm}^2$ (rivet critical)

$P = 3 \times 300 \times 200 \, \text{N} = 180 \, \text{kN}$.

Thus the maximum load which can be sustained by this connection is dictated by plate tension and is equal to 172 kN.

Example 9.2 The same joint as used in Example 9.1, the only difference being the use of grade 4.6 black bolts.

Stage I: Bolt spacings – unaltered from Example 9.1.

Stage II: Plate tension strength.
Again unaltered from Example 9.1

$P = 172 \, \text{kN}$.

Stage III: Bolt shear strength.
Allowed shear stress = $80 \, \text{N/mm}^2$

Bolt area = $(3 \times \pi \times 18^2/4) \times 2$(double shear)
$$= 1526.8 \, \text{mm}^2$$

$P = 122.145 \, \text{kN}.$

Stage IV: Bolt (or plate) bearing.
The end distance is greater than $2 \times$ effective bolt diameter
Allowed bearing stress = $250 \, \text{N/mm}^2$ \qquad (bolt stress critical)

Bearing area = $10 \times 18 = 180 \, \text{mm}^2$

$$P = 3 \times 250 \times 180$$
$$= 135 \, \text{kN}.$$

Thus the maximum tension load which can be carried by this connection is dictated by bolt shear and is equal to 122.145 kN.

Example 9.3 The same joint as used in Example 9.2, the only difference being the use of HSFG bolt of general grade. The bolts used are M 16 in 18 mm clearance holes.

Stage I: Bolt spacings – unaltered from Example 9.1.

Note: from Cl. 3.4 of BS 4604
The outer ply thickness must be increased to 10 mm.

Stage II: Plate tension.

Net area = $(120 - 36) \times 10 = 840 \, \text{mm}^2$

$P = 215 \times 840 = 180.6 \, \text{kN}.$

Stage III: Bolt frictional resistance.

Note: The bolts must be tightened to full proof load
= 92.1 kN

$s_f = 0.45$ \qquad (Cl. 3.1.1 BS 4604: Part 1)

$F = 1.4$

$n_i = 2$

$P = 3 \times (0.45 \times 2 \times 92.1/1.4)$
$$= 177.6 \, \text{kN}.$$

Thus the maximum tension load which may be carried by this connection is dictated by frictional shear and is equal to 177.6 kN. Note that the answer is not much different from that obtained using rivets; however, smaller diameter connectors were used.

Two points should be drawn from the last three examples:

1. The size of connector chosen was largely unrelated to the size of member being joined. If the edge or end distances had been outside the allowed values, simple adjustments to their arrangement within the joint would have solved the problem.
2. The connection was effected by using a top and bottom cover plate. This ensured that the forces were equally disposed between the plates.

Unfortunately, not all connections are so simple to design. The next example illustrates this point.

Example 9.4   The double-angle tension splice shown in Fig. 9.8 is to be designed to carry a load of 135 kN. The $65 \times 50 \times 8$ mm angles are made from grade 43 steel, and the bolts used are of grade 4.6.

Stage I: Determine the bolt size.
Consider Fig. 9.4(a)
e_{min} is determined from Table 21 in BS 449 and is the minimum edge distance to a rolled edge
g_{min} is obtained similarly
f_{req} is obtained from a consideration of the washer size to be used with the given bolt and is obtained from Table 9.6

Hole diameter	Bolt diameter	e_{min} (mm)	f_{req} (mm)	f_{actual} (mm)
12	10	20	10.5	16.0
14	12	22	12.0	14.0
18	16	26	15	10.0

Hole diameter	Bolt diameter	g_{min} (mm)	h_{req} (mm)	h_{actual} (mm)
18	16	26	15.0	25
22	20	30	18.5	21
24	22	32	19.5	19

It is best to use the same diameter of bolt for both the upstand and the base connectors, therefore, 12-mm $\varnothing$ bolts will be used throughout.

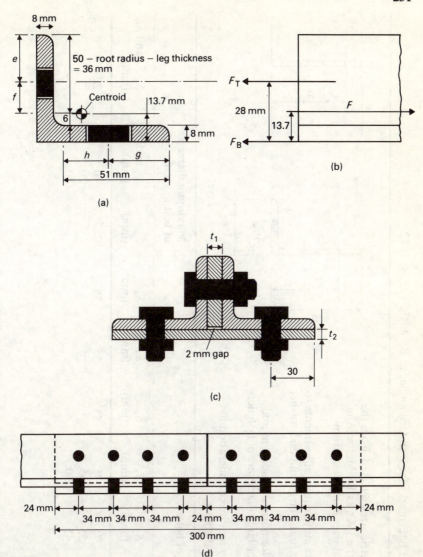

(a)

(b)

(c)

(d)

Fig. 9.8

This is a rather lengthy process, therefore tables such as Table 9.7 are very helpful. These set out the required spacings as 'so called' 'back-marks', with the maximum bolt sizes already calculated. Allied to these tables are Tables 9.8 and 9.9 giving the equivalent for I-sections and channels. However, as will be seen in Example 9.7 they are not to be taken as giving unbreakable rules. In that example a leg

Table 9.7 Recommended backmarks for standard angles to BS 4848: Part 4: 1972

These angles are those metric sizes selected, from the full list recommended by the ISO, as BS Metric Angles. They replaced the Imperial sizes completely from 1 January, 1973.

Note that HSFG bolts may require adjustments to the backmarks shown due to the larger nut and washer dimensions.

Inner gauge lines are for normal conditions and may require adjustment for large diameters of fasteners or thick members.

Outer gauge lines may require consideration in relation to a specified edge distance.

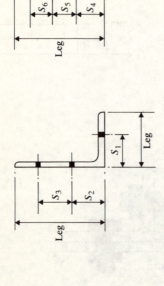

Nominal leg length (mm)	Spacing of holes						Maximum diameter of bolt or rivet		
	S_1 (mm)	S_2 (mm)	S_3 (mm)	S_4 (mm)	S_5 (mm)	S_6 (mm)	S_1 (mm)	S_2 and S_3 (mm)	S_4 S_5 and S_6 (mm)
200	75	75	75	55	55	55	55	30	20
150	55	55	55					20	
125	45	45	50					20	

120			16
100			
90	55		24
80	50		24
75	45		20
70	45		20
65	40	50	20
	35	45	20
60	35		16
50	28		12
45	25		
40	23		
30	20		
25	15		

Note: That with the back marks and cross centres quoted the maximum diameter of fastener stated conforms with Table 21 of BS 449: Part 2: 1969.

This will generally result in the most economical connection.

234

Table 9.8 Recommended spacing of holes in columns, beams and tees to BS 4: Part 1: 1972

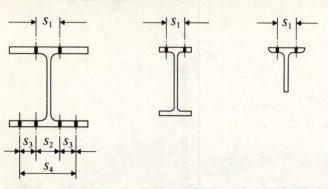

Nominal flange width (mm)	Spacing of holes				Maximum dia. of rivet or bolt (mm)	'b' min (mm)
	S_1 (mm)	S_2 (mm)	S_3 (mm)	S_4 (mm)		
419 to 368	140	140	75	290	24	362
330 and 305	140	120	60	240	24	312
330 and 305	140	120	60	240	20	300
292 to 203	140				24	212
190 to 165	90				24	162
152	90				20	150
146 to 127	70				20	130
102	54				12	98
89	50					
76	40					
64	34					
51	30					

see note opposite

length of 100 mm is used with two bolts in the leg length rather than the one 24 mm bolt described in Table 9.7.

Stage II: Calculate the force on each bolt line.

Consider Fig. 9.4(b)

Taking moments about the shear line of the base bolts

$$F \times 13.7 = F_T \times (50 - 22)$$

$$F_T = \frac{13.7}{28} \times 135 = 66.05 \, kN$$

$$F_B = 68.95 \, kN.$$

Stage III: Check the angle capacity of net area.

$$Net \, area = gross \, area - area \, of \, holes$$

$$= (2 \times 860) - (4 \times 14 \times 8)$$

$$= 1272 \, mm^2$$

$$P = 1272 \times 155 = 197.16 \, kN \qquad O.K.$$

Stage IV: Gusset design (based on tension strength).

1. *Upstand gusset*

Assume gusset dimensions to be $48 \times t_1$

(The 48 mm arises from the need for a 2 mm gap between the upstand gusset and the base gusset)

$$Net \, area = (48 - 14)t_1 = 34t_1$$

$$F_T = 66\,050 = 155 \times 34t_1$$

$$t_1 = 12.53 \, mm \qquad (say \, 13 \, mm)$$

2. *Base gusset*

$$Net \, area = (130 + 13 - 28)t_2 = 115t_2$$

$$F_B = 68\,950 = 155 \times 115t_2$$

$$t_2 = 3.87 \, mm \qquad say \, 5 \, mm \, for \, practical \, reasons$$
$$(such \, as \, protection \, against \, corrosion \, or$$
$$maltreatment \, during \, construction).$$

Note that the actual flange width for a universal section may be less than the nominal size and that the difference may be significant in determining the maximum diameter.

The dimensions S_1 and S_2 have been selected for normal conditions but adjustments may be necessary for relatively large diameter fastenings or particularly heavy masses of serial size.

'b' min. This is the minimum width of flange to comply with Table 21 of BS 449: Part 2: 1969

236

Table 9.9 Recommended back marks for holes in channels to BS 4: Part 1: 1972

Nominal flange width (mm)	S_1 (mm)	Maximum dia. of bolt or rivet (mm)
102	55	20
89	50	20
76	45	20
64	35	16
51	30	10
38	22	

Stage V: Upstand leg bolts.
Assume that the end distances are at least twice the effective bolt diameter $= 2 \times 12 = 24$ mm
Derive the capacity of **one** bolt in double shear

Shear: $P = (2 \times \pi \times 36) \times 80 = 18.095$ kN

Bearing: $P =$ allowed stress on bolt $\times$ least bearing area
or
           $=$ allowed stress on metal $\times$ least bearing area
           whichever is the lesser

The least bearing area is the gusset area in this case, but if the combined thickness of the angle legs had been less than the thickness of the gusset then the least bearing area would have been given by the angle bearing area.
For grade 4.6 bolts passing through grade 43 steel the two allowed stresses in bearing are equal $= 250$ N/mm^2

Least bearing area $= 13 \times 12 = 156$ mm^2

$$P = 250 \times 156 \text{ N}$$
$$= 39 \text{ kN}$$

Therefore, **shear dictates**

Number of bolts required $= \dfrac{F_T}{18.095} = 3.65$ (say 4 bolts).

Stage VI: Base leg bolts.
Assume end distances are at least 24 mm as in *Stage V*
Derive the capacity of **one** bolt in single shear

Shear: $P = (\pi \times 36) \times 80 = 9.048 \, \text{kN}$

Bearing: $P = 250 \times (5 \times 12) = 15 \, \text{kN}$ (5 × 12 being the least bearing area)

Therefore, **shear dictates**

Number of bolts required $= \dfrac{F_B}{9.048} = 7.62$ (say 8 bolts).

Stage VII: Total length of gussets.

Minimum length = 4 × end distance + 6 × minimum bolt pitch

= 276 mm   (say 300 mm).

Thus the arrangement of Fig. 9.8(d) is possible.

Column bases carrying moments fall into two categories: 1. those in which the moment causes stresses of about the same magnitude as the direct load; and 2. those in which the moment predominates. In the first case it is usually possible to use a simple slab base with perhaps column to plate stiffeners. However, in the second case it proves impossible to use this simple plan, and the more complicated base shown in Fig. 9.9 must be used. The combined stiffnesses of the column and the angles are such that the bottom plate may be neglected, its thickness being dictated more by requirements such as the need for it to be of a comparable thickness to the members welded to it than by any structural considerations. The weld design will be considered in a later example. It is therefore assumed (conservatively) that the column pivots about an axis through the centre of one of the bolts. The next example considers the design involved for the bolts in tension.

Example 9.5 Design the bolts required in the column base connection shown in Fig. 9.9. The base of the 305 × 305 mm at 97 kg/m UC is subject to a moment of 150 kNm, and it may be assumed that the axial load on the column is negligible.

Actual value of $D = 307.8 \, \text{mm}$

Tension force on one bolt $= \dfrac{M}{D + 100} = \dfrac{150\,000}{407.8} \, \text{kN}$

$= 367.8 \, \text{kN}$

For a grade 8.8 bolt; allowed tension stress

$= 2.338 \times 120 \, \text{N/mm}^2$

$= 280.6 \, \text{N/mm}^2$

238

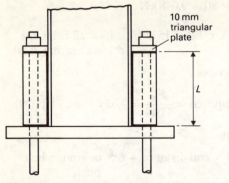 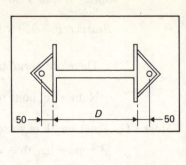

Fig. 9.9

Required stress area $= \dfrac{367\,800}{280.6} = 1310.763\,\text{mm}^2$

Required nominal bolt diameter = 48 mm (cf. Table 9.5)

It will now be found that the angle needed to enclose the bolt without touching is 150×150 mm.

The area of angle required to resist the tension in the bolt

$$= \dfrac{367\,800}{155} = 2373\,\text{mm}^2$$

Thus the angle used is $150 \times 150 \times 10$ mm

$\qquad$ Area $= 2930\,\text{mm}^2$.

Shear produced by eccentricity of loading

Loads on fasteners in an eccentric load connection are usually determined by the method of superposition. An eccentric load can be equated to a force of equal magnitude acting through the centroid of the fastener group provided it is coupled with a moment equal to the value of the load times the perpendicular distance from the centroid to the line of action of the load. Thus the eccentric load shown in Fig. 9.10 is resolved into a direct load P through the centroid O of the fastener group, **plus** a pure moment, Pe, where e is the eccentricity of load P with respect to O. The loads on any fastener due to direct load P and due to pure moment Pe are computed separately and then added vectorially. The computation is usually simplified by resolving the load P into x and y components, P_x and P_y.

$\qquad$ Let $\quad x_i$ and y_i be the coordinates with respect to O of the ith connector

$\qquad\qquad A_i \qquad$ be the cross-sectional area of the ith connector

R'	be the load on a fastener due to direct load P
R''	be the load on a fastener due to moment Pe
R_i	be the resultant of R_i' and R_i'' on the ith fastener

The x and y components of fastener loads are denoted by corresponding subscripts. Then for any fastener with coordinates x_i and y_i and area A_i, the load on the fastener can be computed as follows:

It is assumed that the plate is rigid

Load on fasteners due to direct load, P   The rigidity of the plate ensures that the fasteners are all deformed by an equal amount, resulting in forces which are proportional to the individual fastener area in relation to the total area, $\sum A_i$. Thus:

$$R_{ix}' = \frac{P_x \cdot A_i}{\sum A_i} \qquad \text{and} \qquad R_{iy}' = \frac{P_y \cdot A_i}{\sum A_i}.$$

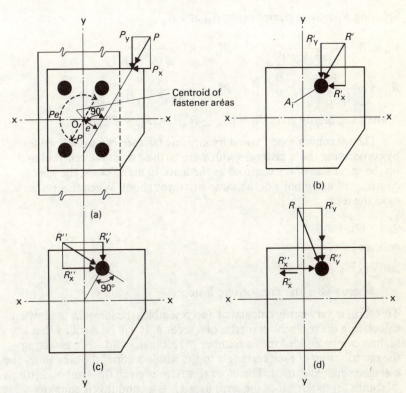

Fig. 9.10   Shear loads on fasteners due to eccentric loads; (a) resultant load on fastener group, (b) load on fastener due to direct load P only, (c) load on fastener due to moment Pe only, and (d) resultant load on fastener

Load on fasteners due to pure moment, Pe   The deformation will be proportional to the distance of each individual fastener from the centroid of the fastener group.

$$\text{Stress} = k \cdot \sqrt{(x_i^2 + y_i^2)}$$

where k is a constant

$$R_i'' = kA_i\sqrt{(x_i^2 + y_i^2)}$$

$$Pe = \sum R_i'' \times \sqrt{(x_i^2 + y_i^2)}$$

$$= k \cdot \sum A_i(x_i^2 + y_i^2)$$

Therefore $$k = \frac{Pe}{\sum A_i(x_i^2 + y_i^2)}$$

and $$R_i'' = \frac{Pe}{\sum A_i(x_i^2 + y_i^2)} \cdot A_i \cdot \sqrt{(x_i^2 + y_i^2)}$$

Splitting R_i'' into its components; R_{ix}'' and R_{iy}''

$$R_{ix}'' = R_i'' \cdot \frac{y_i}{\sqrt{(x^2 + y^2)}}; \qquad R_{iy}'' = R_i'' \cdot \frac{x_i}{\sqrt{(x^2 + y^2)}}$$

$$R_{ix}'' = \frac{Pe}{\sum A_i(x_i^2 + y_i^2)} \cdot A_i y_i; \qquad R_{iy}'' = \frac{Pe}{\sum A_i(x_i^2 + y_i^2)} \cdot A_i x_i$$

Adding vectorially gives: $R_i = (R_{ix}' \pm R_{ix}'')^2 + (R_{iy}' \pm R_{iy}'')^2$.

The maximum load carried by any one fastener can usually be found by considering those fasteners which are farthest from the centroid and on the same side of the centroid as the load. In many cases, the joint is constructed using bolts of the same size throughout. When this is the case, the term

$$\sum A_i \cdot (x_i^2 + y_i^2)$$

reduces to

$$A \cdot \sum(x_i^2 + y_i^2)$$

where A is the area of one fastener

This term is very easily calculated and resembles the formula used when calculating the moment of inertia of a section. It will be recalled that a section can be divided into a number of rectangles and when assessing the contribution of each rectangle to the whole section, the area times the distance squared is used. Therefore this term is given the symbol, I_p (the 'p' stands for 'polar' since the term uses $x_i^2 + y_i^2$ and this is equal to the radius from the centroid to the bolt squared). Thus

$$I_p = \sum(x_i^2 + y_i^2)$$

And

$$R_i'' = \frac{Pe}{I_p} \cdot r_i$$

where r_i is the distance to the bolt at which R_i'' is required

Similarly $$R_{ix}'' = \frac{Pe}{I_p} \cdot x_i$$

And $$R_{iy}'' = \frac{Pe}{I_p} \cdot y_i$$

Also since all the bolts are the same size,

$$R_{ix}' = P_x/N$$

And $$R_{iy}' = P_y/N$$

where N is the total number of bolts.

Example 9.6   In the eccentric joint shown in Fig. 9.11(a) it is required to find the maximum load which can be withstood by using 18-mm $\varnothing$ rivets of grade 43 steel.

Rivet	x_i (mm)	y_i (mm)	$(x_i^2 + y_i^2)$
1	−75	75	11 250
2	0	75	5 625
3	75	75	11 250
4	−75	0	5 625
5	0	0	0
6	75	0	5 625
7	−75	−75	11 250
8	0	−75	5 625
9	75	−75	11 250
			$I_p = 67\,500$

Vertical force due to shear, $R_{iv}' = P/9$ $(R_{ix}' = 0)$
forces due eccentricity (outermost rivets most highly stressed)

$$R_{ix}'' = \frac{175P}{67\,500} \times 75 = 0.1944P$$

$$R_{iy}'' = R_{ix}''$$

Summing the forces, it is found that rivets 3 and 9 are most highly stressed (see Fig. 9.11b)

242

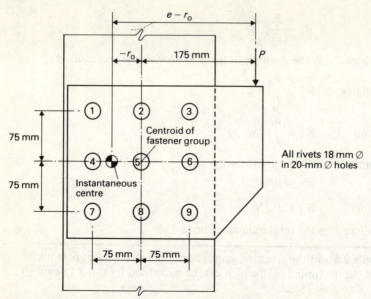

Fig. 9.11(a)

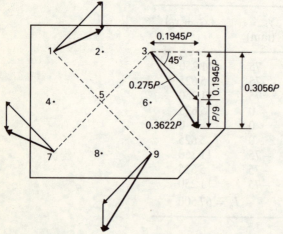

Fig. 9.11(b)

The resultant, $R = \sqrt{(0.1944^2 + (0.1944 + \frac{1}{9})^2)} \times P$

$\qquad\qquad = 0.3622P$

Assuming power-driven rivets:

Capacity of one rivet in shear $=$ allowed stress $\times \dfrac{\pi d^2}{4}$

where d is the effective diameter $=$ hole diameter

Assuming 1 mm clearance all round the rivet $d = 20\,\text{mm}$

Capacity of one rivet in shear $= 100 \times 100\pi = 31.416\,\text{kN}$

$$0.3622P = 31.416$$

Therefore $\hspace{4cm} P = 86.74\,\text{kN}.$

In order to complete the calculation, the plate thickness should be checked to ensure that the bearing stresses are not too high.

Allowed bearing stress $= 300\,\text{N/mm}^2$ (on both the rivets and the steel)

$$300 = \frac{\text{load on one rivet}}{\text{hole diameter} \times t}$$

t being the plate thickness

Thus minimum thickness, $t = \dfrac{314\,16}{300 \times 20} = 5.24\,\text{mm}.$

The combination of rotation and translation of the plate produced by an eccentric load may be reduced to a pure rotation about some point which is known as 'the Instantaneous Centre'. This concept is illustrated in Fig. 9.12; the point is known as 'instantaneous' because it varies, not only with the fastener arrangement but also with the location and direction of the eccentric load. Once this centre is determined for a given loading, the relative displacement between plates at any fastener can be determined, since the magnitude of this displacement is proportional to the distance from the instantaneous centre and its direction is perpendicular to the radius vector from this centre. The unit

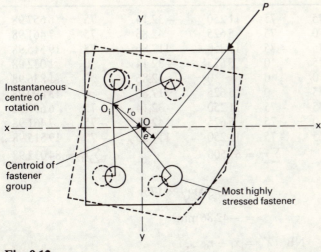

Fig. 9.12

shearing stress on the fasteners is proportional to the displacement and hence the most highly stressed fastener is located the greatest distance away from the instantaneous centre.

The instantaneous centre O_i can be located using the condition that an imaginary fastener located at O_i would carry no load. The load on any fastener consists of components due to direct load and moment; since the load due to direct load is constant in magnitude and direction, it is evident that zero resultant load occurs at the point where the load R'' due to moment exactly counteracts the load R' due to direct load. In order that R'' shall be parallel and opposite to R', the radius vector from the centroid O of the fastener group to O_i must be perpendicular to the line of action of the load P. Also, O_i must be located on the opposite side of O from the line of action of P.

For the load R'' to exactly balance R', the centre must be located at a distance r_0 from the centroid so that:

$$R' = \frac{P}{\sum A_i} \cdot A_0 = -R'' = -\frac{Pe}{\sum A_i r_i^2} \cdot A_0 r_0$$

Thus $\quad r_0 = -\dfrac{\sum A_i r_i^2}{e \cdot \sum A_i}$

When all the fasteners are of equal area:

$$r_0 = -\sum r_i^2 / Ne$$

Therefore, in the last example, the following table would be made:

Rivet	x_i	y_i	r_i^2	X_I	Y_I	r_I^2
1	−75	75	11 250	−32.14	75	6 657.98
2	0	75	5 625	42.86	75	7 461.98
3	75	75	11 250	117.86	75	19 515.98
4	−75	0	5 625	−32.14	0	1 032.98
5	0	0	0	42.86	0	1 836.98
6	75	0	5 625	117.86	0	13 890.98
7	−75	−75	11 250	−32.14	−75	6 657.98
8	0	−75	5 625	42.86	−75	7 461.98
9	75	−75	11 250	117.86	−75	19 515.98
			$\sum r_i^2 = 67\,500$			$\sum r_I^2 = 84\,032.98$

$$\therefore \quad r_0 = -67\,500/(9 \times 175)$$

$$= -42.86\,\text{mm}$$

NB $\quad X_1 = x_i - r_0$

Once the instantaneous centre has been found, the direction of the resultant load on any rivet is located perpendicular to its radius vector r_1 from that centre. The magnitude is given by:

$$R = \frac{P(e - r_0)r_1 A_i}{\sum A_i r_1^2}$$

Or when all the fasteners have the same area

$$R = P(e - r_0)r_1/\sum r_1^2$$

Therefore in the last example

$$R = \frac{(175 + 42.86) \times 139.7}{84\,032.98}$$

$$= 0.3622P \quad \text{as before}$$

However, it can be seen that this is a much more complex route to the answer, but it can prove to be quicker for joints where the fasteners are not distributed in such a regular fashion, or where the load is inclined, or where the fasteners are not all of the same area.

Beam to column connections
These connections may be divided into three categories according to the assumptions made when designing the beam:

1. *Simple.* In this case the beam is not designed to have full moment resistance at its ends. It may be assumed to have no moments at the end, in which case the connections are designed as pure shear joints. On the other hand Cl. 24 of BS 449 allows the beam span to be taken between points 100 mm from the edges of the supporting columns. This imposes small moments on the connections equal to the reactions times 100 mm, making it necessary to design the connections for shear and moment, though shear will usually predominate.
2. *Fixed.* The beam in this case is designed to be fully rigid at its supports, thus attracting large moments and shears to the connections.
3. *Wind.* In this case the beam carries small loads vertically, but is required to resist the horizontal wind forces by providing moment fixity at its ends. Thus the connections will be subjected to rather more moment than shear.

Some of the means of fulfilling these requirements are illustrated in Fig. 9.13. There are many ways of accomplishing these connections and only the more common will be considered in this book. One of the major questions in the designer's mind when designing such a joint is whether to use welding or bolting. The answer to this cannot be given categorically as situations will vary widely, depending on many factors. It may be

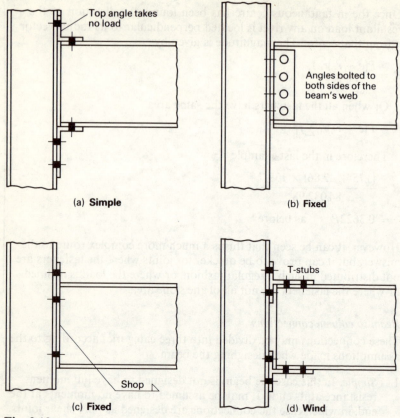

Top angle takes
no load

(a) **Simple**

Angles bolted to
both sides of the
beam's web

(b) **Fixed**

(c) **Fixed**

T-stubs

Shop

(d) **Wind**

Fig. 9.13

that site welding is very expensive, in which case it will probably be best to make the connections using bolts or rivets in some fashion. On the other hand, bolted connections may result in unsightly joints or provide a maintenance problem, therefore site welding might be best. But in all cases the designer must be aware of all the possibilities and decide for himself which method would be more appropriate in his particular circumstances.

Consider Fig. 9.13; the simple connection is subject only to vertical loads, thus in theory, only a bottom connection is needed, unless restraint against torsion is necessary. However, it is always best to include a top connection, thus providing erection stability if nothing else. The bottom connection is the only one requiring full-scale design, the top one being made with the same bolts and an angle of similar size to that used on the bottom. When using this type of connection, the beam must be checked for web bearing and buckling. Example 9.7 considers the design of such a connection.

The connections shown in Figs. 9.13(b) and 9.13(c) may also be used for a simple beam end, especially if Cl. 24 of BS 449 is used. However, they are more commonly used when the connection is of the fixed type. The vertical end reaction is now coupled with the stresses induced by the end moments which in the case of Fig. 9.13(b) can cause tension (column bolts) or shear at right angles to the vertical shear (web bolts). The design of these connections is considered in Example 9.8 and its preamble.

Finally, the wind connection shown in Fig. 9.13(d) experiences far greater moments than it does vertical shear forces, thus these connections are designed as explained in Example 9.9.

Example 9.7 Design a simple beam/column connection similar to that illustrated in Fig. 9.13(a) to carry a reaction of 150 kN. The beam section is 356 × 171 mm @ 51 kg/m, the column section is 254 × 254 mm @ 89 kg/m both in grade 50 steel, with the bolts in grade 8.8 steel.

Stage I: Bolt sizing.

Assume that four bolts are used (two to six is the usual number)

Required shear value = 150/4 = 37.5 kN

Allowed bolt shear stress = 2.338 × 80 = 187 N/mm²

Thus, minimum bolt diameter = $\sqrt{\dfrac{4 \times 37\,500}{187 \times \pi}}$ = 15.98 mm

= 16 mm (nearest size)

Use 16 mm bolts in 18 mm holes.

Stage II: Check bearing and size the angle's thickness.

Allowed bearing stress on the bolts = 2.338 × 250
= 584.5 N/mm²

Allowed bearing stress on the steel = 350 N/mm² (critical)

Actual bearing stress = P/dt

where P is the bolt load
  d is the bolt diameter
  t is the thickness of material passed through

Minimum thickness required = $\dfrac{37\,500}{350 \times 16}$ = 6.7 mm

Column flange thickness = 17.3 mm
therefore *satisfactory*

Whatever angle is chosen, it must be thicker than 6.7 mm.

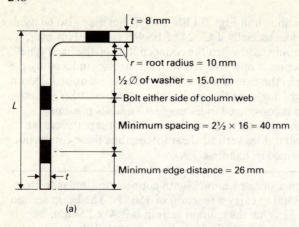

t = 8 mm

r = root radius = 10 mm

½ ∅ of washer = 15.0 mm

Bolt either side of column web

Minimum spacing = 2½ × 16 = 40 mm

Minimum edge distance = 26 mm

(a)

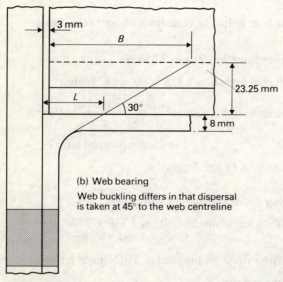

3 mm

B

L

30°

23.25 mm

8 mm

(b) Web bearing

Web buckling differs in that dispersal
is taken at 45° to the web centreline

Fig. 9.14

Stage III: Size the angle's leg length.

As can be seen from the Fig. 9.14(a)

Minimum leg length = $r + t + 15.0 + 40 + 26$
$= r + t + 81 = 99$ mm

Use 100 × 75 × 8 mm unequal angle
($r = 10$ mm of root radius) see tables in Appendix A.

Stage IV: Check the beam's capacity (note the 3 mm clearance between the beam and the column, which allows for inaccuracies in setting out and expansion).

1. *Bearing*
Let the angle of dispersal be $\alpha°$
The student should verify the following formula:

Stiff bearing afforded by the angle $= L$

$$L \tan \alpha = (t + r)(l + \tan \alpha) - r/\cos \alpha$$

in this example: $t = 8$ mm $\qquad r = 10$ mm

for bearing $\alpha = 30°$ $\qquad$ and $L = 29.2$ mm

Thus $\qquad B =$ length of web in bearing
$$= L - 3 + 23.25/\tan 30°$$
$$= 66.45 \text{ mm}$$

Allowed bearing reaction $= 66.45 \times 7.3 \times 260$ N
$$= 126.12 \text{ kN}$$

Thus a stiffener must be designed

2. *Buckling*
Carrying the dispersal at 45° to the centreline of the web

$L = 21.86$ mm

$B = L - 3 + 0.5D = 196.66$ mm

Clear depth between fillets $= 309.1$ mm

Slenderness ratio $= \dfrac{d}{t} \sqrt{3} = \dfrac{309.1}{7.3} \sqrt{3} = 73.34$

From Table 17(b) of BS 449 $\qquad p_c = 143.3$ N/mm^2
Allowed load $= 143.3 \times 7.3 \times 196.66$ N
$$= 205.72 \text{ kN}$$

3. *Shear*
Allowed shear capacity $= D \times t \times p'_q = 355.6 \times 7.3 \times 140$
$$= 363.4 \text{ kN}$$

Thus only bearing capacity needs to be increased.

In the case of the type of connection which has just been considered in Example 9.7, it is interesting to discover the maximum load which can be accommodated in this manner. After all, the beam design may be limited, as in the example, by bending, thus giving a small reaction in comparison to the shear capacity of the beam. On the other hand it may be dictated by shear, and then the cleat will be required to carry much greater loads. The maximum load which can be carried by such a connection is dictated by three considerations:

1. The space available on the column flange.
2. The maximum bolt size available; and
3. The maximum size of angle available.

For the beam and column considered, it will be found that the maximum shear of 363.4 kN can be accommodated easily by using four 27-mm diameter bolts of grade 8.8 and an angle of size 200 × 100 × 10 mm of grade 50 steel. However, this assumes that a suitable web stiffener can be designed, and this may not always be possible. This problem may be overcome by combining the methods used in Figs. 9.13(a) and 9.13(b). If as well as having a bottom flange cleat, a web connection is made, no web bearing or buckling problems will arise. The load is then shared between the flange cleat and the web cleat.

The design of the fixed-end connection is, in comparison to the design of the simple connection, very difficult, not least because the distribution of forces in the bolts is open to many varying analyses, none of which truly represent the actual picture. Only the more modern and very mathematical computer methods of analysis can even approach the truth. Therefore, the methods presented in the following argument must be treated with some degree of restraint and as will be seen, not taken to be too accurate.

Note: **The TEC student should only concern himself with Method II, (p. 259) and he should skip the consideration of Method I.**

Consider the connection of Fig. 9.13(b); the connection of Fig. 9.13(c) may be designed in exactly the same way but the need to check the angle legs bolted to the beam web is omitted.

The first assumption, common to both of the methods here presented, is that the shear force is shared equally between **all** the bolts. This is of course untrue, but it is at least conservative, since part of the shear will be carried by friction induced between the angle flanges and the column flanges due to bending compression.

Method I: *Elastic bending analogy*
(Refer to Fig. 9.15.)

This method assumes that the tension of the bolts is balanced by an elastic compression zone at the base of the connection. Thus the strain diagram of Fig. 9.15(c) may be drawn and the forces in the bolts deduced as set out below.

251

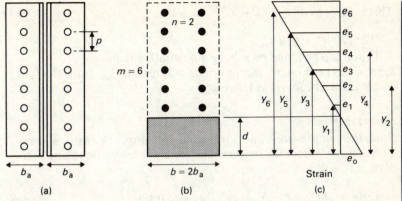

Fig. 9.15

From the geometry of the strain diagram

$$\frac{e_0}{d} = \frac{e_1}{y_1 - d} = \frac{e_2}{y_2 - d} = \cdots = \frac{e_6}{y_6 - d} = \frac{e_i}{y_i - d} = k \text{ (a constant)}$$

Let n = the number of columns of bolts (two in the example)

Let m = the number of rows of bolts **in tension** (six in the example)

Let maximum 'compressive' stress = $f_0 = Ee_0$ (E is the modulus of elasticity of steel)

Thus the bolts stress at row i = $f_i = Ee_i$

Bolt force = $f_i \cdot A_i = E \cdot A_i \cdot e_i$

Equating forces

$$0.5f_0 \cdot bd = \text{sum of bolts' forces} = nE \cdot \sum_{i=1}^{m} A_i \cdot e_i$$

But $\quad e_i = k(y_i - d)$

And $\quad f_0 = Ee_0 = Ekd$

Hence $\quad 0.5Ekd \cdot bd = nEk \cdot \sum_{i=1}^{m} A_i(y_i - d)$

If all the bolts are the same as is the usual case $A_i = A$, remembering that

$$\sum_{i=1}^{m} (y_i - d) = \sum_{i=1}^{m} y_i - md$$

This reduces to $d^2 + mCd - C \sum_{i=1}^{m} y_i = 0$ $\qquad\qquad$ [9.1]

$\qquad$ where $\quad C = 2nA/b$

Thus d can be found by solving a quadratic equation

Once d has been found, the maximum bolt tension f_{max} (equal to f_6 in Fig. 9.15) can be found as follows.

Taking moments

Since the compression zone is triangular in nature, the centre of compression is $d/3$ from the bottom. Therefore, the lever arm to any bolt level is $y_i - d/3$.

$\qquad$ Thus $\qquad M = nEk \cdot \sum_{i=1}^{m} A_i(y_i - d)(y_i - 0.3d)$

$\qquad$ And $\qquad AEk = \dfrac{M}{n \cdot \sum\limits_{i=1}^{m} (y_i - d)(y_i - 0.3d)}$ $\qquad$ [9.2]

This enables the value of k to be found

The maximum bolt force $f_{max} = AEe_{max} = AEk(y_{max} - d)$ $\qquad$ [9.3]

Thus the procedure when using this method is as follows:

1. Guess roughly where the 'neutral axis' lies; thus guess a value for the value of m.
2. Using eqn [9.1] calculate the value of d given by the value of m chosen. If the value of d thus found is not between the rows of bolts originally guessed, then repeat until agreement is obtained.
3. Using eqn [9.2] obtain a value for AEk.
4. Obtain a value for f_{max} and check that the bolt size chosen is adequate. Otherwise, repeat using a different bolt.
5. Through the preceding steps, it must be remembered that when a bolt is subjected to both tension and shear, Cl. 50d of BS 449 states that 'the calculated shear and axial stresses f_s and f_t calculated in accordance with subclause 50a, do not exceed the respective allowable stresses p_s and p_t and that the quantity $f_s/p_s + f_t/p_t$ does not exceed 1.4'.
6. Finally, the bending strength of the angle legs bolted to the beam webs should be checked as shown in the following example.

Example 9.7: Method I A grade 50, 356 × 171 mm @ 45 kg/m UB is designed as a simple beam, but with the assumption that the effective span starts 100 mm out from the column face, viz. Cl. 24 of BS 449 is used. The end reaction is 125 kN, therefore the moment which the connection must simultaneously carry is 125 × 0.1 = 12.5 kNm. Design

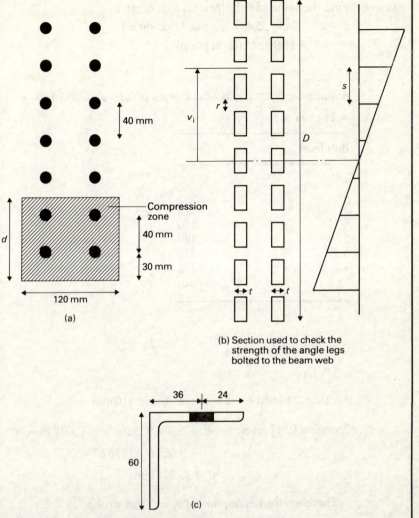

40 mm

v_i

r

D

s

Compression zone

40 mm

30 mm

d

120 mm

(a)

t t

(b) Section used to check the strength of the angle legs bolted to the beam web

36   24

60

(c)

Fig. 9.16 Example 9.7 (Method I)

a suitable connection similar to the type shown in Fig. 9.13(b), using grade 8.8 bolts.

The design is of necessity one of trial and error, therefore:

Try: 60 × 60 mm angles;   10 mm ⌀ bolts;
     fourteen bolts spaced equally at 40 mm

Consider Fig. 9.16(a) which illustrates this trial section.

Stage I: Find the value of d the 'neutral axis depth'.

$n = 2$ $A = 25\pi = 78.5\,\text{mm}^2$ (for shear)

$A_t = 58.0\,\text{mm}^2$ (for tension)

Assume the neutral axis lies between bolt levels 2 and 3

Thus $m = 5$

Bolt level	y_i
1	–
2	–
3	110
4	150
5	190
6	230
7	270
$\sum y_i =$	950

$$C = 2nA/b = \frac{2 \times 2 \times 58}{120}$$

$$= 1.933$$

d is thus assumed to lie between 70 and 110 mm

Equation [9.1] gives $d^2 + 6 \times 1.933d - 1.933 \times 1020 = 0$

$d^2 + 9.667d - 1836.67 = 0$

$d = 38.295\,\text{mm}$

Therefore the assumption of $m = 5$ was wrong

d must lie between levels 1 and 2, making $m = 6$ and $\sum y_i = 1020$

Equation [9.1] gives $d^2 + 6 \times 1.933d - 1.933 \times 1020 = 0$

$d^2 + 11.6d - 1972 = 0$

$d = 38.984\,\text{mm}$

Notice how little difference is made by making the wrong assumption.

Stage II: Find a value for AEk.

Bolt level	y_i	$y_i - d$	$y_i - d/3$	$(y_i - d)(y_i - d/3)$
1	–	–	–	–
2	70	31.02	57.01	1 768.052
3	110	71.02	97.01	6 888.886
4	150	111.02	137.01	15 209.719
5	190	151.02	177.01	26 730.553
6	230	191.02	217.02	41 451.386
7	270	231.02	257.01	59 372.220
				$\sum = 151\,420.816$

From eqn [9.2]
$$AEk = \frac{12\,500\,000}{2 \times 151\,420.816}$$
$$= 41.276\,\text{N/mm}.$$

Stage III: Calculate the maximum bolt force.

From eqn [9.3]
$$f_{max} = 41.276 \times (270 - 38.984)$$
$$= 9.535\,\text{kN}.$$

Stage IV: Check the value of $f_s/p_s + f_t/p_t$.

It is assumed that the shear on each bolt is the same

Thus $f_s = 125/14 = 8.929\,\text{kN}$ $\quad p_s = 187 \times 78.54$
$\qquad\qquad\qquad\qquad\qquad\qquad\qquad = 14.687\,\text{kN}$

$\quad f_t = 9.535\,\text{kN}$ $\qquad\qquad p_t = 280.56 \times 58$
$\qquad\qquad\qquad\qquad\qquad\qquad\qquad = 16.272\,\text{kN}$

$f_s/p_s + f_t/p_t = 1.194$

Therefore the bolts joining the angles to the column are sufficient.

It may, however, be thought that smaller bolts could have been used. If the same configuration, i.e. fourteen bolts 40 mm apart, is used, then the above fraction becomes 1.62, which is too great. But if the spacing is reduced to allow the use of eighteen bolts, 8-mm $\emptyset$ bolts may be used. This exercise is given as Question 9.

Stage V: Check the size of bolt needed on the web of the beam. Since the line of bolts through the beam and angle are set out from the face of the column, the assumed 100 mm distance from reaction to column face is reduced as far as these bolts are concerned. Thus the moment which they have to take is also reduced. As shown in Fig. 9.16(c) the distance is reduced to 36 mm.

Therefore

$$M = 125 \times (0.1 - 0.036) = 8 \, \text{kNm}$$

Since the leg length of the angle being used is only 60 mm, the largest size of bolt which can be used is about M 12, otherwise the edge distance from bolt centre to angle edge will be too small. Thus it is assumed that M 12 bolts are being used.

How many bolts can be accommodated on the length of the angle?

The minimum pitch between bolt centres = 2.5 times bolt diameter

$$= 2.5 \times 12$$
$$= 30 \, \text{mm}$$

From Table 9.3 width across nut corners = 21.9 mm
Therefore if the minimum pitch is used then the clearance between bolts would be 8.1 mm – a little close.
In other words, although nine bolts would fit, a more practical number is eight as shown in Fig. 9.17.

Note: 1. The 22.25 mm edge distance is just greater than the minimum value of 22 mm required for a sawn edge.
      2. The clearance between bolts (to allow access for a box-type spanner) is 14.6 mm – a little more acceptable.

Bolt forces
Shear: (shared equally between eight bolts)
This is a vertically directed force

$$F_v = 125/8 = 15.625 \, \text{kN}$$

Bending: (refer to Fig. 9.16(b))
The neutral axis will be at the centre of the connection since there will be a small gap between beam and column.

If the outermost bolts are a distance of L from the neutral axis and experience force due to bending of F_h – a horizontal force. Other bolts will experience forces which are proportional to their distances from the neutral axis. For a bolt at a distance from the neutral axis of a_i

$$F = F_h \cdot a_i / L$$

$$\text{Moment} = F_h \cdot L + F_h \cdot \frac{a_1}{L} \cdot a_1 + F_h \cdot \frac{a_2}{L} \cdot a_2 + \cdots$$

$$= \frac{F_h}{L} \cdot \sum_{1}^{N} a_i^2$$

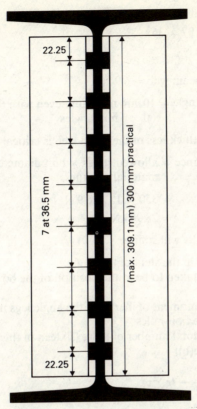

22.25

7 at 36.5 mm

(max. 309.1 mm) 300 mm practical

22.25

Fig. 9.17

For this case
$$a_1 = 3.5s; \qquad a_2 = 2.5s;$$
$$a_3 = 1.5s; \qquad a_4 = 0.5s$$

$$a_5 = 0.5s; \qquad a_6 = 1.5s;$$
$$a_7 = 2.5s; \qquad a_8 = 3.5s$$

$$L = 3.5s \qquad s = 36.5 \text{ mm}$$

$$M = 12sF_h = 8 \times 10^6 \text{ Nmm}$$

Thus $F_h = \dfrac{8 \times 10^6}{12 \times 36.5} = 18.26 \text{ kN}$

The actual force is the vector sum of F_h and $F_v = \sqrt{F_h^2 + F_v^2}$
$$= 24.04 \text{ kN}.$$

Allowed forces
Shear (double shear):

Force = 2 × allowed stress × area of bolt

= 2 × 187 × 36π

= 42.3 kN

Bearing:

Thickness of beam web = 6.9 mm

Thickness of angle = 10 mm minimum even with 60 × 60 × 5 mm angles

Therefore the thickness of the beam web is critical

Bearing resistance = allowed stress × bolt diameter × material thickness

= 350 × 12 × 6.9 N

= 28.98 kN

Thus M 12 bolts will suffice.

Stage VI: Bending stresses in the stem of the angle.
(Critical section taken to be at the position of the bolts in the beam web.)
First obtain the moment of inertia of the angle legs making allowance for the bolt holes.
Let N equal the **total** number of holes (sixteen in this case) (refer to Fig. 9.16(b))

$$I = \frac{2tD^3}{12} - N\frac{tr^3}{12} - tr\sum_1^N v_i^2$$

Try 60 × 60 × 5 mm angles

where $t = 5$ mm $\qquad r = 14$ mm $\qquad D = 309.1$ mm

Bolt level	v_i	v_i^2
1	127.75	16 320.06
2	91.25	8 326.56
3	54.75	2 997.56
4	18.25	333.06
5	18.25	333.06
6	54.75	2 997.56
7	91.25	8 326.56
8	127.75	16 320.06
		55 954.5

$$\sum_1^N v_i^2 = 2 \times 55\,954.5 = 111\,909 \text{ mm}^2$$

$$I = \frac{2 \times 5 \times 300^3}{12} + 16 \times \frac{5 \times 14^3}{12} - 14 \times 5 \times 111\,909$$

$$= 14.648 \times 10^6 \, \text{mm}^4$$

Maximum stress induced in the angles $= \dfrac{M}{I} \times \tfrac{1}{2}D$

$$= \frac{8 \times 10^6}{14.648 \times 10^6} \times 150$$

$$= 81.92 \, \text{N/mm}^2$$

Satisfactory.

Stage VII: Check the stress f_0 between angles and column.

$$f_0 = Ee_0 = Ekd = \frac{AEk}{A}d$$

$$AEk = 43.121$$

$$f_0 = 43.121 \times 44.415/78.5$$
$$= 24.4 \, \text{N/mm}^2$$

Thus the connection is satisfactorily designed.

Finally it should be noted that there is a risk that the web of the column may buckle under the influence of the compression between the angles and the column. This should be checked as though the flange of the column is subjected to a point load equal to the compressive force. A stiffener may thus have to be designed.

Method II: *Pivotal method*

In this method it is assumed that the neutral axis occurs at the level of the lowest line of bolts. (It will be noticed that this is nearly true of the analysis performed in Example 9.7 under the theory of Method I.) The tension in the bolts is then balanced by a compressive region below this level, in which the distribution of stress may not be elastic.

Let u_i = distance to the ith level of bolts

Assume that 1. All bolts are of equal area.
2. The force in each bolt is proportional to its distance from the lowest level, u_i

$$F_i = k \cdot u_i$$

The moment on the connection is given by:

$$M = \sum_1^N F_i u_i = k \cdot \sum_1^N u_i^2$$

where N is the **total** number of bolts

Therefore $\quad k = \dfrac{M}{\sum\limits_{1}^{N} u_i^2}$

The largest bolt force is at the highest level
where $\quad u_i = L$

$$F_{max} = \dfrac{M \cdot L}{\sum\limits_{1}^{N} u_i^2}$$

(Note the similarity between this method and the one presented on pp. 250–252).

Example 9.8: Method II As Example 9.7 but analysed using Method II (refer to Fig. 9.18)

Bolt level	u_i	u_i^2
1	0	0
2	40	1 600
3	80	6 400
4	120	14 400
5	160	25 600
6	200	40 000
7	240	57 600
		145 600

$$\sum_{1}^{14} u_i^2 = 2 \times 145\,600 = 291\,200 \, \text{mm}^2$$

$$F_{max} = \dfrac{12.5 \times 10^6 \times 240}{291\,200}$$

$$= 10.302 \, \text{kN} \quad \text{(compare with 9.535\,kN under Method I).}$$

From this onwards, the analysis is exactly similar to Method I, except that the compression stress is not calculated. It can thus be seen that Method II offers a reduction in the amount of calculation but results in higher bolt forces being deduced. This will rarely, if ever, result in larger bolts being needed, since the increase is so small. In this case, the value of $f_s/p_s + f_t/p_t$ is increased to 1.241 from 1.194, therefore, no larger bolts are needed as a result. The author would therefore suggest that Method II be used as the preferred method. However, he felt the need to introduce Method I since the second method does not conform to established theory with the same ease. In any case, the truth probably lies somewhere in between the two, and one method suggested by the '*Metric*

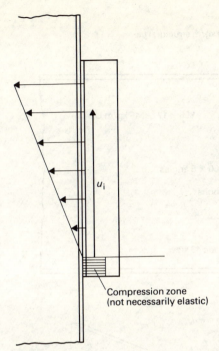

Compression zone
(not necessarily elastic)

Fig. 9.18

Practice for Structural Steelwork' uses a method which is a combination
of the two presented here, in that the compression is elastic in nature,
but the width assumed to be effective in compression is only four times
the nominal bolt diameter. A neutral axis depth is calculated but not on
the assumption that plane sections remain plane.

Example 9.9   Design the wind moment connection shown in Fig. 9.19.
It is assumed that the web angles take all the shear while the Tee-stub
connections on the beam flanges take all the bending moment. This
is obviously an erroneous but conservative and simplifying assumption.
   Use HSFG general grade bolts
   Moment to be resisted = 60 kNm
   Shear to be resisted = 25 kN.

Part I: *Structural tee connections*
These resist **all** the moment – the bolts through the column being in
tension at the top of the beam, while the bolts through the beam flange
take the moment by the shear developed.

   Force on each Tee = M/D

   where M is the moment

262

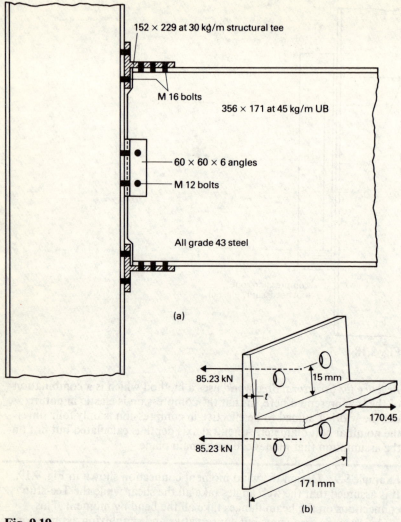

Fig. 9.19

D is the depth of the beam

$$M/D = \frac{60 \times 10^6}{352} N$$

$$= 170.45 \, kN.$$

The maximum size of bolt which may be used is to some extent dictated by Cl. 3.4 of BS 4604, which states:

3.4 Minimum ply thickness *In any joint using high strength friction grip bolts no outer ply, and wherever possible no inner ply, shall be less in* *thickness than half the bolt diameter or 10 mm whichever is smaller, unless it can be demonstrated to the satisfaction of the engineer that any consequent limitations to the performance of the joint owing, for example, to crumpling, tearing, buckling or bending of plies, to corrosion between plies or to limited extensibility of short bolts are adequately catered for.*

Note: *The problem of limited extensibility is not peculiar to the presence of thin plies but may arise with any bolt where the grip is small and where there is only a short length of thread between the bolt head and the face of the nut.*

The thickness of the beam flange = 9.7 mm
This would imply:

1. The web of the Tee should be greater than 9.7 mm.
2. The maximum bolt diameter = 19.4 mm.

However, at joints where neither condensation nor any other damaging effects are likely to be found, most engineers can be persuaded that the ply thickness may be reduced to 80 per cent of that called for in Cl. 3.4 above.

    Thus, if M 20 bolts are used, the Tee section may have a web thickness of just 8 mm and still meet with the engineer's approval.
Try M 16 bolts as shown in Fig. 9.19.

Stage I: Shear.

        Shear value of one bolt

$$= s_f \times p/F \quad \text{(terms as used on pp. 220)}$$

$$= \frac{0.45}{1.2} \times 92.1 \, \text{kN}$$

        (1.2, since wind forces)

$$= 34.54 \, \text{kN}$$

Number of bolts needed = 170.45/34.54 = 4.935
(say six – three either side of the beam web).

Stage II: Tension between Tee and column.
From the requirements of geometry, the minimum number of bolts which may be used is four – two either side of the column web and two either side of the Tee's web.

$$\text{Tension capacity} = 4 \times 0.6p = 4 \times 0.6 \times 92.1$$
$$= 221.04 \, \text{kN}$$

Thus M 16 bolts will suffice provided a Tee section can be found which will accommodate the spacing of the bolts.

Stage III: Design the Tee-section.
Web: Should be thicker than 8 mm (Cl. 3.4 of BS 4604).

Should be thick enough to take the tension on the net area.

Should be long enough to accommodate the three bolts.

Minimum thickness – assume 8 mm and check the tension strength

Net area = (width − 2 × diameter of one hole) × 8 mm²

= (171 − 2 × 18) × 8

(171 mm derives from the beam width)

= 1228.5 mm²

Tension strength = 165 × 1.25 × 1228.5   (1.25 is the wind factor from Cl. 13 of BS 449)

= 253.38 kN

Thus the tension strength is satisfactory

Minimum length = allowed end distance

+ minimum bolt spacing

+ clearance to the flange of the Tee

Hole diameter = 2 mm in excess of the bolt diameter
= 18 mm

Referring to Table 21 of BS 449

Distance to a rolled edge = 26 mm

Minimum bolt spacing = 2.5 × 2 × 16 = 80 mm

Clearance to Tee flange consists of allowing a washer to lie flat on the web without having to be cut. This can be found from Table 9.11, where the washer size is given as 37 mm (unclipped).

One half this distance is the required clearance distance

Minimum length of Tee web = 26 + 80 + 18.5 = 124.5 mm

Flange: (refer to Fig. 9.19(b))

Distance from centre of bolt to root of web = 15 mm (clipped washers are needed)

Assuming that the portion of the flange between bolt and web acts as a cantilever, the moment to be resisted is given by

$$M = \frac{170.45}{2} \times 0.015 = 1.278 \text{ kNm}$$

Section modulus = $171t^2/6$

where t is the flange thickness **at the root of the web**

$$\text{Bending stress caused} = 165 \times 1.25 = M/Z = \cfrac{1.278 \times 10^6}{\cfrac{171}{6} \times t^2}$$

Thus
$$165 \times 1.25 \times \frac{171}{6} t^2 = 1.278 \times 10^6$$

$$t = 14.75 \text{ mm minimum}$$

Note: The flange thickness will be specified at the centre of the flange outstand, therefore it will be necessary to allow for the 5 per cent inside slope on some Tee sections.

Summary of section constraints:

1. Web thickness – greater than 8 mm.
2. Web length – greater than 124.5 mm.
3. Flange thickness – greater than 14.75 mm at the root of the web.

Try Tee Section 152 × 229 mm at 34 kg/m which is cut from
$$457 \times 152 \text{ mm at } 67 \text{ kg/m UB}$$

Web thickness   = 8.0 mm   satisfactory

Flange thickness = 15.0 mm measured at the centre of the flange outstand

However, since this Tee section is cut from a UB which has parallel flanges, this is also the thickness at the root of the web.

Web length $= A - 15$ (approx)

A is given in the tables as the overall depth of the Tee
$$= 227.3 - 15$$
$$= 212 \text{ mm}$$

Thus, this tee is satisfactory
Alternately it will be found that a long stalk tee-bar of size 102 × 203 mm at 25 kg/m is also satisfactory.

Part II: *Angle connections to the beam web*

Shear per angle   = 12.5 kN

Length of angle possible = 352 − (152.9 − 8)
$$= 207.1 \quad \text{(say 200 mm)}$$

In theory one M 12 bolt will have a shear capacity of 18.525 kN. However, two bolts are better from a practical viewpoint, and therefore will be adopted in this case. The length of angle used is 100 mm which is adequate for the spacing of two such bolts. No account need to be taken of

Table 9.10 High strength friction grip bolts and nuts BS 4395 Parts 1 and 2 – Dimensions

Nominal dia. D	Dia. of unthreaded shank B		Pitch (course pitch) series	Width across flats A		Depth of washer face C	Thickness of hexagon head F	
(mm)	Max. (mm)	Min (mm)	(mm)	Max. (mm)	Min. (mm)	(mm)	Max. (mm)	Min. (mm)
(M 12)	12.70	11.30	1.75	22	21.16	0.4	8.45	7.55
M 16	16.70	15.30	2.0	27	26.16	0·4	10.45	9.55
M 20	20.84	19.16	2.5	32	31.00	0.4	13.90	12.10
M 22	22.84	21.16	2.5	36	35.00	0.4	14.90	13.10
M 24	24.84	23.16	3.0	41	40.00	0.5	15.90	14·10
M 27	27.84	26.16	3.0	46	45.00	0.5	17.90	16.10
M 30	30.84	29.16	3.5	50	49.00	0.5	20.05	17.95
M 33	34.00	32.00	3.5	55	53.80	0.5	22.05	19.95
M 36	37.00	35.00	4.0	60	58.80	0.5	24.05	21.95

Not recommended diameter in brackets ( )

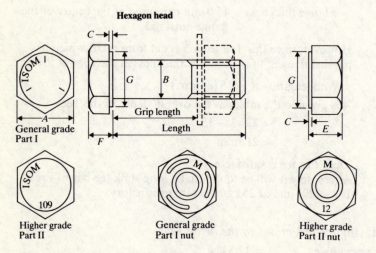

Hexagon head

General grade
Part I

Higher grade
Part II

General grade
Part I nut

Higher grade
Part II nut

The symbol 'M' may be used as an alternative to 'ISOM' on bolt heads

Table 9.10 *cont.*

Dia. of csk. head J (mm)	Dia of washer face G Max. (mm)	 Min. (mm)	Depth of flash H (mm)	Thickness of nuts E Max. (mm)	 Min. (mm)	Addition to grip length to give length of bolt* required (mm)
24	22	19.91	2.0	11.55	10.45	22
32	27	24.91	2.0	15.55	14.45	26
40	32	29.75	3.0	18.55	17.45	30
44	36	33.75	3.0	19.65	18.35	34
48	41	38.75	4.0	22.65	21.35	36
54	46	43.75	4.0	24.65	23.35	39
60	50	47.75	4.5	26.65	25.35	42
66	55	52.55	5.0	29.65	28.35	45
72	60	57.75	5.0	31.80	30.20	48

Countersunk head dimensions are to BS 4933
*Allows for nut, one flat round washer and sufficient thread protrusion beyond nut.

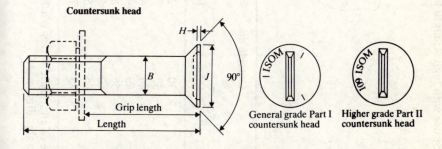

Countersunk head

General grade Part I countersunk head

Higher grade Part II countersunk head

Thread lengths

Nominal length of bolt	Length of thread*	
	Part 1	Part 2
Up to and including 125 mm	$2d + 6$ mm	$2d + 12$ mm
Over 125 mm up to and including 200 mm	$2d + 12$ mm	$2d + 18$ mm
Over 200 mm	$2d + 25$ mm	$2d + 30$ mm

$d*$ = thread diameter

Table 9.11 Flat round washers for use with high strength friction grip bolts (dimensions in mm)

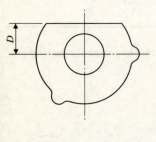

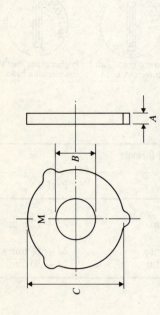

Nominal size	Inside dia. B		Outside dia. C		Thickness A		$D^{\dagger}$
	(Max.)	(Min.)	(Max.)	(Min.)	(Max.)	(Min.)	
M 12	13.8	13.4	30	29	2.8	2.4	11.5
M 16	17.8	17.4	37	36	3.4	3.0	14.0
M 20	21.5	21.1	44	43	3.7	3.3	17.5
M 22	23.4	23.0	50	48.5	4.2	3.8	19.0
M 24	26.4	26.0	56	54.5	4.2	3.8	21.0
M 27	29.4	29.0	60	58.5	4.2	3.8	22.5
M 30	32.8	32.4	66	64.5	4.2	3.8	26.0
M 33	35.8	35.4	75	73.5	4.6	4.2	29.0

The symbol 'M' appears on the face of all Metric Series Washers.
† When required washers clipped to this dimension.

the eccentricity of the bolts in a situation such as this since the top and bottom Tee stubs serve to force the beam to deflect vertically without sufficient rotation to cause bending stresses in the angle bolts.

Angle size required = $60 \times 60 \times 6\,\text{mm}$.

Example 9.10 An example to demonstrate a method for coping with prying forces. Design the connections for Fig. 9.20 if,

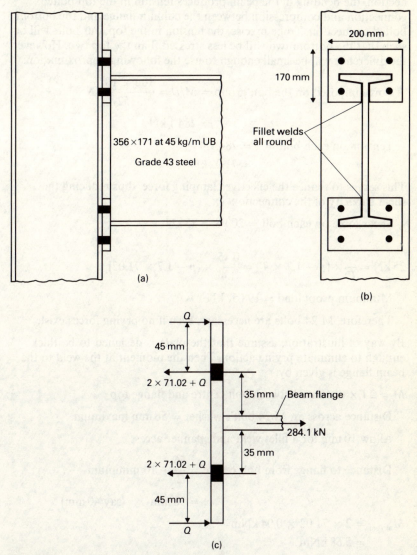

Fig. 9.20

Moment $= 100\,\text{kNm}$

Shear $= 200\,\text{kN}$

Use HSFG general bolts.

Part I: *Design assuming NO prying forces*

This is done to size the required bolts roughly. For this particular connection, the bending of the beam produces tension in the top bolted connection and compression between the column flange and the bottom bolted connection. To be precise, the tension in the top four bolts will be such that the bottom two will be less stressed than the top two. However the difference will be small enough to use the following approximation:

$$\text{Force imposed on the bolt group} = M/D = \frac{100 \times 10^6}{352}\,\text{N}$$

$$= 284.1\,\text{kN}$$

Tension on each bolt $= T = 284.1/4$

$$= 71.02\,\text{kN}$$

This serves to reduce the effective clamping force, thus reducing the shear capacity of the connection

Shear force on each bolt $= 200/8 = 25\,\text{kN}$

$$25\,\text{kN} = \frac{s_f}{F} \times (p - 1.7 \times T) = \frac{0.45}{1.4} \times (p - 1.7 \times 71.02)$$

Minimum proof load $= 198.52\,\text{kN}$

Therefore, M 24 bolts are necessary, even if no prying forces exist.

By way of illustration, assume that the plate is designed to be thick enough to eliminate prying action. Then the moment at the weld to the beam flange is given by:

$M = 2T \times$ distance between bolt centre and flange top

Distance across an M 24 bolt's washer $= 56\,\text{mm}$ maximum.

Allow 10 mm for a fillet weld and spanner access.

$$\text{Distance to flange from bolt centre} = \frac{56}{2} + 10 \text{ minimum}$$

$$= 38\,\text{mm} \qquad (\text{say } 40\,\text{mm})$$

$M_{\text{at root}} = 2 \times 71.02 \times 0.04\,\text{kNm}$

$\qquad = 5.68\,\text{kNm}$

Plate width $= 200\,\text{mm}$

$$Z = 200t^2/6$$

t = plate thickness

Thus $\dfrac{200}{6}t^2 = \dfrac{5.68 \times 10^6}{165}$

$t = 32.1\,\text{mm}$ (say 33 mm).

Part II: *Allow for the prying forces*

These are assumed to be concentrated right at the edge of the plate. The European Recommendations permit **any** value for Q to be adopted, provided that:

1. The bolts are capable of accepting the summation of the forces due to the applied loading and the prying action

   i.e. bolt force = $71.02 + 0.5Q$ for this example

2. The end plate is capable of resisting the double curvature bending.

Therefore, the method adopted here is to choose a bolt size larger than the M 24 minimum and *make* the prying forces such that the bolts are fully stressed.

Choose M 27 bolts

Shear capacity = $25\,\text{kN} = \dfrac{0.45}{1.4} \times (p - 1.7T)$

$$= \dfrac{0.45}{1.4} \times (234 - 1.7T)$$

Therefore $T = 91.89\,\text{kN}$

$$0.5Q = 91.89 - 71.02 = 20.88\,\text{kN}$$

Minimum edge distance = $44\,\text{mm}$ (say 45 mm)

(hole = 30 mm; to a machine flame cut edge)

Distance from bolt to beam flange = $\dfrac{60}{2} + 10 = 40\,\text{mm}$

Moment in plate at bolt centre lines = $0.045Q\,\text{kNm}$

$$= 0.045 \times 2 \times 20.88$$

$$= 1.88\,\text{kNm}$$

$$Z = (200 - 2 \times 30)t^2/6$$

$\dfrac{140}{6}t^2 = \dfrac{1.88 \times 10^6}{165}$

$t = 22.1\,\text{mm}$

$$\text{Moment in plate at beam flange} = 2T \times 0.04 - Q \times 0.085$$
$$= 2 \times 91.89 \times 0.04 - 2 \times 20.88$$
$$\times 0.085$$
$$= 3.802 \text{ kNm}$$
$$Z = 200t^2/6$$
$$\frac{200}{6}t^2 = \frac{3.802 \times 10^6}{165}$$
$$t = 26.3 \text{ mm}.$$

Thus a plate thickness of at least 27 mm would be chosen with a maximum distance of 85 mm from the surface of the beam flange. Thus by taking one bolt size bigger than used in Part I, the plate thickness has been reduced from 33 mm to 27 mm – a very worthwhile saving.

Example 9.11 This example sets out to demonstrate the difference between HSFG bolts and ordinary bolts when used in a beam to column moment connection. Consider the connection shown in Fig. 9.21, which consists of an end plate welded to the end of a 356 × 171 mm at 67 kg/m UB and then bolted to the column by HSFG bolts. What is the maximum moment which can be taken by this arrangement?

First the student must understand how the connection reacts under load.

Due to the initial prestress of the bolts, no part of the end plate will lose contact with the surface of the column. Thus the neutral axis must lie at the neutral axis of the beam, in other words, at mid depth.

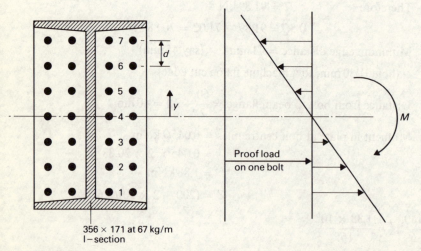

356 × 171 at 67 kg/m
I−section

Fig. 9.21

Let the force in any one bolt be F

F is proportional to the distance of the bolt from the neutral axis.

Therefore $F = k \cdot y$

The external moment, $M = \sum F \cdot y = k \cdot \sum y^2$

If the maximum value of $y = y_0$ and the maximum value of $F = F_0$

$$F_0 = k \cdot y_0 = \frac{M}{\sum y^2} \cdot y_0$$

The student will recognise a familiar formula.

F_0 must be less than 0.6 times the proof load of the bolt being used.

Thus $M_{max} = 0.6p \times \sum y^2/y_0$

Hence, when the maximum allowable size of bolt has been determined, the maximum value of external moment may be determined.

Consider one side of the end plate only:

Width = 0.5 (beam width − web width)

= 82.05 mm.

In the configuration shown in Fig. 9.21, two bolts must be accommodated in this width, and thus it will be found that M 16 bolts will be the largest possible:

Width necessary for M 16 bolts = edge distance + minimum spacing
+ clearance to beam web (washer width)

= 26 + 40 + 18.5

= 84.5 mm

Therefore, using clipped washers, M 16 bolts may be used. However, it would probably be better to widen the end plate by about 5 mm and use unclipped washers.

The depth of end plate available = beam depth − 2 × flange thickness

= 332.6 mm

Using seven rows of bolts as shown, requires a depth (with M 16 bolts) of:

$D = 6 \times$ minimum spacing + 2 × clearance to flange

= 6 × 40 + 2 × 18.5

= 277 mm.

Thus M 16 bolts are possible.

The value of d (Fig. 9.21) chosen = 45 mm. This gives a clearance to the flanges of 31.3 mm as against the 18.5 needed and is thus satisfactory.

Bolt level	y	y^2
1	-135	18 225
2	-90	8 100
3	-45	2 025
4	0	0
5	45	2 025
6	90	8 100
7	135	18 225
		56 700

Remembering that there are four bolts at each level

$\sum y^2 = 4 \times 56\,700 = 226\,800\,mm^2$

$M_{max} = 0.6 \times 92.1 \times 226\,800/135\,kNmm$

$\qquad = 92.84\,kNm$

The maximum moment of the section $= 176.7\,kNm$.

Shear capacity

Although those bolts above the neutral axis have a reduced shear capacity because the external moment imposes tension on them, this is fully compensated for by the increased pressure on the column below the neutral axis.

Therefore, the shear capacity is unimpaired

Shear capacity $= 28 \times \dfrac{0.45}{1.4} \times 92.1 = 828.9\,kN$.

Final note on beam to column connections

It will have been noticed that in all the beam to column examples which have been considered, the distribution of the bolts in the joints has been of a symmetrical nature.

There are two reasons for this:

1. It results in an easier analysis of the bolt forces.
2. It gives greater stability and doesn't tend to produce out-of-balance moments in the weaker, perpendicular direction.

Welded Connections

When consideration is given to the design of welded connections it is important to make a clear distinction between the following two terms:

1. The joint type, and
2. The weld type.

The first refers to the geometry of the plates of metal being joined, while the latter refers to the geometry of the metal used in the weld. For any

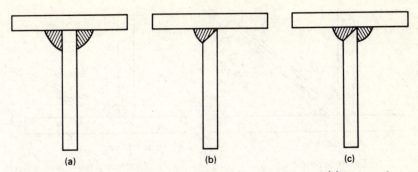

Fig. 9.22 Types of weld: (a) fillet weld, (b) butt weld, and (c) butt and fillet weld

given joint type, it is usually possible to make the connection using at least two types of weld. First, consider the classes of weld which are in common use.

Types of weld

The two most common types of weld are the 'butt weld' and the 'fillet weld'. These are illustrated in Fig. 9.22, and are shown as they would be used in a 'Tee joint'.

As can be seen, the butt weld involves shaping one of the plates in such a fashion that the weld metal can be placed into the plate, whereas the fillet weld only affects the surface of the plates. The butt weld is said to have full penetration throughout the plate, while the fillet weld has markedly less penetration. There are a number of factors which affect the choice between one weld type and the other, and as shown in Fig. 9.22(c) they are sometimes used in conjunction with one another. Some of the factors are listed in the following paragraphs; however, the list is by no means a complete one.

1. *The type of stress to be resisted*   As will be seen, the two types of weld have different mechanical properties; this leads to one being considered in one situation and the other in a different situation.

2. *The size and thickness of the parts to be connected*   A butt weld is practically impossible when the thickness of parent metal – the metal of the part to be joined – is less than 2 or 3 mm when a Tee joint is to be made, because the electrode used to deposit the weld metal will tend to melt the parent metal rather than cause proper fusion. Again, especially when working with RHSs of small dimension, a fillet weld is easier because the preparation needed is less than with a butt weld, and the weld itself is simpler to fabricate.

3. *Cost*   With the extra preparation needed with the butt weld, it can prove more expensive than the fillet weld. However, the equivalent fillet

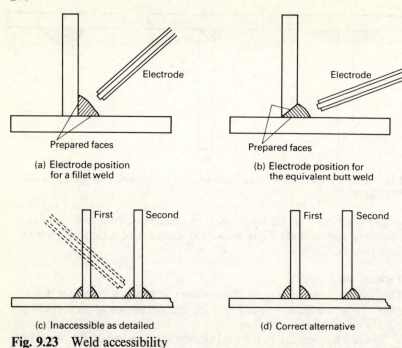

(a) Electrode position
   for a fillet weld

(b) Electrode position for
   the equivalent butt weld

(c) Inaccessible as detailed

(d) Correct alternative

Fig. 9.23 Weld accessibility

weld may need more weld metal which may make the butt weld cheaper;
each case must be judged on its merits.

4. *Accessibility* (consider Fig. 9.23)   In order to fabricate any weld, the
welder must be able to place the electrode in the correct position, as
shown in Fig. 9.23(a) for a fillet weld, and as in Fig. 9.23(b) for the
equivalent butt weld. As a general rule, the tip of the electrode should
bisect the included angle between the surfaces of the prepared parts.
Figure 9.23(c) shows a typical situation where the designer has detailed
both upstands to be welded using fillet welds. However, the position of
the electrode for the internal weld in the upstand welded second is impossible to obtain. Therefore, the configuration of Fig. 9.23(d) must
be used in this situation.

5. *Type of welding process being used*   There are many methods of
placing the weld metal in the joint, but the three most common classes
are:

1.   Oxy-acetylene gas welding.
2.   Manual arc welding.
3.   Automatic arc welding.

Of these three, the first is less used in structural work for two reasons:
(a) it is slower than electric arc welding; and (b) because of this it subjects

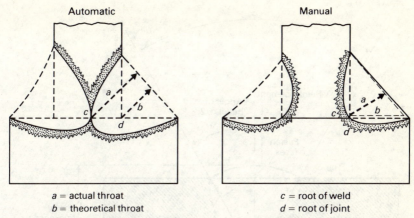

a = actual throat   c = root of weld
b = theoretical throat   d = root of joint

Fig. 9.24 A comparison of penetration in automatic and manual fillet welds

the structure to a greater amount of heat, which in some cases can be detrimental.

The main difference between the second and third methods lies in the degree of penetration which can be achieved, as shown in Fig. 9.24. Automatic welding is also faster and easier to control. However, it necessitates special equipment and factory conditions, thus it is not very practical for much welding on site. It will also be noted from Fig. 9.24, that although the automatic fillet weld achieves a penetration almost equal to that of the butt weld, it is still classified as a fillet weld with equal strength as the manual equivalent. This is obviously conservative.

However, it is not the intention of this work to dwell any longer on the reasons for choosing one type of weld in favour of the other. This may be obtained from any textbook on the welding process. Our concern in this chapter is to give the student the ability to calculate the size of weld required, regardless of its type. However, some knowledge of the foregoing at least makes the student aware that the mathematics of the design process is not everything.

The anatomy of a weld (and its idealisation for design purposes)
Figures 9.25 and 9.26 show the various parts of first a butt weld and then a fillet weld. The student should become familiar with the following terms:

Fusion faces – the original surfaces of the parent metal into which the weld metal penetrates

Root – the vertex of the weld, where defects can easily form, such as included slag (the product of the weld flux and oxygen and metal oxides)

Toe – the line formed between the weld metal and the parent metal

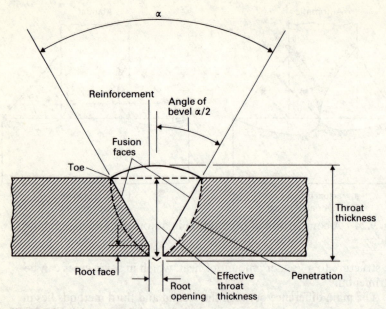

Fig. 9.25 A butt weld

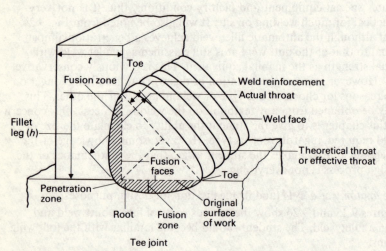

Fig. 9.26 A fillet weld

Throat – the central portion of the weld. It is at this section of the weld that the greatest stresses develop

Weld reinforcement – the part of the weld metal which lies above the surface joining the two toes of the weld. As will be seen later this is discounted in analysing the stresses in the weld. It is of course possible that instead of having a convex surface to the welds as is illustrated, the surfaces might be concave. In this case, the thickness of throat taken by the designer will be as shown in Figs. 9.27 and 9.28. Thus the weld is designed using an **effective throat thickness**.

For a right angled Tee joint, with a fillet weld having equal leg lengths, the effective throat thickness = 0.71 *L*.

The strength of a weld
Allowable stresses in welds

53. a. **General.** *When electrodes complying with Sections 1 and 2 of BS 639* are used for the welding of grade 43 steel, or with Sections 1 and 4 of BS 639* are used for the welding of grade 50 steel, or with Sections 1 and 4 of BS 639* are used for the welding of grade 55 steel[†] and the yield stress of an all-weld tensile test specimen is not less than 430 N/mm² when tested in accordance with Appendix D of BS 639*, the following shall apply:*

(i) *Butt welds. Butt welds shall be treated as parent metal with a thickness equal to the throat thickness (or a reduced throat thickness as specified in Clause 54 for certain butt welds) and the stresses shall not exceed those allowed for the parent metal.*

(ii) *Fillet welds. The allowable stress in fillet welds, based on a thickness equal to the throat thickness, shall be 115 N/mm² for grade 43 steel or 160 N/mm² for grade 50 steel or 195 N/mm² for grade 55 steel.*

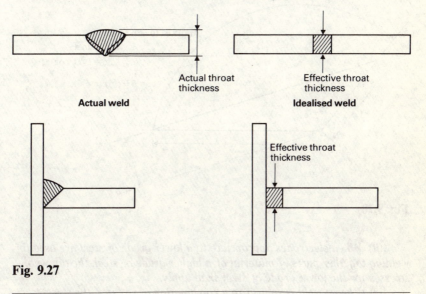

Fig. 9.27

* *BS 639, 'Covered electrodes for the manual metal-arc welding of mild steel and medium-tensile steel'.*
[†] *See 4.1.1. of BS 639.*

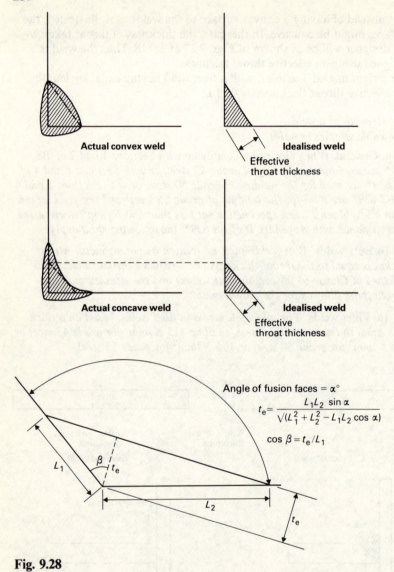

Actual convex weld

Idealised weld

Effective throat thickness

Actual concave weld

Idealised weld

Effective throat thickness

Angle of fusion faces = $\alpha°$

$$t_e = \frac{L_1 L_2 \sin \alpha}{\sqrt{(L_1^2 + L_2^2 - L_1 L_2 \cos \alpha)}}$$

$$\cos \beta = t_e / L_1$$

Fig. 9.28

(iii) *When electrodes appropriate to a lower grade of steel are used for welding together parts of material of a higher grade of steel, the allowable stresses for the lower grade of steel shall apply.*

(iv) *When a weld is subject to a combination of stresses, the stresses shall be combined as required in Subclause 14 c and d, the value of the equivalent stress f_e being not greater than that permitted for the parent metal.*

Figure 9.29 illustrates the various types of stress which can be experienced by a weld. It will be noted that the overriding assumption which is made is that the critical section of the weld is the throat; therefore, every calculation is based upon this thickness.

ABCD is the throat plane on which it is assumed that the maximum stresses occur

f_n is the stress normal to the throat plane
f_h is the horizontal shear stress
f_v is the vertical shear stress

Both f_n and f_h would be caused by a direct tension between the connected parts. f_v, on the other hand, would be caused by a vertical shear force. When the three stresses exist together, Cl. 14c of BS 449 must be used with a slight modification. This states that, the equivalent stress is obtained from:

$$f_e = \sqrt{(f_n^2 + 3f_q^2)}$$

where f_q is the shear stress. This is obtained from f_v and f_h by adding them vectorially, thus:

$$f_q = \sqrt{(f_v^2 + f_h^2)}$$

Therefore $f_e = \sqrt{(f_n^2 + 3(f_v^2 + f_h^2))}$ [9.4]

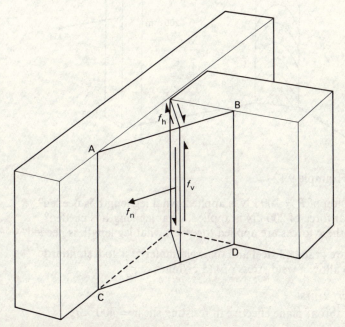

Fig. 9.29 Stresses in a weld

This should not be less than the stresses found from Table 1 of BS 449.

However, this formula is not often used since a simpler and more conservative assumption is used, as shall be seen later. A comparison of the two will be presented.

The following three examples introduce the student to the fundamental methods of calculation.

Example 9.12   Figure 9.30 shows a Tee joint formed by using fillet welds either side of the stalk.

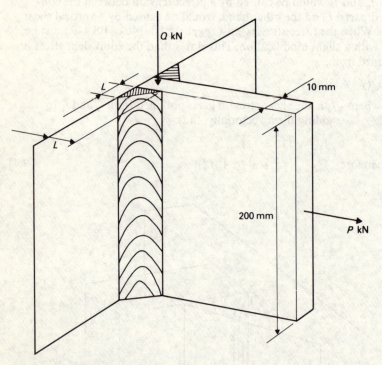

Fig. 9.30   Example 9.12

1. If a tension of $P = 200$ kN is applied, what leg length is needed?
2. If a shear force of 200 kN is applied, what leg length is needed?
3. If both these forces are applied together, what leg length is needed?

The plates are grade 43 steel and the weld material is to a standard sufficient to allow a weld stress of 115 N/mm².

Consider part 2 first

The area of throat plane effective in resisting shear $= 400 \times t_e$

$$t_e = 0.71L$$

Shear capacity of weld $= 115 \times 400 \times 0.71\,LN$

$$= 32.66\,L\,\text{kN}$$

$$= 200\,\text{kN}$$

Therefore $L = 6.1$ mm (say 7 mm)

Now consider part 1

Applying the same assumption as in part 2. above

Area of throat $= 400t_e = 284\,L\,\text{mm}^2$

Tension strength $= 115 \times 284\,L = 32.66\,L\,\text{kN}$ as before

$L = 6.1$ mm as before.

This is the simple, yet conservative method of calculating stresses on welds. However, looking at Fig. 9.31, it will be seen that the true stresses are given by resolving P into its components P_h and P_n for a weld in which the leg lengths are equal,

where β = one half the angle between the fusion faces

= 45° in this example

Thus $P_h = P_n = P/\sqrt{2}$

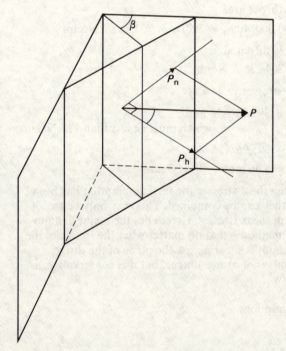

Fig. 9.31

$$f_n = P_n/284L = \frac{200\,000}{\sqrt{2} \times 284L} = 497.96/L\,\text{N/mm}^2$$

$$f_h = f_n = 497.96/L\,\text{N/mm}^2$$

$$f_e = \sqrt{(f_n^2 + 3f_h^2)} = \sqrt{[(497.96/L)^2 + 3 \times (497.96/L)^2]}$$
$$= 2 \times 497.96/L = 995.92/L\,\text{N/mm}^2$$

The allowed effective stress, $p_e = 230\,\text{N/mm}^2$ (Table 1 of BS 449)

$$995.62/L = 230$$

$$L = \frac{995.62}{230}$$

$$= 4.33\,\text{mm}   (\text{say 5 mm})$$

Thus a saving has been made, but at the cost of extra calculation effort. Most designers use the simpler method for speed, since the saving is not too great in many instances, especially where a constant stress is being resisted.

3. Both forces present

(a)   Conservative assumption – simple method:

Shear stress = Q/throat area

$$= 200\,000/400t_e = 500/t_e = 704.225/L\,\text{N/mm}^2$$

Tension stress = P/throat area

$$= 704.225/L\,\text{N/mm}^2$$

Combining these stresses vectorially

Combined stress = $995.96/L\,\text{N/mm}^2$

which must be less than $115\,\text{N/mm}^2$

$$L = 995.96/115$$

$$= 8.66\,\text{mm}   (\text{say 9 mm})$$

Note: In combining these stresses the tacit assumption has been made that they can be combined. This may only be true if they are **both** shear stresses. Herein lies the major assumption of this method – that no matter what the source of the stress, the result is shear across the plane of the throat. However, this is of course untrue, but it is conservative, as shown below.

(b)   More 'exact' assumption:

Stresses due to P:

$$f_h = f_n = 497.96/L\,\text{N/mm}^2   \text{as before}$$

Stresses due to Q:

$f_v = Q/\text{throat area}$

$\quad = 704.225/L \, \text{N/mm}^2$

$f_e = \sqrt{[(497.96/L)^2 + 3 \times ((497.96/L)^2 + (704.225/L)^2)]}$

$\quad = \sqrt{(2\,479\,655.2/L^2)}$

$\quad = 1574.7/L \, \text{N/mm}^2$

This must be less than 230 N/mm² the stress from Table 1 of BS 449

Hence $\quad L = 1574.7/230$

$\qquad\quad = 6.85 \, \text{mm} \qquad$ (say 7 mm).

For most normal design purposes, the simpler method will prove satisfactory. Throughout the rest of the examples in this section this is the assumption which will be adopted, but the inquisitive student may like to recheck them on the basis of the 'more exact' assumption. In any case, most butt welds do not suffer from these difficulties.

It should be noted that the International Institute of Welding proposes a method similar to that of 3(b) above, but with one important difference. The allowable stresses they quote are based on the tension strength of the parent metal divided by, in the case of grade 43 metal, 0.7, and 0.85 in the case of grade 50 metal. This results in a weld size in between the two here calculated; however, the stress given in Table 1 of BS 449 should be used for designs to BS 449.

Example 9.13   Figure 9.32 shows essentially the same joint as in Example 9.12, except that the weld is a butt weld. Will a butt weld be sufficient in this case?

1. *P alone*

Effective throat = 10 mm

Weld stress = $P/(200 \times 10) = 200\,000/2000$

$\qquad\qquad\quad = 100 \, \text{N/mm}^2$

Allowed stress is 155 N/mm² obtained from Table 19 of BS 449.

2. *Q alone*

Weld stress = $Q/2000 = 100 \, \text{N/mm}^2$

Allowed stress = 100 N/mm²

Note the use of the 'average' allowed shear stress obtained from Table 11 of BS 449.

3. *P and Q together*

This necessitates the use of Cl. 14c of BS 449

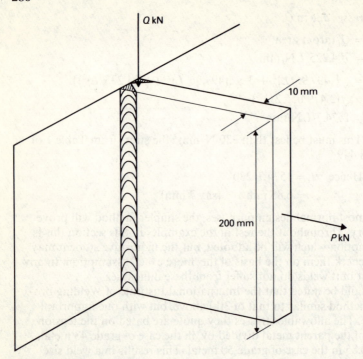

Fig. 9.32 Example 9.13

$f_e = \sqrt{(100^2 + 3 \times 100^2)}$
$\quad = 200 \text{ N/mm}^2$

$p_e = 230 \text{ N/mm}^2$

Thus the butt weld is satisfactory.

Example 9.14 In Fig. 9.33, a Tee joint is again shown, the weld being the same as in Example 9.12. The difference is that the shear force Q, is set at an eccentricity of e from the face of the weld. This induces a bending stress in the weld as well as the shear stress.

Welds subject to bending are analysed in a manner similar to that used for the bending of any structural section, in that a moment of inertia is calculated. In this example, the eccentric force is not in the plane of the welds, therefore the calculation of this inertia is simple. The next example covers the case where the force is in the plane of the welds, giving rise to a small degree of extra calculation.

The moment of inertia of the weld is given by:

$I_w = \text{total throat width} \times d^3/12$

This is, of course, exactly similar to $I = bd^3/12$

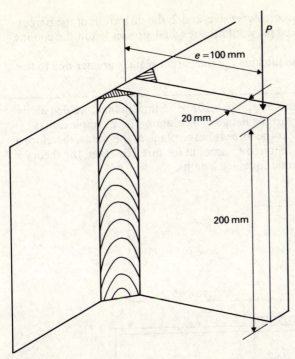

Fig. 9.33

Thus in this case $I_w = 2 \times 0.71L \times 200^3/12 = 942\,809.0\,L\,\text{mm}^4$

Weld stress caused by bending $= \dfrac{Pe}{I_w}\cdot\dfrac{d}{2}\left(\text{ cf. stress} = \dfrac{M}{I}\cdot y\right)$

$$= \frac{200\,000 \times 100}{942\,809.0L}\cdot 100$$

$$= 2121.3/L\,\text{N/mm}^2$$

Weld shear stress due to $P = P/(2 \times 200 \times t_e)$

$$= 200\,000/(400 \times 0.71L)$$

$$= 707.1/L\,\text{N/mm}^2$$

Combining these vectorially gives $f_q = \sqrt{(2121.3^2 + 707.1^2)}/L$

$$= 2236.1/L\,\text{N/mm}^2$$

$$= 115\,\text{N/mm}^2$$

Thus $L = 19.44\,\text{mm}$ (say 20 mm)

It should be noted that a weld of this size would prove impossible to deposit in only one pass with the welding rod, therefore this weld would have to be a multi-layered weld. This would result in two consequences:

288

1. the weld would be more expensive, and 2. the distortion of the parent metal would be reduced (as would the residual stresses left in the cooling of the plates).

Note also that the thickness of the parent plate is greater due to the bending stresses.

Example 9.15 Figure 9.34 shows a 203 × 76 mm channel welded along the edges of its flanges to the flange of a column. This example differs from the last one in that the force is in the plane of the welds, therefore a slight modification is required to account for this. However, the theory is not too different. Assume equal leg lengths.

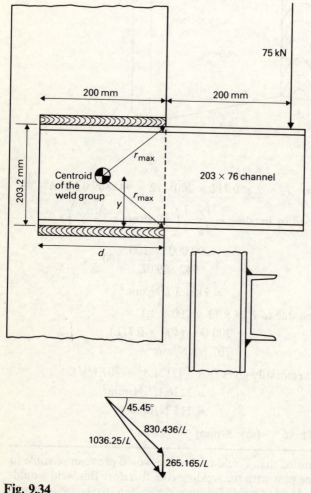

Fig. 9.34

As with the last example, the load induces both bending and shear stresses (both assumed to cause shear along the throat plane). The bending stress is given by:

$$f_b = \frac{Mr}{I_p}$$

which is similar to the last example

where M is the moment of the load about the centroid of the weld group
r is the distance from the centroid to the point at which the stress due to bending is needed
I_p is the 'polar' moment of inertia

In order to calculate the value of I_p the following formula is used:

$$I_p = 2\left(\frac{t_e d^3}{12} + t_e d \cdot y^2\right) = 2t_e d(d^2/12 + y^2)$$

The proof of this formula is given at the end of this example. It can be seen that the formula differs from the one used in the last example in having one extra term which allows for the offsetting of the weld centroid from the line of the weld itself.

The factor of two is required because of the presence of two lines of weld. In the more general case of two lines of weld in which one of the lines has a smaller throat than the other, the procedure is similar to the above, but the designer is first required to find the centroid and then calculate the contribution of each weld separately. This will be demonstrated later. For this example:

$$I_p = \frac{2L}{\sqrt{2}} \cdot 200\,(200^2/12 + 101.6^2)\ \text{mm}^4$$

$$= 3.862L \times 10^6\ \text{mm}^4$$

$$r_{max} = \sqrt{(100^2 + 101.6^2)} = 142.557\ \text{mm}$$

Thus the stress due to bending $= \dfrac{75\,000 \times 300 \times 142.557}{3.862L \times 10^6}$

$$= 830.436/L\,\text{N/mm}^2$$

Stress due the shear effect of the load $= \dfrac{75\,000}{400 \times L/\sqrt{2}}$

$$= 265.165/L\,\text{N/mm}^2$$

Both these stresses *are* shear stresses

Therefore maximum stress on the weld $= 1036.25/L = 115\ \text{N/mm}^2$

Thus $L = 1036.25/115$

$$= 9.01\ \text{mm} \quad \text{(say 10 mm)}$$

There now follows a derivation of the general formula for I_p. The TEC student should not regard this as central to his studies.

Consider the small elemental area in Fig. 9.35.

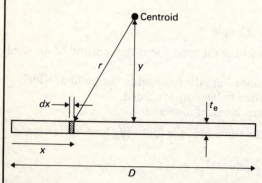

Fig. 9.35

The stress due to external moment $= f = kr$

In other the words, it is assumed that the stress is proportional to the radial distance from the centroid. In turn this assumes, as did the method for bolts on p. 240, that rotation is about the centroid, and that the plate is rigid.

Let the contribution of this weld to the resistance of moment $= M_o$

$$M_o = \int_0^L (\text{stress} \times \text{area} \times r) = \int_0^L (kr \cdot t_e dx \cdot r)$$

$$r^2 = y^2 + (0.5D - x)^2 = x^2 + y^2 + 0.25D^2 - Dx$$

$$M_o = kt_e \cdot \int_0^L (x^2 + y^2 + 0.25D^2 - Dx) \, dx$$

$$= kDt_e \left(D^2/12 + y^2 \right) = kI_p$$

Thus the stress $= f = kr = M_o r / I_p$

Thus $\quad I_p = Dt_e \left(D^2/12 + y^2 \right)$

Applying this to Fig. 9.36 gives

$$I_p = D_1 t_1 \left(D_1^2/12 + y_1^2 \right) + D_2 t_2 \left(D_2^2/12 + y_2^2 \right)$$

where y_1 and y_2 are found by a centre of gravity calculation as follows:

$$y_2 = \frac{D_1 t_1}{D_1 t_1 + D_2 t_2} \cdot (y_1 + y_2) \qquad y_1 = \frac{D_2 t_2}{D_1 t_1 + D_2 t_2} \cdot (y_1 + y_2)$$

The distance between the welds, $(y_1 + y_2)$ is, of course, known.

On closer examination of the formula for I_p it can be seen that it consists of two distinct parts:

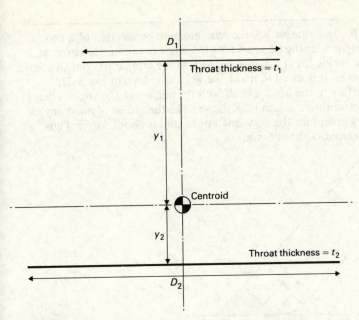

Fig. 9.36

$$I_p = I_{xx} + I_{yy}$$

where $I_{xx} = t_e \cdot \dfrac{D^3}{12}$ and $I_{yy} = t_e D \cdot y^2$

This should remind the student of the parallel axis theorem used in the calculation of the moment of inertia of a shape consisting of a number of rectangles, where

$I_{\text{about the centroidal axis}} = I_{xx} + \text{area of rectangle} \times y^2$
$$= bd^3/12 + bd \cdot y^2$$

where I_{xx} is the moment of inertia about the centre of the rectangle

and

y is the distance from the centroid of the rectangle to the centroid of the complete section

Thus the general rule for calculating the value of I_p for any weld group is as follows:

1. Divide the group into a number of straight lines.
2. Find the overal centroid and the distances (y) to the weld (the perpendicular distances).
3. Calculate $I_p = I_{xx} + I_{yy}$ for each straight section of the group.
4. Sum the various I_p's to obtain the total value of I_p.

Example 9.16 In Example 9.5, the base moment connection of a column was designed for the bolts and for the angle required. However, at that stage the welds could not be designed. This example sets out to complete the process. It is usual to make the weld as shown in Fig. 9.37, where the surface of the weld is flush with the surface of the angle. Therefore, the effective throat width is as shown. The toe radius of most angles does not vary except for the very small ones, and is about 5 mm. Thus the effective throat is about 6 mm.

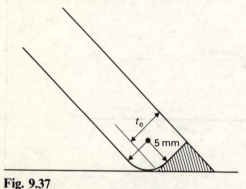

Fig. 9.37

Area of throat = $6L$ mm^2

Weld strength = $115 \times 2 \times 6L = 1380L\,N = 367\,800\,N$

Length of angle required = $L = 367\,800/1380$

$$= 266.5 \text{ mm} \quad \text{(say 275 mm)}$$

Example 9.17 Redesign the angle seating cleat of Example 9.7 using a welded connection.

Stage I: Determine the thickness of angle needed
This is dictated by the bearing on the metal of the angle

Load = 150 kN

Allowed bearing stress = 260 N/mm^2 (grade 50 steel)

Required bearing area = $150\,000/260 = 576.9$ mm^2

Assuming that the angle length equals the beam flange width

Actual bearing area = $171.5 \times$ angle thickness

Minimum angle thickness = 576.9/171.5

$$= 3.36 \text{ mm}$$

Thus a $25 \times 25 \times 4$ mm equal angle may be suitable; however, this may not give an adequate amount of weld, and this must also be checked.

Stage II: Design the welding

Assuming, based on Stage I above, that the leg length is the maximum 4 mm allowed by the angle thickness, and that the welding is to the two sides and bottom of the angle, but not to the top surface on which the beam rests.

Throat area $= 0.71 \times 4 \times (2 \times 25 + 171.5)$

$\qquad\qquad = 626.5 \text{ mm}^2$

Weld strength $= 626.5 \times 160 \text{ N}$ (welding suitable for grade 50 steel)

$\qquad\qquad = 100 \text{ kN}$

Thus the angle is not large enough
Try $50 \times 50 \times 5$ equal angle using 5 mm welds

Throat area $= 0.71 \times 5 \times (2 \times 50 + 171.5)$

$\qquad\qquad = 963.8 \text{ mm}^2$

Weld strength $= 154.2 \text{ kN}$
Satisfactory

Example 9.18   Figure 9.38 shows a UB section welded onto the face of a column as a bracket to support a crane rail imposing a load of 250 kN on the bracket at a distance from the column face of 300 mm. Design the required fillet welds; the steel is grade 50.

The major controversy over the design of a connection such as this lies with determining the position of the neutral axis. However, the author feels it reasonable to assume that the neutral axis lies at the mid depth of the beam section.

The welds are subjected to two types of stress:

1.   Bending stress – Above the neutral axis, this gives rise to tension in the weld. Below the neutral axis, the weld is unstressed, since the compression is resisted by the beam pressing on the column flange.
2.   Shear stress.

Therefore, the weld above the neutral axis is more heavily stressed than that below the neutral axis. Thus the upper portion of fillet weld should be larger than the lower portion. However, most of the bending tension is taken by the welds close to the flange of the beam. A compromise is thus made by allowing the welds above the three-quarter mark to take *all* the tension, while the rest of the welds take *all* the shear.

1. Tension welds
   the eccentricity of the load causes a moment at the
   column flange $= M = 250 \times 0.3 \text{ kNm}$
   $\qquad\qquad\qquad = 75 \text{ kNm}.$

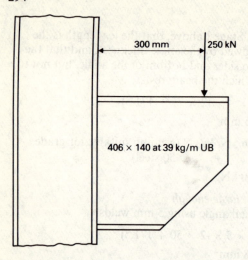

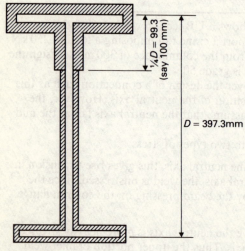

Fig. 9.38

The tension **force** in the welds is found by dividing this moment by the lever arm of the section.

For a rectangular section this would be equal to two thirds of the total depth, since the linear stress gives rise to a linearly varying force, whose centre of action is one sixth of the depth down from the compression flange and similarly, one sixth of the total depth up from the tension edge. However, for an I-section, although the stress varies linearly across the section, the force is mostly concentrated in the flanges, therefore, the lever arm will be greater than two thirds of the total depth. A suitable compromise is to use the depth between the fillets as the lever arm, which for this section = 357.4 mm.

Tension in welds = 75/0.3574

$$= 209.85 \, \text{kN}$$

Length of welds effective in tension
= perimeter of the beam in contact with the weld

$$= 2B - t + 2 \times 100 - 2 \times T$$

where the symbols have the significance given in Appendix B

$$= 2 \times 141.8 - 6.3 + 200 - 17.2$$
$$= 460.1 \, \text{mm}$$

Throat area = $0.71L \times 460.1 \, \text{mm}^2$
(L being the leg length)

Thus $0.71L \times 460.1 \times 160 = 209\,850 \, \text{N}$

$$L = 4.01 \, \text{mm} \qquad (\text{say 4 mm})$$

2. *Welds in shear*

Length of welds effective in shear $= 2B - t + 2 \times (397.3 - 100)$
$$- 2 \times T$$

$$= 283.6 - 6.3 + 594.6 - 17.2$$

$$= 854.7 \, \text{mm}$$

Throat area = $0.71L \times 854.7 \, \text{mm}^2$

Thus $0.71L \times 854.7 \times 160 = 250\,000$

$$L = 2.57 \, \text{mm} \qquad (\text{say 3 mm})$$

Thus theoretically, the weld size would be changed from 4 mm leg length to 3 mm leg length in the lower section of the connection.

However, the difference between the welds is not very great; thus it is probably more efficient to continue the 4 mm weld over the whole connection. In any case, with a beam of this thickness, **the minimum practical weld size**, is about 6 mm leg length. However, the method will hold for other more heavily loaded sections, as will be demonstrated in the questions at the end of this chapter.

This same method may be used to design the welds necessary when attaching an end plate which will subsequently be bolted to the supporting column. However, in both these cases the method used in the following example may be adopted. It is based on the calculation of a weld moment of inertia.

Example 9.19 Design the welds needed (fillet welds) for the end plate of Fig. 9.39 (grade 43 steel).

Maximum bending moment $= \dfrac{200 \times 5}{8}\,\text{kNm} = 125\,\text{kNm}$

Maximum shear force $= 100\,\text{kN}$ (end reaction).

Stage I: Calculate the moment of inertia of the welds.
The welds divide up into twelve separate straights:

$$I_{AB} = 171.5t_e \times \left(\frac{355.6}{2}\right)^2 = 5.422 \times 10^6 t_e \,\text{mm}^4$$

$$I_{GH} = I_{AB} = 5.422 \times 10^6 t_e \,\text{mm}^4$$

$$I_{AM} = 11.5t_e \times \left(\frac{355.6}{2} - \frac{11.5}{2}\right)^2 + t_e \times \frac{11.5^3}{12}$$

$$= 0.341 \times 10^6 t_e \,\text{mm}^4$$

$$I_{BC} = I_{JH} = I_{FG} = I_{AM} = 0.341 \times 10^6 t_e \,\text{mm}^4$$

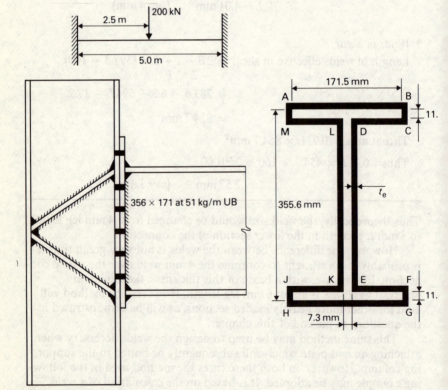

Fig. 9.39

$$I_{ML} = 82.1t_e \times \left(\frac{355.6}{2} - 11.5\right)^2 = 2.271 \times 10^6 t_e\,\text{mm}^4$$

$$I_{CD} = I_{EF} = I_{JK} = I_{ML} = 2.271 \times 10^6 t_e\,\text{mm}^4$$

$$I_{DE} = I_{KL} = t_e \times \frac{332.6^3}{12} = 3.066 \times 10^6 t_e\,\text{mm}^4$$

$$I = 27.42 \times 10^6 t_e\,\text{mm}^4.$$

Stage II: Calculate the maximum bending and shear stresses.

$$\text{Maximum bending stress} = f_b = \frac{M}{I} \times \frac{355.6}{2}$$

$$= \frac{125 \times 10^6 \times 177.8}{27.42 \times 10^6 t_e}$$

$$= 810.55/t_e$$

$$\text{Maximum shear stress} = \frac{\text{end reaction}}{\text{total throat area}}$$

$$\text{Total throat area} = (\text{weld perimeter}) \times t_e = 1382.6 t_e\,\text{mm}^2$$

$$\text{Maximum shear stress} = \frac{100\,000}{1382.6 t_e} = 72.33/t_e$$

$$\text{Maximum weld stress} = \sqrt{(810.55^2 + 72.33^2)}/t_e$$

$$= 813.77/t_e\,\text{N/mm}^2$$

$$= 115\,\text{N/mm}^2$$

Minimum $t_e = 813.77/115 = 7.07\,\text{mm}$

Minimum leg length = 10 mm (exactly)

Example 9.20 In the double channel bracket support shown in Fig. 9.40, the weld nearest to the point load – TU – is made 1.5 times larger than all the other welds. This is in an effort to equalise the stresses experienced, at points S, T, U and V. If the steel is of grade 43, design the welds.

Stage I: Find the centroid of the weld group.
From symmetry, the centroid lies mid-way between the upper and lower horizontal welds.
Taking area moments about the vertical line TU:

$$\text{Area of SV} \times 300 = (\text{area of SV} + \text{area of TU}) \times y$$

$$y = \frac{300}{2.5} = 120\,\text{mm}.$$

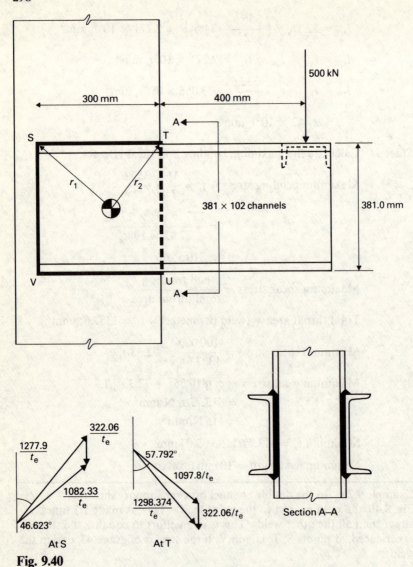

298

Fig. 9.40

Stage II: Calculate the value of I_p the polar moment of inertia.

$$I_{ST} = I_{VU} = t_e \cdot \frac{300^3}{12} + 300t_e \cdot 190.5^2 = 13.137 \times 10^6 t_e \, \text{mm}^4$$

$$I_{SV} = t_e \cdot \frac{381^3}{12} + 381t_e \cdot 180^2 = 16.953 \times 10^6 t_e \, \text{mm}^4$$

$$I_{TU} = 1.5 \times t_e \times \frac{381^3}{12} + 1.5 \times t_e \times 381 \times 120^2$$

$$= 15.143 \times 10^6 t_e \, \text{mm}^4$$

$$I_p = 58.37 \times 10^6 t_e \, \text{mm}^4.$$

Stage III: Calculate the stresses at S, T, U, and V.

Bending: moment $= M = 250 \times (0.4 + 0.12) = 130 \, \text{kNm}$

$$f_{bs} = Mr_1/I_p \qquad\qquad r_1 = \sqrt{(190.5^2 + 180^2)}$$

$$= \frac{130 \times 10^6 \times 262.088}{58.37 \times 10^6 t_e} \qquad = 262.088 \, \text{mm}$$

$$= 583.72/t_e \, \text{N/mm}^2$$

$$f_{bt} = Mr_2/I_p \qquad\qquad r_2 = \sqrt{(190.5^2 + 120^2)}$$

$$= \frac{130 \times 10^6 \times 225.145}{58.37 \times 10^6 t_e} \qquad = 225.145 \, \text{mm}$$

$$= 501.441/t_e \, \text{N/mm}^2$$

Shear: $f_q = \dfrac{\text{force}}{\text{area}} = \dfrac{250\,000}{1552.5 t_e} = 161.03/t_e \, \text{N/mm}^2$

Referring to Fig. 9.40

Adding vectorially: at S and V

$$f_S = \frac{\sqrt{(f_{bs} \sin 46.623°)^2 + (f_{bs} \cos 46.623° - 161.03)^2}}{t_e}$$

$$= 487.4/t_e \, \text{N/mm}^2$$

at T and U

$$f_T = \frac{\sqrt{(f_{bt} \sin 57.792°)^2 + (f_{bt} \cos 57.792° + 161.03)}}{t_e}$$

$$= 602.87/t_e \, \text{N/mm}^2$$

Note: It is worth comparing these values against those obtained by allowing all the welds to be of the same size, when:

$$f_S = 506.25/t_e \, \text{N/mm}^2 \qquad f_T = 726.8/t_e \, \text{N/mm}^2$$

Thus the difference between the two stresses is greatly reduced. Of the two stresses, f_T is the larger, and must be limited to $115 \, \text{N/mm}^2$

Thus $\quad t_e = 602.87/115$

$$= 5.24 \, \text{mm}$$

and

Leg length = $\sqrt{2} \times t_e = 7.4$ mm   (say 8 mm)

This makes the leg length of the front weld = 12 mm.
If all the welds were of the same size, the weld would need to
be of leg length = 9 mm.

The limiting factor in all of this must be the thickness of
the column flange, however, most of the sections have a
thickness greater than the 12 mm required.

The foregoing should be compared with Question 9 which brings the
stresses at T and S even closer.

Questions

1. Calculate the maximum tensile load which can be taken by the butt-
   splice joint of Fig. 9.7, if 16-mm ⌀ grade 8.8 black bolts are used.
   Plates of grade 50 steel.
2. A circular steel scaffold tube of outside diameter 48.3 mm and wall
   thickness 4 mm is to be joined in tension by means of a bolted splice
   as shown in Fig. 9.41. If the axial tension capacity of the tube is
   70 kN, what percentage of this is possible using the proposed splice?
   Assume grade 43 material stresses divided by 1.23. Try bolt diameters ranging from 8 to 16 mm.

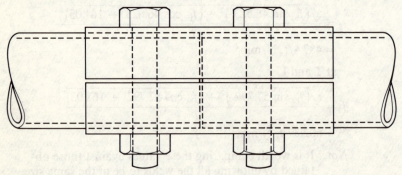

Fig. 9.41

3. Design the semi-circular cover plates needed in the last example for
   the optimum bolt size. Assume the cover plate material to be the
   same as the tube material.
4. If the connection of Fig. 9.8 is made using M 10 grade 8.8 bolts in
   place of M 12 grade 4.6 bolts, design the connection.
5. A tension member consists of two 120 × 120 × 10 mm angles placed
   back to back. A splice similar to Fig. 9.8 is to be made to carry a

load of 500 kN. Using HSFG bolts design the connection. All plates and angles to grade 43 steel.

6. Redesign the connection of Question 5 using grade 8.8 black bolts. Which of the two designs is the more practical?

7. Some of the large box girders now required in power stations are so deep that they cannot be transported from the works as one entire member. They are therefore fabricated as two separate box sections which are connected together on site by the use of HSFG bolts. Details of such a girder are given below and in Fig. 9.42. The girder

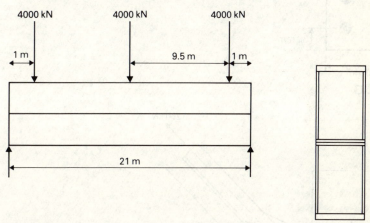

Fig. 9.42

has an effective span of 21 m, a depth of 4.875 m and a width of 1.25 m. The flanges are 90 mm thick, the webs are 20 mm thick and the division plate holding the bolts, at the position of the beam's neutral axis, is 25 mm thick. Design the bolted connection.

8. The beam/column connection shown in Fig. 9.43 is constructed from a steel plate welded to the end of the beam. It uses twelve bolts, each of 16 mm ⌀. Calculate the maximum values of moment and shear that can be transmitted by this joint simultaneously, if the connection is made in turn with
1. Grade 4.6 bolts.
2. Grade 8.8 bolts and the maximum tensile force is induced.

9. Check the beam to column connection of Example 9.7 (use Method II) using eighteen 8-mm ⌀ bolts in the column web. Assume the web connection remains unchanged. Angle length = 300 mm.

10. A tie carrying a load of 250 kN is composed of two 60 × 60 × 6 mm angles welded to either side of the stalk of a cutting from a 533 × 210 mm at 82 kg/m UB, which in turn is bolted to the flange face of a universal column. If the slope of the tie is 45° as shown in Fig. 9.44, design the bolts required.

302

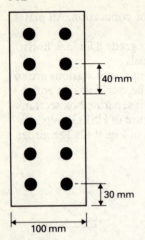

40 mm

30 mm

100 mm

Fig. 9.43

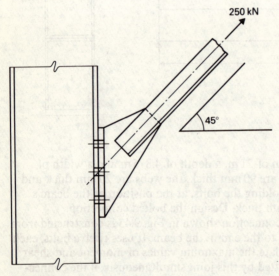

250 kN

45°

Fig. 9.44

(Note that the central two bolts are designed in such a fashion that the centroid of the angle lies on a line through the centre of the bolts. This reduces any tendency to put more load on some bolts than others.)

11. In the refurbishment of a warehouse, it is necessary to strengthen an existing 457 × 152 mm @ 52 kg/m UB spanning 5 m to carry a UDL of 100 kN/m.

It is proposed to effect this by bolting a 381 × 152 mm @ 52 kg/m UB to the tension flange. If HSFG bolts are used and the surfaces suitably prepared to allow the use of 0.45 as a slip factor, design the bolts needed.

12. Figure 9.45 shows a 50 × 50 × 6mm, grade 43 steel angle welded to the gusset of a roof truss. If the tension of the angle is 75 kN design the welds in such a way that the **force** in AB is equal to the **force** in CD.

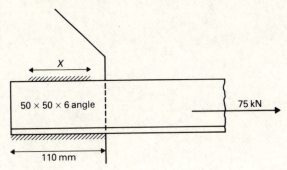

Fig. 9.45

13. A 305 × 102 mm @ 28 kg/m UB is supported on an angle cleat welded to the face of the supporting column. Design the connection if the load to be supported is 350 kN and the steel is of grade 43.

14. A 305 × 102 mm @ 28 kg/m UB is welded directly onto the face of a universal column. Thus it forms a bracket supporting a crane gantry girder, the centre of which is 150 mm from the face of the column. Assuming grade 50 steel and welding to an equivalent grade, design the weld to withstand 175 kN load.

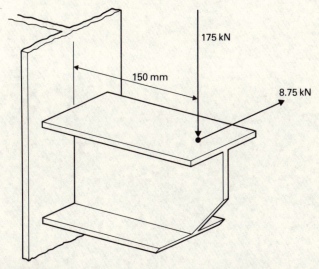

Fig. 9.46

15. The bracket of the previous question is subjected to a horizontal load (as shown in Fig. 9.46) of 8.75 kN at the same time as the 175 kN vertical load. Redesign the welds on this basis.
    (Note that this horizontal load causes both moment about the minor axis and produces torsion.)
16. Redesign the connection of Example 9.20 making TU 2.5 times as large as the others.

Chapter 10

Designing with angles

So far the chapters in this book have been concerned with the design of sections which are symmetrical about the vertical axis, or the axis through the web of the section. When this type of section is used, the simple theories of bending may be applied and stresses found using them. However, when a section is not found to be symmetrical in other words, when it is asymmetrical, these simpler theories cannot be used without some modification. This chapter is concerned with one of these asymmetrical sections – the rolled steel angle. The reader will recall that this section has already been introduced as a tension member. However, when in tension, it behaves without regard to its asymmetricality and thus requires no special attention. It is only when the section is subjected to bending or compression (when it buckles in bending) that the design must be modified.

The main uses for angles are in trusses and as purlins, but many designers would incline to the view that they have been superseded in these uses by the more modern sections. For example the 'zed-purlin' is very often used because of its lightness and ease of fixing; the RHS is increasingly used in the fabrication of trusses because of its inherent strength in compression, and because of the ease with which it may be welded together. It might therefore be argued that there is no point in considering its design. The author would argue against this if for no other reason than that it gives an insight into problems involved with the design of asymmetrical sections in a clear and simple fashion.

Example 10.1 Consider the $80 \times 60 \times 7$ mm grade 43 angle shown in Fig. 10.1(a) which is subjected to a bending moment of 1 kNm about the x–x axis. Calculate the bending stresses which result.

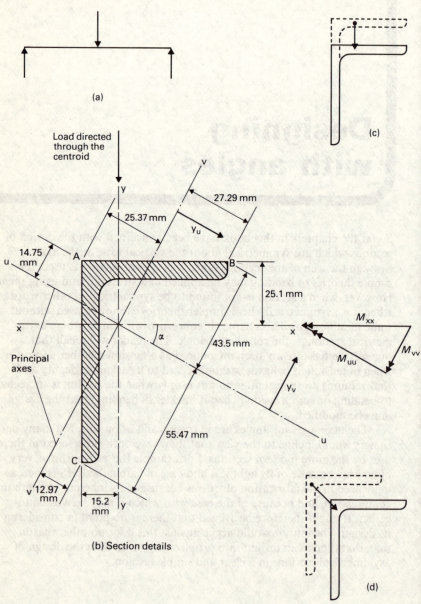

(a)

Load directed through the centroid

(c)

(b) Section details

(d)

Fig. 10.1

It is assumed that the load is directed through the centroid of the section. The first impulse may very well be to use the well-tried formula

$$\frac{M}{I} = \frac{f}{y}$$

Therefore consulting the section tables in Appendix B gives

$I_{xx} = 59\,\text{cm}^4 = 0.59 \times 10^6\,\text{mm}^4$

For corners A and B $y = c_x = 25.1\,\text{mm}$

For corner C $y = 80 - c_x = 54.9\,\text{mm}$

Thus $f_A = f_B = \dfrac{1 \times 10^6 \times 25.1}{0.59 \times 10^6} = 42.5\,\text{N/mm}^2$

and

$$f_C = \frac{1 \times 10^6 \times 54.9}{0.59 \times 10^6} = 93.1\,\text{N/mm}^2$$

However This assumes that the angle will deflect in a manner illustrated by Fig. 10.1(c), i.e. The deflection will be parallel to the y–y axis.

If the angle is unrestrained from twisting, the natural inclination of the section is to deflect as shown in Fig. 10.1(d). Therefore, it is easily seen that this simple approach is not sufficient. But why should such a section as this choose to deflect in such a fashion? The classic bending formula is based upon a very important assumption: that the section is symmetrical about the vertical axis. In other words it is a mirror image of itself if the mirror is placed on the vertical y–y axis. Obviously this is not true for the angle; it is known as an asymmetrical section.

The theory of bending for such a section may be found in most degree-level textbooks on structural analysis, and they may be consulted if the student wants to pursue the subject further. It will suit our present purposes to use this theory without derivation. It is found that any asymmetrical section can be analysed in a similar manner to the simple theory of bending, if all moments are resolved into components about the two 'Principal Axes' of the section. The directions of these sections may be found by consulting the tables in Appendix B. The only difficulty now lies in obtaining the values of y for use in the bending formula. These must be taken as the perpendicular distances from u–u and v–v axes shown in Fig. 10.1(b). The student should verify the values which have been given before proceeding further in this study. He should then be able to prove that the following formulae for these values apply:

Point	y_v	y_u
A	$c_x \cos \alpha - c_y \sin \alpha$	$c_x \sin \alpha + c_y \cos \alpha$
B	$c_x \cos \alpha + (B - c_y) \sin \alpha;$	$-c_x \sin \alpha + (B - c_y) \cos \alpha$
C	$(A - c_x) \cos \alpha + c_y \sin \alpha;$	$(A - c_x) \sin \alpha - c_y \cos \alpha$

Resolving the moment along the principle axes gives:

$$M_{uu} = M_{xx} \cos \alpha \qquad M_{vv} = M_{xx} \sin \alpha$$

for this particular case where the load is directed along the y–y axis.

M_{uu} causes compression at A and B

tension at C

M_{vv} causes compression at A

tension at B and C

Thus $f_A = -\dfrac{M_{uu}}{I_{uu}} \cdot 14.75 - \dfrac{M_{vv}}{I_{vv}} \cdot 25.37$

$f_B = -\dfrac{M_{uu}}{I_{uu}} \times 43.5 + \dfrac{M_{vv}}{I_{vv}} \times 27.29 \qquad$ *compression* $- ve$

$f_C = \dfrac{M_{uu}}{I_{uu}} \times 55.47 + \dfrac{M_{vv}}{I_{vv}} \times 12.97$

From the tables $\quad \alpha = \tan^{-1}(0.546) = 28.635°$

$$I_{uu} = 72\,\text{cm}^4 \qquad I_{vv} = 15.4\,\text{cm}^4$$

Therefore $\quad M_{uu} = 0.878\,\text{kNm} \qquad M_{vv} = 0.479\,\text{kNm}$

and $\quad f_A = 96.93\,\text{N/mm}^2 \qquad$ compression

$f_B = 31.89\,\text{N/mm}^2 \qquad$ tension

$f_C = 107.98\,\text{N/mm}^2 \qquad$ tension

If the allowable stress had been $165\,\text{N/mm}^2$, the maximum moment would have been $1.528\,\text{kNm}$. Thus the mistake of using the simple approach and forgetting the asymmetrical nature of the section resulted in an error of more than 13 per cent.

Example 10.2 The last example explained the procedure involved in the calculation of stresses for an angle in bending. It was also shown that the value of the maximum moment which can be taken by the section may be found. However, this was accomplished for a load directed along the vertical y–y axis. It is informative to determine the maximum value of moment which can be taken if the load is inclined to the vertical axis as shown in Fig. 10.2.

If the load is in the direction indicated, this induces a moment at right angles to this direction, equal to $M\,\text{kNm}$. As before, this can be resolved into components M_{uu} and M_{vv}. Assuming that the angle of inclination is 45° and that the same angle as was used in the last example is still being considered:

$$M_{uu} = M \cos 73.635° = 0.2818\,M$$

$$M_{vv} = M \sin 73.635° = 0.9595\,M$$

$$f_A = -\frac{M_{uu}}{I_{uu}} \times 14.75 - \frac{M_{vv}}{I_{vv}} \times 25.37 = -163.84 \times 10^{-6}\,MN/mm^2$$

Similarly

$$f_B = 153.0 \times 10^{-6}\,MN/mm^2 \qquad f_C = 102.51 \times 10^{-6}\,MN/mm^2$$

Maximum allowable stress = 165 N/mm² and will occur at A first

$$M = 1.007\,kNm \qquad \left(= \frac{165}{163.84 \times 10^{-6}}\,Nmm \right)$$

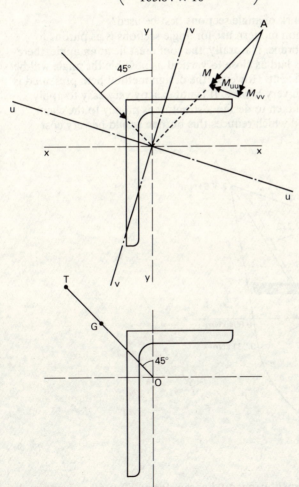

Fig. 10.2

This may now be plotted in a manner shown in Fig. 10.2(b). The direction of the load is drawn as OT. Along this line a point G is marked. The length of OG is scaled to represent the value of the maximum moment which may be taken by the section with the load in this direction. If this same exercise is now repeated for many angles of load inclination, the polar graph of Fig. 10.3 may be drawn.

The maximum moment of 2.306 kNm occurs when the direction of the load lies very close to the v–v axis (only 4° different). The minimum moment occurs when the load is directed along the x–x axis and is equal to 0.894 kNm.

Thus it can be seen that it is approximately true to say that the angle is strongest when loaded along the v–v axis, or when the moment is about u–u axis.

How can this quirk of angle sections best be used?

The most common modern use for angle sections is as purlins to support a roof membrane. Naturally, the roof must lie at an angle, therefore, if the v–v axis is laid as close to vertical as possible the angle will be acting at its most efficient. However, the design method here presented is rather long, although very precise. It is not exactly very easy to apply. Moreover, the time taken to design a structure is money to the client, therefore any method which reduces this burden would be very cost

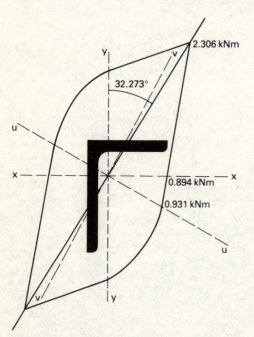

Fig. 10.3 Polar moment capacity diagram for an 80 × 60 × 7 mm angle in grade 43 steel

effective, especially since the purlin is only a secondary structural member. The next example illustrates two such methods.

Example 10.3 The roof of a building is shown in Fig. 10.4, and is to be supported by angle purlins. Design the purlins required. Main members are 5 m apart.

Method I:

The **vertical** load on one purlin $= 1.5 \times 5 \times 1 \times \cos 25°$
$$= 6.797 \, \text{kN}$$

The **component perpendicular** to roof $= 6.797 \times \cos 25°$
$$= 6.16 \, \text{kN}$$

Thus
$$M_{xx} = \frac{6.16 \times 5}{8} = 3.85 \, \text{kNm}$$

Allowable stress $= 165 \, \text{N/mm}^2$

Therefore required $Z_{xx} = \dfrac{3\,850\,000}{165} \, \text{mm}^3$

$$= 23.33 \, \text{cm}^3$$

Thus the following sections will suffice

$$100 \times 75 \times 10 \quad \text{or} \quad 125 \times 75 \times 8$$

the second of these sections being the lighter.

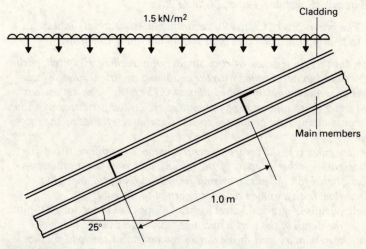

Fig. 10.4

Method II: BS 449 Cl. 45.

Total load on the purlin = 6.797 kN

L = span of purlin   = 5000 mm

$L/60$ = 83.3 mm

$L/45$ = 111.1 mm

$$Z_{xx} = \frac{WL}{1.8} \times 10^{-3} = \frac{6.797 \times 5000}{1800}$$

$$= 18.88 \text{ cm}^3$$

To fully meet the dimensional requirements a $150 \times 90 \times 10$ mm section would have to be used, however, the Z_{xx} value of the $125 \times 75 \times 8$ mm angle chosen as one of the alternatives in Method I is 29.6 cm³. This is very much greater than the 18.88 needed, and the author would argue that the shortfall on the width of the angle may be disregarded.

Obviously Method II is the easier to use and the author advises its use in all cases. However, the student is no longer using it without realising its theoretical basis.

As a final note on this example, if the $125 \times 75 \times 8$ mm angle is used the stress found in a manner similar to that of the previous example is equal to 129.11 N/mm² (compression) at corner B. It is therefore arguable that the smaller angle given by Method I would still be sufficient since the actual stress in this large angle is so very much less than the 165 N/mm² allowed stress.

Purlins

45. *All purlins shall be designed in accordance with the requirements for uncased beams (Subclause 19a and Clause 25).*

Note: *The provisions of Clause 26 regarding lateral stability and of Clause 15 limiting the deflection of beams do not apply to purlins.*

Alternatively, in the case of roof slopes not exceeding 30° pitch, purlins may be designed in accordance with the following empirical rules, which are based on a minimum imposed loading of 0.75 kN/m². The deflections obtained under these rules may be found to exceed those permitted by Clause 15 but should in any case be limited to suit the characteristics of the roof covering.

In these rules L is the centre-to-centre distance in millimetres of the steel principals or other support of the purlins, and W is the total distributed load, in kN, on the purlin arising from dead load and snow, but excluding wind, both assumed as acting normal to the roof.

Angle purlins of grade 43 steel for roof slopes not exceeding 30° pitch. The leg or the depth of the purlin taken approximately in the plane of action of the maximum load or maximum component of the load shall be not less than L/45; the other leg or width of the purlin shall be not less than

L/60, *and the numerical value of the section modulus of the purlin in centi-metre units shall be not less than WL/1.8 × 10⁻³.*

Tubular purlins of grade 43 steel for roof slopes up to and including 30° pitch. *The diameter of the purlins shall be not less than L/64 and the numerical value of the section modulus of the purlins in centimetre units shall be not less than WL/2 × 10⁻³.*

Note: *Purlins designed in accordance with these rules shall be deemed to be capable of carrying the forces resulting from the requirements of Sub-clause 26e (ii).*

In cases where the slope of the roof is greater than 30°, the author would advocate the use of Method I in the design of the purlins with the possible addition of tie rods at the third points of the purlin span. The reason for the seemingly arbitrary 30° as the slope beyond which Cl. 45 of BS 449 may not be used may be understood upon consulting Fig. 10.3.

For most angles the angle at which the maximum moment is obtained is just greater than 30°. Beyond this angle the maximum moment reduces very quickly, therefore it would be unsafe to use an empirical method in such cases. Even Method I presented in this last example will prove unreliable for slopes greater than about 45°, and in these cases the 'exact' method presented in the previous two examples should be used.

Angles as struts

Angles are used as compression members in trusses when they are placed in the top chords. As members of roof trusses they are still in regular use, but the RHS has tended to supersede their use for the larger lattice girders and bridge trusses. The reason for this decline lies with the inherent weakness of angles in compression.

The student will remember that in Chapter 8 is the important parameter when obtaining the allowable stress in compression from Table 17 of BS 449 was the stanchion's 'slenderness ratio' l/r_y. The procedure for the design of angles in compression is similar, in that once a slenderness ratio has been obtained, the allowable stress may be obtained from Table 17 as before. However, calculating the correct value for slenderness ratio is of necessity rather empirical, even though the code gives it the air of theory. The clause which is of importance is Cl. 31c (although Cl. 37 should not be neglected).

c. *Angles as struts. (i) For single-angle discontinuous struts connected to gussets or to a section either by riveting or bolting by not less than two rivets or bolts in line along the angle at each end, or by their equivalent in welding, the eccentricity of the connection with respect to the centroid of the strut may be ignored and the strut designed as an axially-loaded member provided that the calculated average stress does not exceed the allowable stresses given in Table* 17, *in which l is taken as* 0.85 *times the length of the strut, centre-to-centre of intersections at each end, and* r *is the minimum radius of gyration.*

Single angle struts with single-bolted or riveted connections shall be treated similarly, but the calculated stress shall not exceed 80 per cent of the values obtained from Table 17, *centre-to-centre of intersections shall be taken. In no case, however, shall the ratio of slenderness for such single angle struts exceed* 180.

(ii) For double angle discontinuous struts, back-to-back connected to both sides of a gusset or section by not less than two bolts or rivets in line along the angles at each end, or by the equivalent in welding, the load may be regarded as applied axially. The effective length l shall be taken as between 0.7 *and* 0.85 *times the distance between intersections, depending on the degree of restraint, and the calculated average stress shall not exceed the values obtained from Table* 17 *for the ratio of slenderness based on the minimum radius of gyration about a rectangular axis of the strut. The angles shall be connected together in their length so as to satisfy the requirements of Clause* 37.

(iii) Double angle discontinuous struts back to back, connected to one side of a gusset or section by one or more bolts or rivets in each angle, or by the equivalent in welding, shall be designed as for single angles in accordance with c *(i) and the angles shall be connected together in their length so as to satisfy the requirements of Clause* 37.

(iv) The provisions in this clause are not intended to apply to continuous angle struts such as those forming the rafters of trusses, the flanges of trussed girders or the legs of towers which shall be designed in accordance with Clause 26 *and Table* 17.

This chapter shall not concern itself with Class (iv) – the continuous angle – the design of which is rather complicated. Of the remaining three classes, the first deals with single angles according to their end support conditions, while the other two classes are concerned with a strut composed from two angles placed back to back. The remaining examples in this chapter deal with these three classes.

Example 10.4 Design a single-angle strut to take an axial compression load of 50 kN. The length of the strut between fasteners is 2.5 m and the grade of steel is 43.

1. *End connections – single bolts (bolts not designed)*
   Effective length = actual length = 2.5 m

   Try a 90 × 90 × 8 mm angle

   Minimum radius of gyration = r_v = 17.6 mm

   $$l/r_v = \frac{2500}{17.6} = 142$$

From Table 17 of BS 449 $\quad p_c = 45\,\text{N/mm}^2$

$$0.8p_c = 36\,\text{N/mm}^2$$

Allowed axial load on this angle $= 36 \times$ area of section

$$= 36 \times 1390\,\text{N}$$
$$= 50.04\,\text{kN}.$$

2. *End connections – two bolts in line at each end*
Effective length $= 0.85 \times 2.5 = 2.125\,\text{m}$

Try a $80 \times 80 \times 8\,\text{mm}$ angle

$$l/r_v = \frac{2125}{15.6} = 136.2$$

$$p_c = 48\,\text{N/mm}^2$$

Allowed axial load $= 48 \times 1230\,\text{N}$

$$= 59.04\,\text{kN}.$$

Example 10.5 Design a double-angle, discontinuous strut connected to both sides of a gusset by two bolts either end of the strut, to carry a load of 100 kN if the actual length of the strut is 2.5 m.

The main difficulty with this type of strut lies with the determination of the effective length, since the code is not precise about its requirements. In this example, the factor of 0.85 will be used since two bolts only are used.

The value to be used for the radius of gyration is taken about the axis perpendicular to the plane of the gusset. Since it is desirable to achieve the maximum allowable stress, the radius of gyration should be as large as possible; therefore, if possible, the longer legs of unequal leg angles should be placed next to the gusset. The student may like to prove that this does indeed give the best results.

Try two $65 \times 50 \times 8\,\text{mm}$ angles with the 65 mm leg placed on the gusset.

Effective length $= 0.85 \times 2500 = 2125\,\text{mm}$

$$r_x = 20.1\,\text{mm}$$

$$l/r_x = 105.7$$

From Table 17 of BS 449 $p_c = 73.5\,\text{N/mm}^2$

Allowed axial load $= 73.5 \times 2 \times 860\,\text{N}$

$$= 126.4\,\text{kN}$$

Section area $= 1720\,\text{mm}^2$

Satisfactory.

It will be found that if equal angles are used the following will also suffice:

Angles	Load allowed	Section area
Two $60 \times 60 \times 8$	110.2 kN	1806 mm^2
Two $70 \times 70 \times 6$	128.4 kN	1626 mm^2

However, for a double-angle strut such as this, the strength of the strut relies upon the joint action of the two angles. Therefore they must be joined together in such a manner as to ensure that this is guaranteed. This is ensured by complying with the requirements of Cl. 37 of BS 449 which states:

Compression members composed of two components back-to-back
37. *Compression members composed of two angles, channels or tees, back-to-back in contact or separated by a small distance shall be connected together by riveting, bolting or welding so that the maximum ratio of slenderness l/r of each member between the connections is not greater than 40 or greater than 0.6 times the most unfavourable ratio of slenderness of the strut as a whole, whichever is the less.*

In no case shall the ends of the strut be connected together with less than two rivets or bolts or their equivalent in welding, and there shall be not less than two additional connections spaced equidistant in the length of the strut. Where the members are separated back-to-back the rivets or bolts through these connections shall pass through solid washers or packings, and where the legs of the connected angles or tables of the connected tees are 125 mm wide or over, or where webs of channels are 150 mm wide or over not less than two rivets or bolts shall be used in each connection, one on the line of each gauge mark.

Where these connections are made by welding, solid packings shall be used to effect the jointing unless the members are sufficiently close together to permit welding, and the members shall be connected by welding along both pairs of edges of the main components.

The rivets, bolts or welds in these connections shall be sufficient to carry the shear forces and moments (if any) specified for battened struts, and in no case shall the rivets or bolts be less than 16 mm diameter for members up to and including 10 mm thick; 20 mm diameter for members up to and including 16 mm thick; and 22 mm diameter for members over 16 mm thick.

Compression members connected by such riveting, bolting or welding shall not be subjected to transverse loading in a plane perpendicular to the washer-riveted, bolted or welded surfaces.

Where the components are in contact back-to-back, the spacing of the rivets, bolts or intermittent welds shall not exceed the maximum spacing for compression members as given in Subclauses 51c (i) and 54c.

For this example: 65 × 50 mm angle $r_v = 10.5$ mm

Maximum bolt spacing = $40r_v = 420$ mm   (say 415 mm to fit into the length more easily)

60 × 60 mm angle $r_v = 11.6$

Maximum bolt spacing = 464 mm   (again 415 mm)

70 × 70 mm angle $r_v = 13.7$ mm

Maximum bolt spacing = 548 mm   (say 500 mm).

Finally, since the two angles are physically separated by the thickness of the gusset, packing washers must be placed between the angles letting the bolts pass through them. On tightening the bolts, these packers become compressed against the backs of the angles and prevent the angles becoming distorted.

Questions

1. The roof of a building slopes at an angle of 20° to the horizontal. The main frames are spaced 4.5 m apart; the purlins are spaced 1.2 m apart (measured on the slope). Using Cl. 45 of BS 449, design the purlins if the roof loading is 1.4 kN/m^2.
2. Design a single-angle strut 5 m long carrying a load of 100 kN if the ends are supported by two bolts in line at either end.
3. Design the same strut as above if the strut is made from two angles bolted to either side of the supporting gusset by two bolts at either support.
4. Construct a moment capacity similar to that of Fig. 10.3 for a 100 × 100 × 12 mm grade 43 angle.

Appendix A

Answers

Chapter 2

1. (a) $A_s = 434.3\,\text{mm}^2$; $x = 78.08\,\text{mm}$; $f_{yd2} = 217.4\,\text{N/mm}^2$
   (b) $A_s = 834\,\text{mm}^2$; $M_u = 42.21\,\text{kNm}$; $x = 150\,\text{mm}$;
   $f_{yd2} = 217.4\,\text{N/mm}^2$.

2. $M_u = 252.4\,\text{kNm}$; $x = 192.2\,\text{mm}$; $f_{yd2} = 434.8\,\text{N/mm}^2$;
   $\epsilon_{st} = 0.0056$.

3. $A_s = 1285.8\,\text{mm}^2$ but x/d exceeds 0.5, therefore it contravenes the requirements of CP 110.

4. Fig. A.1 indicates two points at which the lines 'kink'. It is at these points that the steel just yields. For lower values of either A_s/bd or M/bd^2 the steel is fully yielded.

5. $M_u = 53.29\,\text{kNm}$; $A_s = 456.8\,\text{mm}^2$.

6. *Worked solution*

$$M_u = 0.25\,PL = \frac{100 \times 3}{4} = 75\,\text{kNm}$$

For any given section of known dimensions, the maximum singly reinforced moment allowed by CP 110 occurs when $x/d = 0.5$. Conversely, if M_u and b are known then when $x/d = 0.5$, d will be a minimum

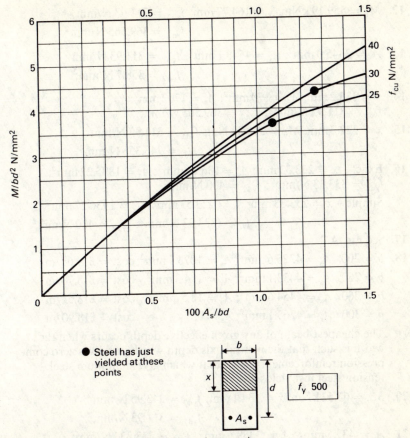

Fig. A.1 Singly reinforced beams (Chart 6)

referring to chart 30 (or any of 27, 28, 29) for no compression steel:

$M/bd^2 = 6.172$ Thus $d = 348.6$ mm

$100 A_s/bd = 2.285$ Thus $A_s = 796.7$ mm^2.

7. $b = 100.24$ mm; $A_s = 1069.7$ mm^2; $x = 240.3$ mm.

8. (a) $b = 100$ mm; $A_s = 1040.7$ mm^2; cost/m $= £4.37$
   (b) $b = 125$ mm; $A_s = 966.7$ mm^2; cost/m $= £5.34$
   (c) $b = 150$ mm; $A_s = 931.7$ mm^2; cost/m $= £6.33$.

9. $f_{cu} = 25$ N/mm^2: $A_s = 289.75$ mm^2
   $f_{cu} = 40$ N/mm^2: $A_s = 268.4$ mm^2.

10. $M_u = 527.36$ kNm; $x = 280.12$ mm; $f_{yd2} = 369.6$ N/mm^2.

11. $A'_s = 1075.3$ mm^2; $A_s = 4763.5$ mm^2; $f_{yd1} = 311.93$ N/mm^2
    $f_{yd2} = 369.565$ N/mm^2.

12. $M_u = 559.19\,\text{kNm}$; $x = 164.77\,\text{mm}$; $f_{yd1} = 311.93\,\text{N/mm}^2$
$$f_{yd2} = 369.565\,\text{N/mm}^2.$$

13. $A_s' = 2345.0\,\text{mm}^2$; $A_s = 4548.1\,\text{mm}^2$; $f_{yd1} = 311.93\,\text{N/mm}^2$
$$f_{yd2} = 358.7\,\text{N/mm}^2.$$

14. $x = 238.2\,\text{mm}$; $A_s' = 170\,\text{mm}^2$; $A_s = 1702\,\text{mm}^2$;
$f_{yd1} = 311.93\,\text{N/mm}^2$; $f_{yd2} = 369.565\,\text{N/mm}^2.$

15. $x = 322.7\,\text{mm}$; $M_u = 396.8\,\text{kNm}$; $f_{yd1} = 311.93\,\text{N/mm}^2$
$$f_{yd2} = 343.353\,\text{N/mm}^2.$$

16. Exact $- x = 233.33\,\text{mm}$; $A_s' = 136.76\,\text{mm}^2$; $A_s = 1895.22\,\text{mm}^2$;
$f_{yd1} = 333.33\,\text{N/mm}^2$; $f_{yd2} = 400\,\text{N/mm}^2.$
Simple $- x = 233.33\,\text{mm}$; $A_s' = 223.53\,\text{mm}^2$; $A_s = 1936.27\,\text{mm}^2$
$$f_{yd1} = 333.33\,\text{N/mm}^2;\ f_{yd2} = 400\,\text{N/mm}^2.$$

17. See Fig. A.2.

18. $b = 200$: $A_s = 4325.6\,\text{mm}^2$; $A_s' = 1073.5\,\text{mm}^2$; cost $= £19.62/\text{m}$
$b = 250$: $A_s = 4433.1\,\text{mm}^2$; $A_s' = 194.3\,\text{mm}^2$; cost $= £18.70/\text{m}$
$b = 260$: $A_s = 4454.6\,\text{mm}^2$; $A_s' = 18.5\,\text{mm}^2$; cost $= £18.52/\text{m}$
$b = 300$: $A_s = 4263.1\,\text{mm}^2$; $A_s' = 0$; cost $= £18.90/\text{m}$
The cheapest beam of any given effective depth occurs when the width is such that the neutral axis depth which just gives zero compression reinforcement is also that which gives minimum steel – tension steel just yields.

20. $M_u = 65.52\,\text{kNm}$; $x = 134.1\,\text{mm}$; $f_{yd2} = 343.95\,\text{N/mm}^2$;
$$f_{yd1} = 311.93\,\text{N/mm}^2.$$

21. $A_s' = 437.5\,\text{mm}^2$, $A_s = 3489.6\,\text{mm}^2$, $f_{yd1} = 333.333\,\text{N/mm}^2$
$$f_{yd2} = 400\,\text{N/mm}^2.$$

22. $A_s' = 735.8\,\text{mm}^2$, $A_s = 3023.9\,\text{mm}^2$, $f_{yd1} = 333.333\,\text{N/mm}^2$
$$f_{yd2} = 400\,\text{N/mm}^2.$$

23. (a) No shear steel is required for the central 0.9 m
    (b) Nominal shear steel is required for the central 2.66 m. It takes the form of 8 mm links at 375 mm spacing (maximum spacing allowed)
    (c) The steel near the supports must satisfy
        $A_{sv}/S_v = 0.9$. Therefore, use either of
            12-mm $\emptyset$ links at 250 mm spacing or
            10 mm $\emptyset$ links at 175 mm spacing.

25. When the two bars are bent up, the shear resistance V_c must be based on the area of the continuing bars.
    $A_s = 1608.5\,\text{mm}^2$ $\qquad 100A_s/bd = 0.835$
    From Table 5 of CP 110: $v_c = 0.60\,\text{N/mm}^2$

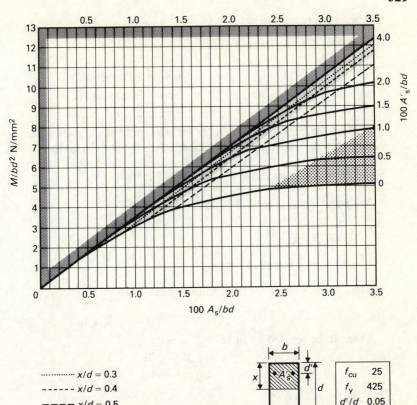

.... $x/d = 0.3$
- - - - $x/d = 0.4$
- - - $x/d = 0.5$

f_{cu}	25
f_y	425
d'/d	0.05

▨ Impossible to meet the requirements of CP110
▨ Tension steel is plastic
▨ Tension steel is elastic

Fig. A.2 Doubly reinforced beams (Chart 31)

Shear force at ends of beam $= V = 400\,\text{kN}$

Nominal shear stress $= v = V/bd = 2.08\,\text{N/mm}^2$

Shear stress to be taken by shear steel $= 2.08 - 0.6$
$$= 1.48\,\text{N/mm}^2$$

The tension bars will be bent up as near to the ends of the beam as possible since this is where the shear force is largest. Therefore the bars are bent up at the position where their effective pitch just meets the support.

$$P = (1 + \cot \alpha)d = (1 + \cot 45°) \times 550$$
$$= \underline{1100\,\text{mm}}$$

Resistance of these bent-up bars $= 0.87 f_{yv} A_{sv} \sin 45°$

Therefore, reduction in

shear stress to be taken by links $= \dfrac{0.87 f_{yv} A_{sv}}{bd\sqrt{2}}$

$$= \dfrac{0.87 \times 425 \times 1608.5}{350 \times 550 \times 1.4142}$$

$$= 2.18 \, \text{N/mm}^2$$

Therefore the bent-up bars can take all the shear at the ends of the beam. However, at least half the shear must be taken by links.

Shear to be taken by links $= \dfrac{1.48}{2} = 0.74 \, \text{N/mm}^2$

Minimum spacing $= 0.75 \times 550 = 412.5 \, \text{mm}$

$\dfrac{A_{sv}}{S_v} = \dfrac{350 \times 0.74}{0.87 \times 425} = 0.7$

Use 10 mm links at 225 spacing

Beyond the bent-up bars the nominal shear stress $= \dfrac{262\,500}{350 \times 550}$

$$= 1.36$$

Shear to be resisted by links $= 1.36 - v_c$

$$= 1.36 - 0.78$$

$$= 0.58$$

Thus $A_{sv}/S_v = 0.55$

Minimum shear steel given by $A_{sv}/S_v = 0.0012 \times 350$

$$= 0.42$$

Therefore use 10 mm links at 275 mm spacing for rest of beam

Note: v_c for centre section of beam was based on four bars.

26. Load to nearest support – 8 mm links at 216 mm (say 210 mm)
    Load to farthest support – 8 mm links at 375 mm (max. spacing allowed).
27. Load is less than $2d$ from end of beam.
    For 1 m from support to load: $A_{sv}/S_v = 0.358$, which is less than the nominal requirement of $A_{sv}/S_v = 0.36$
    Therefore, use 8 mm $\varnothing$ links at 275 mm spacing.
    The shear stress in the rest of the beam is less than $0.5v_c$. Thus no shear steel is needed in this section.
28. Local bond: $f_{bs} = 2.12 \, \text{N/mm}^2$ against 3.36 allowed
    Anchorage bond: anchorage length $= 807.6 \, \text{mm}$ from point load on *both* sides.
    Therefore it is O.K. so long as the end anchorage complies with Section 3.11.7 of CP 110.

29. Local bond: $f_{bs} = 2.45\,\text{N/mm}^2$ <u>O.K.</u>

Anchorage bond: *Tension steel*
$$L = 1398.6\,\text{mm} \qquad \text{too long as shown}$$

Remedy – either use hook at free end or change from two 40 mm to four 25 mm + two 20 mm

then $L_{25} = 874.1\,\text{mm}$ <u>O.K.</u>

Compression steel
$L_{10} = 349.7\text{mm}$ <u>O.K.</u>

Note: Compression bars *cannot* be anchored with a hook. If the anchorage length had been too great the only remedy would have been to change the bar sizes.

30. Compression steel = three 16-mm $\varnothing$ bars = $603.2\,\text{mm}^2$

Tension steel   = four 40-mm + one 25-mm $\varnothing$ bars
              = $5517.4\,\text{mm}^2$

Shear steel   = use nominal steel throughout except where no shear reinforcement is needed

Local bond: $f_{bs} = 1.27\,\text{N/mm}^2 < \text{allowed}$

Anchorage bond: Tension steel $L_{40} = 1398.6\,\text{mm}$
                    $L_{25} = 874.1\,\text{mm}$

Compression steel $L_{16} = 380\,\text{mm}$

Deflection: $d_{min} = 756\,\text{mm} < \text{actual }775\,\text{mm}.$

31. Compression steel = three 16-mm $\varnothing$ bars = $603.2\,\text{mm}^2$

Tension steel = four 40 mm + one 25 mm $\varnothing$ bars = $5517.4\,\text{mm}^2$

Note that the exact formulae give the same areas of steel for the shallower section that the simplified method gives for the deeper section.

Shear steel   Nominal as in Example 2.13

Local bond $f_{bs} = 1.31 < \text{allowed}$

Anchorage bond as in the Example 2.13

Deflection: $d_{min} = 754.9\,\text{mm}$

This exceeds the actual depth by only 0.65 per cent and is judged close enough for practical purposes.

Chapter 3

1. (Worked)   Assuming that load width = 0

Effective width of slab strip = b

$$b = 2.4 \times (1 - x/L)$$
$$M = Px(1 - x/L)$$

Therefore $M/bd^2 = P/2.4d^2$

Thus the value of M/bd^2 does not depend on the position of the load on the slab span. Therefore, the value of A_s is also independent of position. However, the closer to a support a load is the less is the effective width, thus the constant area of steel must be fitted into a smaller portion of concrete. Apart from this when the load is close to the support, the shear is larger. Therefore, the critical position of a point load is as close to the support as possible.

2. For the 2.64 m under the press

   $A_s = 3696\,\text{mm}^2$

   For the rest of the slab

   $A_s = 800\,\text{mm}^2/\text{m}$

   Secondary steel

   $A_s = 348\,\text{mm}^2/\text{m}$

   Both types of shear satisfactory. Deflection $d_{min} = 209\,\text{mm}$.

3. Assuming $100\,A_s/bd = 1$ gives $d_{min} = 225.7\,\text{mm}$

   Using $d = 250\,\text{mm}$ and overall depth $= 290\,\text{mm}$ gives:

   $A_s = 2088\,\text{mm}^2/\text{m}$ of slab width – 20 mm bars at 150 mm spacing.

   For this arrangement, shear and deflection are satisfactory

4. Use $d = 200\,\text{mm}$ $M_u = 67.64\,\text{kNm}$

   $A_s = 1034\,\text{mm}^2/\text{m}$ – 20 mm bars at 300 mm spacing

   $v = 0.3\,\text{N/mm}^2$ whereas $v_c = 0.5$ **No shear steel**

   $d_{min} = 185\,\text{mm}$, hence the slab is properly designed

   Note: In these last two questions, the unit weight of reinforced concrete is taken as $24\,\text{kN/m}^3$.

Chapter 4

1. x–x axis: lower section $l_e = 0.754l_0 = 3016.0\,\text{mm}$

   upper section $l_e = 0.716l_0 = 2864.0\,\text{mm}$

   y–y axis: lower section $l_e = 0.752l_0 = 2632.0\,\text{mm}$

   upper section $l_e = 0.711l_0 = 2488.5\,\text{mm}$.

2. Upper section: $l_{ex}/h = 2864/200 = 14.32$

   $l_{ey}/b = 2488.5/300 = 8.30$

   Lower section: $l_{ex}/h = 3016/200 = 15.10$

   $l_{ey}/b = 2632/300 = 8.77$

   Therefore in the y direction the column is short but in the x direction the dimension should be increased to 251.3, say 260 mm.

3. Using the modified size of 260×300 mm

Upper storey:
Total design load $= 1.4 \times 125 + 1.6 \times 190 = 479$ kN
Thus $\quad 479\,000 = 0.35 \times 30 \times 260 \times 300 + 0.6A_{sc} \times 425$
$$A_{sc} = -ve$$
Therefore theoretically no steel is needed. However, Cl. 3.11.4.1 of CP 110 insists that a minimum area of steel be inserted $=$ 1 per cent $= 0.01 \times 260 \times 300 = 780$ mm^2
Use four 16-mm $\varnothing$ bars $= 804$ mm^2

Lower storey:
Total design load $= 479 + 1.4 \times 160 + 1.6 \times 500$
$$= 1503 \text{ kN}$$
Thus $1\,503\,000 = 0.35 \times 30 \times 260 \times 300 + 0.6A_{sc} \times 425$
$A_{sc} = 2682.4$ mm^2
Use four 32-mm $\varnothing$ bars $= 3220$ mm^2.

4. $N = 2505$ kN.
5. There is no one solution to this problem, but there are limits. First, there is a minimum percentage of steel allowed; this will result in the largest column. Second, Cl. 3.11.5 of CP 110 states that the maximum area of steel is 6 per cent; this will result in the smallest column.

Minimum area of steel solution:
$A_{sc} = 0.01 \times A_c$
$N = 3\,000\,000 = 0.4 \times 50 \times A_c + 0.67 \times (0.01\,A_c) \times 250$
$A_c = 138\,408.3$ mm$^2 \quad$ dimension $= 372$ mm
$$A_{sc} = 1384 \text{ mm}^2$$
The nearest to this is four 25-mm $\varnothing$ bars $= 1963$ mm^2 and the size of the column $= 365.5$ square $\quad$ (say 370 mm)

Maximum area of steel solution:
Use four 40-mm $\varnothing$ bars in a column of side $= 330$ mm
(Note that this is a maximum area of steel since bars larger than 40 mm cannot be obtained, and not because they give 6 per cent. In fact the percentage is less than 6 per cent by quite a margin.)
Percentage of steel $= 4.6$
Alternatively use eight 32-mm $\varnothing$ bars in a column of side $= 330$ mm
Percentage of steel $= 5.9$ but this is less economic.

6. Use eight 32-mm bars

Chapter 5

1. (a) $f_A = 0$; $f_B = f_D = 0.37\,\text{N/mm}^2$; $f_C = 0.74\,\text{N/mm}^2$
    (b) $f_A = 0.36\,\text{N/mm}^2$; $f_B = 0.57\,\text{N/mm}^2$; $f_C = 0.24\,\text{N/mm}^2$
    $$f_D = 0.02\,\text{N/mm}^2$$
    (c) Load is outside the kern, therefore the maximum stress occurs along edge BC = $0.948\,\text{N/mm}^2$
    (d) $f_A = 0.107\,\text{N/mm}^2$; $f_B = 0.64\,\text{N/mm}^2$; $f_C = 0.56\,\text{N/mm}^2$
    $$f_D = 0.027\,\text{N/mm}^2$$
    (e) $f_A = f_B = 0.34\,\text{N/mm}^2$; $f_C = f_D = 0.32\,\text{N/mm}^2$.

2. (a) $u = 166.67\,\text{kN/m}$ $\quad B_{min} = 416.67\,\text{mm}$ (420 would be used)
    (b) $u_1 = u_3 = u_5 = u_7 = 250\,\text{kN/m}$
    $u_2 = u_4 = u_6 = 100\,\text{kN/m}$ $\quad B_{min} = 625\,\text{mm}$
    (c) Between 1 and 2: $u_1 = 250\,\text{kN/m}$; $u_2 = 100\,\text{kN/m}$
    Between 2 and 3: $u = 150\,\text{kN/m}$
    Between 3 and 4: $u_3 = 100\,\text{kN/m}$; $u_4 = 250\,\text{kN/m}$
    Between 4 and 5: $u_4 = 250\,\text{kN/m}$; $u_5 = 100\,\text{kN/m}$
    Between 5 and 6: $u_5 = u_6 = 150\,\text{kN/m}$
    Between 6 and 7: $u_6 = 100\,\text{kN/m}$; $u_7 = 250\,\text{kN/m}$

3. (Worked)
    Stage I: Depth of base.

    $$\text{Anchorage length} = \frac{\varnothing f_y}{4.6 f_{bs}} = \frac{16 \times 250}{4.6 \times 2.7} = 322.1\,\text{mm}$$

    Cover needed = 30 mm
    Overall depth = 352.1 (say 360 mm)
    $d = 320\,\text{mm}$ for the steel parallel to the long side of the column. It will be less by one bar diameter in the other direction.

    Stage II: Length of base.
    Since f_1/f_2 must be less than 3
    L must be greater than twelve times the effective eccentricity

    $$e = \frac{425 \times 10^6}{1600 \times 10^3} = 265.625\,\text{mm}$$

    $L_{min} = 3187.5$

    Therefore the length of the base is sufficient.

Stage III: Width of base.

$$f = \frac{N}{A} \pm \frac{M}{Z}$$

$$f_{max} = \frac{1.6 \times 10^6}{3200B} + \frac{6 \times 425 \times 10^6}{3200^2 \times B} < = 0.24$$

$$= 0.234 \, \text{N/mm}^2 \quad \text{therefore width is satisfactory}$$

$$f_{min} = 0.078 \, \text{N/mm}^2.$$

Stage IV: Bending moments and areas of tension steel.
Refer to Fig. 5.13

$$f_3 = 0.174 \, \text{N/mm}^2 \qquad V_L = 0.2042 \times 1225 \times 3200$$
$$= 800.231 \, \text{kN}$$

$$M_L = V_L \times C \qquad C = \frac{0.174 + 2 \times 0.234}{3(0.174 + 0.234)} \times 1225$$
$$= 642.5 \, \text{mm}$$

$$M_L = 514.17 \, \text{kNm} \qquad \text{from Chart 1 of CP 110: Part 2}$$

$$100 \, A_s/bd = 0.77$$

$$A_s = 7884.8 \, \text{mm}^2$$

Use 20-mm bars at 125 mm spacing

Thus the effective depth in the other direction = 300 mm

$$V_B = 0.5(0.234 + 0.078) \times 3200 \times 1425 = 711.36 \, \text{kN}$$

$$M_B = 506.844 \, \text{kNm} \qquad 100 \, A_s/bd = 0.87$$

$$A_s = 8352 \, \text{mm}^2$$

Use 16-mm bars at 75 mm spacing.

Stage V: Flexural shear stress.

Section $f_4 = 0.198 \, \text{N/mm}^2$

$$V = 514.56 \, \text{kN} \qquad v = 0.503 \, \text{N/mm}^2$$

From Table 5 of CP 110 $\qquad v_c = 0.631 \, \text{N/mm}^2$

Section
$$V = 471.74 \, \text{kN} \qquad v = 0.491 \, \text{N/mm}^2$$

$$v_c = 0.66 \, \text{N/mm}^2$$

Therefore the base is safe in flexural shear.

Stage VI: Local bond stress.

Section ABCD: $f_{bs} = \dfrac{800\,231}{320 \times 1306.9}$ $\qquad$ twenty-six bars

$$= 1.91 \, \text{N/mm}^2$$

Section STUV: $f_{bs} = \dfrac{711\,360}{300 \times 2161.4}$ forty-three bars

$$= 1.1 \text{ N/mm}^2$$

From Table 21 of CP 110 allowed $f_{bs} = 2.2 \text{ N/mm}^2$ (plain bar).

Stage VII: Punching shear.

Effective perimeter $= 2(750 + 350) + 2\pi \times 1.5 \times 360$

$$= 5592.9 \text{ mm}$$

Enclosed area $= (750 \times 350) + 2(750 \times 1.5 \times 360$

$$+ 350 \times 1.5 \times 360)$$

$$+ \pi \times (1.5 \times 360)^2$$

$$= 2.367 \times 10^6 \text{ mm}^2$$

Load from earth on the enclosed area

$$= \text{average stress} \times \text{enclosed area}$$

$$= 0.156 \times 2.367 \times 10^6 \text{ N}$$

$$= 369.2 \text{ kN}$$

Punching load $= 1600 - 369.2 = 1230.8 \text{ kN}$

Punching shear stress $= \dfrac{1\,230\,800}{310 \times 5592.9} = 0.71 \text{ N/mm}^2$

Note the use of average effective depth

Average value of $100\,A_s/bd = 0.82$

Hence from Table 5 of CP 110 $v_c = 0.646 \text{ N/mm}^2$

(no allowance can be made for depth since the depth is greater than 300 mm)

Thus some modification must be made if no shear steel is to be added. One of the following will suffice:

1. Increase the overall depth of the slab.
2. Increase the area of steel to just over 1 per cent.
3. Introduce a plinth over the section at which punching shear is checked.

4. (Partially worked)

Stages I, II and III are the same as in the previous worked solution.

Stage IV: Bending moments and shear forces.

This time there are four 'critical sections'

1. *Section* ABCD width $= 750 \text{ mm}$

   effective depth $= 250 \text{ mm}$

   $M_L = 514.17 \text{ kNm}$ as in last example

   thus $A_s = 6176.3 \text{ mm}^2$ <u>use five 40-mm $\emptyset$ bars</u>

2. *Section* A′B′C′D′ width = 3200 mm
   effective depth = 250 mm

   $f'_3 = 0.184 \, \text{N/mm}^2$ $V'_L = 685.99 \, \text{kN}$
   $M'_L = 365.54 \, \text{kNm}$ $A_s = 7200 \, \text{mm}^2$

   However this is to be spread equally over the whole
   of the 3200 mm width. Since the amount of steel
   needed in the section under the plinth is already in
   excess of this amount, only the outer sections will
   have this area of steel in proportion to the width of
   the outer sections.

   Width of outer sections = 3200 − 750 = 2450 mm

   $$\text{Area of steel} = \frac{2450}{3200} \times 7200 = 5512.5 \, \text{mm}^2$$

   Use eighteen 20-mm $\emptyset$ bars = 5654.9 mm²

3. *Section* STUV width = 1150 mm
   effective depth = 410 mm

   $M_B = 506.844$ as in last example
   $A_s = 6412.4 \, \text{mm}^2$ use eight 32-mm $\emptyset$ bars

4. *Section* S′T′U′V′ width = 3200
   effective depth = 230 mm

   $M'_B = 374.56 \, \text{kNm}$
   $A_s = 8266 \, \text{mm}^2$ over 3200 mm width
   $= 5296 \, \text{mm}^2$ over 2050 mm width

   Use eighteen 20-mm $\emptyset$ bars = 5655 mm².

Stage V: Flexural shear.
There are only two critical sections since the plinth is
small enough for 1.5 times the effective depth at section
ABCD to be greater than the width of the plinth top.
On checking shear at the two sections it is found that
the actual stresses are less than the allowed stress v_c
(based on the 100 A_s/bd for the outer sections).

Stage VI: Local bond.
No problem.

Stage VII: Punching shear (not considered on plinth).

Effective perimeter = 6627.4 mm

Enclosed area = $3.209 \times 10^6 \, \text{mm}^2$

Uplift force on the enclosed area = 500.55 kN

Punching stress = $(1600 - 500.55)/(6627.4 \times 230)$
                = $0.721 \, \text{N/mm}^2$

The only difficulty in this stage is in obtaining the value of $100 A_s/bd$ to be used in Table 5 of CP 110. Cl. 3.4.5.2 states that the average of the values in the two directions at right angles shall be taken. However this does not mean that **all** the steel in both directions is counted. Only that steel which is a distance of three times the slab depth on either side of the loaded area is to be taken. This is illustrated in Fig. A.3(d), and the bars affected in Fig. A.3(a) and A.3(c).

Area of steel in major direction = 6176.3 + 4398.2
$$= 10\,574.5$$

Area of steel in minor direction = 6412.4 + 5026.5
$$= 11\,438.9$$

major axis: $100 A_s/bd = 1.057 \times 10^6/(250 \times 2550) = 1.88$

minor axis: $100 A_s/bd = 1.144 \times 10^6/(230 \times 2950) = \underline{1.69}$

$$\text{Average} = \overline{1.785}$$

Thus $v_c = 0.857 \, \text{N/mm}^2$

against actual stress of $0.721 \, \text{N/mm}^2$

5. $d = 890 \, \text{mm}$ (overall depth = 920 mm)

Earth pressure = $0.675 \, \text{N/mm}^2$ and pressured length = 2.667 m

Major axis steel: $A_s = 3239 \, \text{mm}^2$ – eleven 20-mm $\varnothing$ bars at 190 mm

Minor axis steel: A_s used is the minimum allowed = 0.15 per cent evenly distributed across width = $4005 \, \text{mm}^2$ – twenty 16-mm bars at 150 mm

Shear and punching shear: not relevant since both $1.5d$ and $1.5h$ lie outside the base

Local bond: allowed stress = $2.7 \, \text{N/mm}^2$ (plain bar)

major axis actual = $1.85 \, \text{N/mm}^2$

minor axis actual = $0.82 \, \text{N/mm}^2$.

6. Width = 400 mm (width dictated by the column width)

Effective depth: anchorage requirements suggest

$$d = 222.8 \, \text{mm}$$

Clause 3.11.4.3 of CP 110 suggests that

$$0.75d_{\text{min}} = \text{width} - \text{cover} - \text{one link diameter}$$

Assume cover = 25 mm (Table 19 of CP 110)

Assume links are 8-mm $\varnothing$ bars

$$d = 456 \quad \text{(say 460 mm)}$$

$M = 156.25 \, \text{kNm}$ $100 A_s/bd = 0.547$ $A_s = 1006.5 \, \text{mm}^2$

Use five 16-mm $\varnothing$ bars

All steel between these arrows counted for punching shear

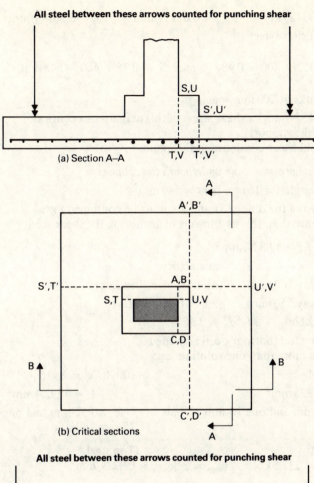

(a) Section A–A

(b) Critical sections

All steel between these arrows counted for punching shear

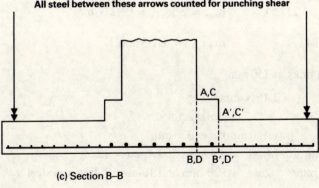

(c) Section B–B

Fig. A.3

Shear: $V_{max} = 125\,\text{kN}$ $\qquad v_{max} = 0.679\,\text{N/mm}^2$ $\qquad v_c = 0.569\,\text{N/mm}^2$

Use nominal shear steel

$$\frac{A_{sv}}{S_v} > = 0.0012 \times 400 = 0.48 \qquad \text{max. } S_v = 0.75 \times 460 = 345\,\text{mm}$$

Use 8-mm links at 200 mm spacing

$0.5v_c = 0.2845\,\text{N/mm}^2$ a shear stress of this magnitude occurs at 1.05 m from the supports
hence nominal shear continues to this point.

Steel in the compression zone underneath the columns

Anchorage length for '16 mm bars = 549 mm

However, two of the outer 16 mm bars must be continued for a greater distance than this as a means of anchoring the shear steel.

Local bond: $f_{bs} = 1.08\,\text{N/mm}^2$

satisfactory.

7. $b = 158.3$ (say 160 mm) (no transverse bending)
    $d = 547.5$ (say 550 mm) (anchorage dictates)
    $M = 356.25\,\text{kNm}$ $\qquad M/bd^2 = 7.36$

Compression steel (bottom steel) is needed
Thus there is more than one solution, e.g.

$100\,A'_s/bd = 1$ $\qquad\qquad\qquad\qquad\qquad 100\,A_s/bd = 2.415$

$\qquad A'_s = 825\,\text{mm}^2$ $\qquad\qquad\qquad\qquad\qquad A_s = 1992.4\,\text{mm}^2$

Use two 20-mm and one 16-mm $\varnothing$ bars $\qquad$ Use two 32-mm and one 25-mm $\varnothing$ bars

Bottom steel $\qquad\qquad\qquad\qquad\qquad\qquad$ **Top steel**

Shear: $V_{max} = 237.5\,\text{kN}$ $\qquad v_{max} = 2.88; v_c = 0.92\,\text{N/mm}^2$

$$\frac{A_{sv}}{S_v} > = 0.847 \qquad\qquad \text{max } S_v = 412.5\,\text{mm}$$

Use 12 mm links at 250 mm

Local bond: $f_{bs} = 2.45\,\text{N/mm}^2$

satisfactory.

8. $b = 148.1$ $\quad$ (say 150 mm); $d = 330\,\text{mm}$
    $M = 75\,\text{kNm}$ $\qquad M/bd^2 = 4.59$ $\qquad 100\,A_s/bd = 1.66$
    $A_s = 821.7\,\text{mm}^2$ $\quad$ use two 20-mm and 16-mm $\varnothing$ bars **top steel**

Shear: $V_{max} = 100\,\text{kN}$ $\qquad v_{max} = 2.02\,\text{N/mm}^2$ $\qquad v_c = 0.832$

$$\frac{A_{sv}}{S_v} > = 0.5 \qquad \text{max } S_v = 247.5$$

Use 8 mm links at 200 mm spacing along the length of the beam.

Under the columns use two 20-mm and one 16-mm bars for 810 mm either side of the column centreline. Then 8-mm bars must be lapped with the outer 20-mm bars in order to give proper anchorage to the links. Lapping to be in accordance with Cl. 3.11.6.5.

Local bond: $f_{bs} = 1.72 \, \text{N/mm}^2$

satisfactory.

Chapter 7

1. Using 203×133 mm at 30 kg/m UB

   Average shear stress $= 98.09 \, \text{N/mm}^2$

   Maximum shear stress $= 110.27 \, \text{N/mm}^2$

   Maximum bending stress $= 137.68 \, \text{N/mm}^2$

   Maximum stress allowed $= 165 \, \text{N/mm}^2$

   Deflection $= 1.8$ mm at centre

   allowed $= 5.56$ mm

   Stiff bearing needed $= 74.81$ mm (bearing dictates).

2. Grade 43: 914×419 mm at 340 kg/m (bending critical)

   bearing $= 5.85$ mm (bearing critical)

   Grade 50: 914×305 mm at 253 kg/m UB (bending)

   bearing $= 19$ mm (bearing)

   Grade 55: 914×305 mm at 201 kg/m (bending)

   bearing $= 23.6$ mm (bearing).

3. Deflection criterion exceeded

   actual $= 22.83$ mm

   allowed $= 22.22$ mm

   However, most designers would allow this.

4. *Shear:* Average shear stress is exceeded **but** maximum shear stress is not   (O.K.)

   Bending: $p_{bc} = 165 \, \text{N/mm}^2$

   $f_{bc} = 114.6 \, \text{N/mm}^2$

   Bearing and buckling exceeded under load and over support

   Deflection $= 2$ mm

   allowed $= 11.11$ mm.

5. Under load: use 20-mm thick stiffeners to full width of beam – dictated by bearing on stiffeners

At support: use 18-mm thick stiffeners to full width of beam – again dictated by bearing on stiffeners.

6. (a) 7.416 m; (b) 2.9625 m; (c) 2.646 m;
   (d) 3.08 m; (e) 3.096 m.

7. Shear gives UDL = 1752.3 kN

   Bending gives UDL = 454 kN

   Deflection gives UDL = 447.6 kN

   Web buckling is satisfactory

   Web bearing length needed = 16.33 mm

   Web stiffeners of thickness 5 mm to full width of beam.

8. Bending $p_{bc} = 165\,\text{N/mm}^2$; $f_{bc} = 162\,\text{N/mm}^2$

   Shear: average shear stress = $12.37\,\text{N/mm}^2$

   Bearing: no stiff bearing required

   Deflection:

      actual = 10.06 mm

      allowed = 18.06 mm

9. Deflection = $0.2386P$ mm   (P in kN)

10. Bending $p_{bc} = 89.37\,\text{N/mm}^2$ $P_{\text{allowed}} = \underline{50.09\,\text{kN}}$

   Shear: $P_{\text{allowed}} = 318\,\text{kN}$

   Bearing: $P_{\text{allowed}} = 147.44\,\text{kN}$

   Deflection: $P_{\text{allowed}} = 93.14\,\text{kN}$

   Bending dictates the maximum load which may be supported.

11. Since the load is applied to the lower flange, web bearing and buckling problems do not arise.

   Live load moments

      Maximum cantilever moment = 50 kNm

      Maximum span moment = 112.5 kNm.

   Live load shear forces

   The maximum shear force occurs when the crab is close to a support

      Maximum shear force = 75 kN.

   Effective lengths

   Cantilever effective length = 2 m

   for the span $k_1 = 1$; $k_2 = 0.7$

   Effective length = $0.7 \times 6 = 4.2$ m.

Try a 457 × 152 mm at 60 kg/m UB:

$r_y = 32.3$ mm $Z = 1.12 \times 10^6$ mm^3 $D/T = 34.2$ $l/r_y = 130$

$p_{bc} = 126$ N/mm^2

Maximum moment (span) = 115.15 kNm $f_{bc} = 102.8$ N/mm^2

Safe in shear.

Try a 457 × 152 mm at 52 kg/m UB:

$r_y = 36.7$ mm $Z = 0.949 \times 10^6$ mm^3 $D/T = 41.3$

$l/r_y = 135$ $p_{bc} = 120.5$ N/mm^2

Maximum moment (span) = 114.8 kNm $f_{bc} = 120.96$ N/mm^2

Safe in shear.

Try a 406 × 178 mm at 54 kg/m UB:

$r_y = 36.7$ mm $Z = 0.9228 \times 10^6$ mm^3 $D/T = 36.9$

$l/r_y = 114.4$ $p_{bc} = 141$ N/mm^2

Maximum moment = 114.9 kNm $f_{bc} = 124.5$ N/mm^2

Safe in shear.

Deflection (live load only):

Section	Deflection (mm)	
	span	cantilever
457 × 152 mm at 60 kg/m	6.31	2.18
457 × 152 mm at 52 kg/m	7.53	2.60
406 × 178 mm at 54 kg/m	8.65	3.00
Allowed	16.67	2.78

Thus the third section is just too flexible however, most designers would allow this to pass since it does not involve damage to another part of the structure.

Chapter 8

1. Grade 43: 356 × 368 mm at 129 kg/m
   305 × 305 mm at 137 kg/m
   254 × 254 mm at 167 kg/m

Grade 50: 305 × 305 mm at 118 kg/m
(305 × 305 mm at 97 kg/m just fails)
254 × 254 mm at 132 kg/m

Grade 55: 305 × 305 mm at 97 kg/m
254 × 254 mm at 107 kg/m.

2.	Section	Cased size
Grade 43	305 × 305 mm at 97 kg/m	410 × 410 mm
	254 × 254 mm at 107 kg/m	370 × 360 mm
Grade 50	254 × 254 mm at 89 kg/m	370 × 360 mm
Grade 55	254 × 254 mm at 73 kg/m	370 × 360 mm

3. 254 × 254 mm at 132 kg/m UC   or   305 × 305 mm at 118 kg/m.

4. 305 × 305 mm at 240 kg/m UC   or
356 × 368 mm at 202 kg/m UC   or
356 × 406 mm at 235 kg/m UC.

5. Details for a 305 × 305 mm at 198 kg/m UC:

$B = 314.1$ mm; $D = 339.9$ mm; $r_y = 80.2$ mm area $= 25\,230$ mm^2

$Z_{xx} = 2.991 \times 10^6$ mm^3; $Z_{yy} = 1.034 \times 10^6$ mm^3

Details for the column cased:

$r_y = 0.2(B + 100) = 0.2 \times 414.1 = 82.82$ mm $l/r_y = 60.37$

$p_c = 168.6$ N/mm^2 stress on concrete $= \dfrac{p_c}{0.19 p_{bc}} = 3.736$ N/mm^2

Permissible load on steel $= 168.6 \times 25\,230\,N = 4253.8$ kN

Permissible load on concrete $= 3.736 \times A_c$

A_c is obtained as the product of breadth times depth where

Breadth $= 470$ mm or $B + 150$ whichever is the lesser
$= \underline{464.1\text{ mm}}$

Depth $= 490$ mm or $D + 150$ whichever is the lesser
$= \underline{489.9\text{ mm}}$

Permissible load on the concrete $= 3.736 \times 464.1 \times 489.9$
$= 849.47$ kN

Total permissible axial load = 5103.27 kN

Actual axial load = 3150 kN

Bending stresses: $M_{xx} = 500 \times 10^3 \times 269.95/2 = 67.49 \, \text{kNm}$

$M_{yy} = 350 \times 10^3 \times 109.6/2 = 19.18 \, \text{kNm}$

$f_{bc} = 41.11 \, \text{N/mm}^2 \qquad p_{bc} = 230 \, \text{N/mm}^2$

$$\frac{f_{bc}}{p_{bc}} + \frac{f_c}{p_c} = \frac{41.11}{230} + \frac{3150}{5103.27} = 0.797 \qquad \text{O.K.}$$

6. **(a)** Reduce the size of casing to give just in excess of minimum cover, viz. to 440 × 415 mm.

$$\frac{f_{bc}}{p_{bc}} + \frac{f_c}{p_c} = 0.814$$

(b) Reduce steel section to 305 × 305 mm at 158 kg/m and the casing to 480 × 460 mm.

$$\frac{f_{bc}}{p_{bc}} + \frac{f_c}{p_c} = 0.975.$$

7. The section is satisfactory $\dfrac{f_{bc}}{p_{bc}} + \dfrac{f_c}{p_c} = 0.877.$

8. **(a)** Reduce casing to 540 × 515 mm $\dfrac{f_{bc}}{p_{bc}} + \dfrac{f_c}{p_c} = 0.89.$

(b) Use 356 × 406 mm at 393 kg/m universal column of grade 55. steel and reduce the casing to 520 × 510

$$\frac{f_{bc}}{p_{bc}} + \frac{f_c}{p_c} = 0.886.$$

9. *Effective lengths:* AB $-\, l = 0.85 \times 5 = 4.25 \, \text{m}$

BC $-\, l = 0.7 \times 5 = 3.5 \, \text{m}$

Weight of AB and BC = 5 × 137 × 9.81/1000 kN = 6.72 kN

Check the column at roof level:

Axial load = 750 kN $\qquad M_{xx} = 750 \times 10^3 \times 260.25$

$= 195.19 \, \text{kNm}$

$p_{bc} = 230 \, \text{N/mm}^2 \qquad f_{bc} = M_{xx}/2.049 \times 10^6$

$= 95.26 \, \text{N/mm}^2$

$l/r_y = \dfrac{4250}{78.2} = 54.35 \qquad p_c = 178.3 \, \text{N/mm}^2$

Allowed axial load $= 178.3 \times 17\,460 = 3113.2\,\text{kN}$

$$\frac{f_{bc}}{p_{bc}} + \frac{f_c}{p_c} = \frac{750}{3113.2} + \frac{95.26}{230} = 0.655$$

Satisfactory

Check the column at level B:
The question here arises as to which of the sections of the column, the upper or the lower, should be checked. Since the effective length of BC is less than the effective length of AB, it can carry a greater axial load. Therefore the check is performed on the upper section where the allowed axial load is less.

Axial load $= 750 + 6.72 + 1250 = 2006.72$

Moments: the moments generated by the eccentricity of the beams are divided equally between the upper and lower sections:

$$M_{xx} = 0.5 \times 550 \times 10^3 \times 260.25 = 71.569\,\text{kNm}$$

$$M_{yy} = 0.5 \times 200 \times 10^3 \times 106.9 = 10.69\,\text{kNm}$$

$$\frac{f_{bc}}{p_{bc}} + \frac{f_c}{p_c} = \frac{2006.72}{3113.2} + \frac{50.39}{230} = 0.864$$

Satisfactory

At the base of the column, there is no moment and virtually the same axial load as at level B. Therefore, it is obvious that the column is satisfactory at this level also. **Thus the stanchion is satisfactory.**

Chapter 9

1. *Spacings: Satisfactory*
   Plate tension: 180.6 kN
   Bolt shear: 225.6 kN
   Bolt bearing: 189 kN (material critical)
   Thus maximum load $= 180.6\,\text{kN}$.

2. (Worked)
   Consider Fig. 9.41 which will be used to derive:

   (a) The bearing area on the tube
   (b) The net tension area of the tube

   (a) If the diameter of bolt hole $= d\,\text{mm} = $ bolt diameter $+ 2\,\text{mm}$
   If the internal radius of the tube $= r\,\text{mm}$

Then the angle subtended by the hole at the centre $= \theta$

where $\quad \theta = 2\sin^{-1}\dfrac{d}{2r} = 2\sin^{-1}\dfrac{d}{40.3}$ deg.

Gross area of section of tube $= 556.7\,\text{mm}^2$

Net area of section for tension $= \dfrac{180 - \theta}{180} \times 556.7\,\text{mm}^2$

(b) If bolt diameter $= D\,\text{mm}$

Angle of bearing area subtended at centre $= \alpha = 2\sin^{-1}\dfrac{D}{2r}$

Bearing area $= \dfrac{\alpha}{180} \times 556.7$

Allowed stresses: tension $= 155/1.23 = 126\,\text{N/mm}^2$ (Table 19 of
BS 449)

$\qquad$ bearing $= 250/1.23 = 203.25\,\text{N/mm}^2$ (material)

$\qquad$ shear $\quad= 187\,\text{N/mm}^2$ (grade 8.8 bolts) (double
shear)

Try the following bolts:

			Tube strength in:		
Bolt dia. (mm)	Hole dia. (mm)	Shear area (mm^2)	tension (kN)	bolt shear (kN)	bearing (kN)
6	8	56.55	61.1	10.57	10.76
8	10	100.5	58.8	18.8	14.4
10	12	157.1	56.5	29.4	18.1
12	14	226.2	54.2	42.3	21.8
16	18	402.1	49.4	75.2	29.4
20	22	628.3	44.3	117.5	37.4
24	26	904.8	38.75	169.2	46.0

Therefore, the largest force which can be accommodated by this splice occurs when the bolt diameter is 24 mm

Maximum tension transmitted $= 38.75\,\text{kN}$

$\qquad\qquad\qquad\qquad\quad = 55.4$ per cent of full strength

Note: If the connection is redesigned to take six bolts instead of two, the optimum size of bolt is 10 mm and the maximum tension strength $= 54.3\,\text{kN}$ (bearing)

$\qquad\qquad\quad = 77.6$ per cent

3. Allowed end distance from bolt to join in tubes = 36 mm
   Therefore minimum length of cover plate = 36 × 4 = 144 mm
   (say 150 mm)
   A thickness of 3 mm would give the same net area as the tube but a thickness of 4.15 mm is needed to give adequate bearing. **Thus it is suggested that a thickness of 5 mm be used.**

4. Angle capacity = 207.08 kN (smaller holes)
   Upstand gusset needs 11.83 mm (say 12 mm thickness)
   Base gusset needs 3.77 mm (say 5 mm thickness)
   The arrangement of bolts shown in Fig. A.4 will suffice.

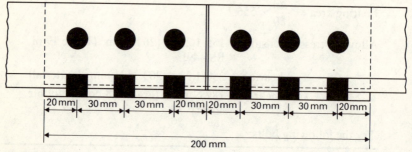

| 20 mm | 30 mm | 30 mm | 20 mm | 20 mm | 30 mm | 30 mm | 20 mm |

200 mm

Fig. A.4

5. Bolt size is dictated by Cl. 3.4 of BS 4604
   Maximum size allowed = M 12
   Assume the distance between bolt centreline and edge of angle is 50 mm. (This does not accord with the back marks of Table 9.7 but is necessary since only one bolt per leg length is being used. However, it should be noted that two bolts could be used in the space afforded by the leg length. This would reduce the length of the splice.)

   Angle capacity = 632.4 kN
   Upstand gusset needs 14.67 mm thickness   (say 15 mm)
   Base gusset needs 7.49 mm thickness   (say 8 mm)
   Upstand gusset needs 7.44   (say eight) bolts either side of the splice
   Base gusset needs 16.60   (say eighteen) bolts either side of the splice

   The minimum length of gusset = 336 mm   (say 350 mm)

   Thus it can be seen that it would be best to have two bolts in the space of the leg length, which would reduce the length of the splice to 216 mm   (say 250 mm).

341

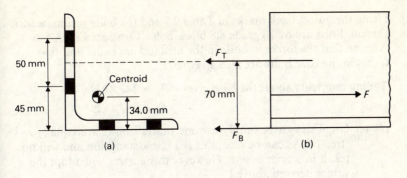

(a)   (b)

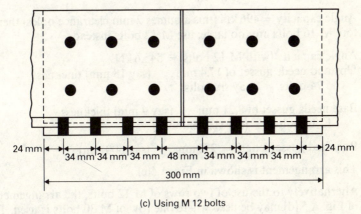

24 mm ← → 34 mm 34 mm 34 mm 48 mm 34 mm 34 mm 34 mm → ← 24 mm

300 mm

(c) Using M 12 bolts

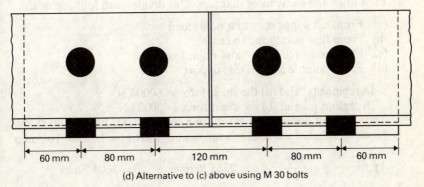

60 mm   80 mm   120 mm   80 mm   60 mm

(d) Alternative to (c) above using M 30 bolts

Fig. A.5

6. Using the gusset back marks of Table 9.7 and the bolts recommended therein. Bolts are M 16 grade 8.8 black bolts. Consider Fig. A.5. Assume that the forces in bolts in the upstand are equal and that forces in the base bolts are also all equal.

Taking moments about the base gives $F_T = 242.8\,\text{kN}$

$$F_B = 257.1\,\text{kN}$$

(*Note:* Often designers simply assume that **all** bolts are equally stressed. As can be seen, this is a fair assumption and will not result in a poor design. However this answer will adopt the values derived above.)

Angle capacity = 496 kN (this assumes 2 mm clearance holes) therefore M 16 bolts are too large; use M 12 bolts instead

Angle capacity (with M 12 bolts) = 545.6 kN
Upstand needs gusset of 17.4 mm   (say 18 mm) thickness
        5.74 bolts   (say six bolts)

Base needs gusset of 8.21 mm   (say 9 mm) thickness
        12.16 bolts   (say sixteen bolts)

Required gusset length = 276 mm   (say 300 mm)

This arrangement is shown in Fig. A.5(c).

Alternatively to the use of two rows of M 12 bolts, the arrangement of Fig. A.5 (d) may be used where one row of M 30 bolts is used. The required sizes of gussets is the same, but the length is now 400 mm.

7. First draw the shear force diagram. This divides into four regions:

(a) From left support to first point load
(b) From first point load to centre
(c) From centre to right-hand point load
(d) From point load to right support

    In regions (a) and (d) the shear force = 6000 kN
    In regions (c) and (b) the shear force = 2000 kN
Thus the bolt design divides into two sections 6000 kN and 2000 kN
Find the value of the section's moment of inertia = $1.633 \times 10^{12}\,\text{mm}^4$

Apply $q = \dfrac{Va\bar{y}}{I}$ $a = 236\,650\,\text{mm}^2$ $\bar{y} = 1604.7\,\text{mm}$

In the 6000 kN regions this gives a shear of 1395.2 N/mm
In the 2000 kN regions this gives a shear of 465.1 N/mm.

A number of bolt distributions will cope with this, one of which is:
regions (a) and (d) 20 M 27 bolts as in Fig. A.6
regions (b) and (c) 190 M 16 bolts as in Fig. A.6.

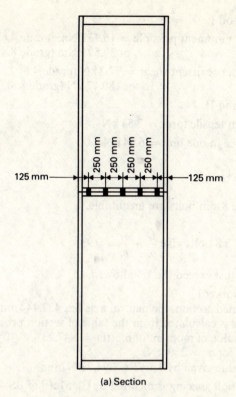

125 mm → ← 250 mm | 250 mm | 250 mm | 250 mm → ← 125 mm

(a) Section

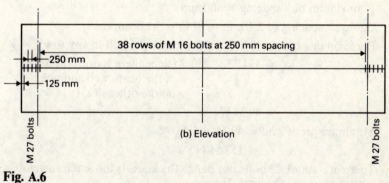

38 rows of M 16 bolts at 250 mm spacing

250 mm

125 mm

M 27 bolts

M 27 bolts

(b) Elevation

Fig. A.6

Note: Clause 51c(i) of BS 449 limits the pitch of the bolts to 300 mm.
Clause 51d(ii) of BS 449 limits the edge distance to no more
than 140 mm.
Therefore this is the smallest number of bolts which may be
used, even if larger bolts are used.

8. Using Method I:

    Maximum moment possible = 16.93 kNm (grade 4.6)
                               or 39.59 kNm (grade 8.8)

    Maximum coexistent shear = 77.2 kN (grade 4.6)
                              or 180.47 kN (grade 8.8).

9. Using Method II:

    Maximum tensile force = 7.54 kN

    Shear force in one bolt = 6.94 kN

$$\frac{f_s}{p_s} + \frac{f_t}{p_t} = 1.47$$

    Therefore 8 mm bolts are unsuitable.

10. M 20 grade 8.8 bolts give $\dfrac{f_s}{p_s} + \dfrac{f_t}{p_t} = 0.93$

    M 16 bolts just exceed the 1.4 allowed.

11. (Partially worked)

    As a combined section, the neutral axis lies 432.444 mm down from the top (this is calculated from the table of section properties) the combined value of moment of inertia = 947.231×10^6 mm^4 maximum shear force = 250 kN.

    Thus the value given by $Va\bar{y}/I = 394.73$ N/mm

    Maximum bolt spacing is dictated by Cl. 51c(i) of BS 449

        maximum bolt spacing = 300 mm

    Choose a spacing of 250 mm to fit into the beam easily

    Force on one bolt = 394.73 × 250/number of bolt in any row

                  = 394.73 × 250/2 (since there is one on one side
                                    of the beam web and one
                                    on the other side)

                  = 49.34 kN

    Minimum proof load = 49.34 × 1.4/0.45

                      = 153.5 kN

    Therefore use M 22 bolts but flange thickness is too small therefore reduce spacing of bolts to 75 mm

    Force on one bolt = 394.73 × 75/2 = 14.8 kN

    Minimum proof load = 46.1 kN

    Thus use M 12 bolts throughout the length of the beam.

12. Centroid lies 14.5 mm above the bottom leg.

    Force in AB = 75 × 14.5/50 = 21.75 kN

    Force in CD = 53.25 kN

CD is designed first:

Throat area $= 110t_e$

$$115 \times 110t_e = 53\,250$$
$$t_e = 4.21\,\text{mm}$$

Minimum leg length $= 4.21 \times \sqrt{2} = 5.95\,\text{mm}$ (say 6 mm)

AB is now designed:

(a) using the same 6 mm leg length

$$X = \frac{21.75}{53.25} \times 110 = 44.9\,\text{mm} \qquad \text{(say 45 mm)}$$

(b) using full length of 110 mm

Minimum leg length $= \dfrac{21.75}{53.25} \times 6 = 2.45\,\text{mm}$ (say 3 mm) or

anywhere in between these two extremes.

13. Length of angle = width of beam flange $= 101.9\,\text{mm}$

Bearing area $= 101.9 \times$ leg thickness $= 101.9t$

$$101.9t \times 190 = 100\,000$$
$$t = 5.165\,\text{mm} \qquad \text{(say 6 mm)}$$

Using 6 mm fillet welds to sides and bottom

Need to use a $60 \times 60 \times 6$ angle

However, if the length of the angle is increased slightly to 105 mm a $50 \times 50 \times 6$ angle may be used.

14. $M_{xx} = 175 \times 0.15 = 26.25\,\text{kNm}$

Area of weld throat $= 1013.2t_e$

I_{xx} (for weld) $= 13.833 \times 10^6\,t_e\,\text{mm}^4$

Bending stress $= f_{bx} = 293.09/t_e\,\text{N/mm}^2$

Shear stress $= f_{qx} = 172.72/t_e\,\text{N/mm}^2$

Resultant stress $= 340.197/t_e\,\text{N/mm}^2$

$t_e = 2.96$ leg length $= 4.18$ (say 5 mm).

15. I_{yy} for weld $= 0.45 \times 10^6\,t_e\,\text{mm}^4$

$I_p = I_{xx} + I_{yy} = 14.283 \times 10^6\,t_e\,\text{mm}^4$

$M_{yy} = 1.3125\,\text{kNm} \qquad f_{by} = 148.604/t_e\,\text{N/mm}^2$

Torque $= 1.3514\,\text{kNm} \qquad f_t = 15.388/t_e\,\text{N/mm}^2$

$$f_{qy} = 8.636/t_e\,\text{N/mm}^2$$

Vector resultant

$$= \sqrt{(f_{qx} + f_t \cos 71.743)^2 + (f_{qy} + f_t \sin 71.743)^2 + (f_{bx} + f_{by})^2}$$
$$= 476.6/t_e \qquad t_e = 4.144\,\text{mm} \qquad \text{leg length} = 5.86 \qquad \text{(say 6 mm)}.$$

16. Centroid is located at 85.714 mm from TU

$I_p = 66.898 \times 10^6 \, t_e \, \text{mm}^4$

Stress at R.H. corners $= 448.02/t_e \, \text{N/mm}^2$

Stress at L.H. corners $= 432.42/t_e \, \text{N/mm}^2$

Thus leg length of TU $= 13.75$ mm   (say 14 mm)

Leg length elsewhere $= 5.5$ mm   (say 6 mm).

Chapter 10

1. $L = 4500$ mm $L/60 = 75$ mm

$\qquad\qquad\qquad\qquad L/45 = 100$ mm

$W = 1.4 \times 1.2 \times 4.5$ kN

$\quad = 7.56$ kN

Minimum section (elastic) modulus $= \dfrac{7.56 \times 4500}{1800} \text{cm}^3$

$$= 18.9 \, \text{cm}^3$$

Thus suitable angles would be

$100 \times 75 \times 8$ $Z = 19.3 \, \text{cm}^3$

or bigger angles

2. Effective length $= 0.85 \times 5 = 4.25$ m

The following angles will suffice

$150 \times 150 \times 10$ Allowed load $= 128.9$ kN

or larger equal angles

$200 \times 100 \times 15$ Allowed load $= 103.2$ kN

or larger unequal angles

3. Two $100 \times 65 \times 7$ mm angles give a compression resistance of 109.76 kN if the longer leg is bolted to the gusset.

The intermediate stitch blots must be spaced apart at no more than 560 mm, therefore make the spacing 555 mm in order to better fit them into the strut length.

4. See Fig. A.7. (opposite)

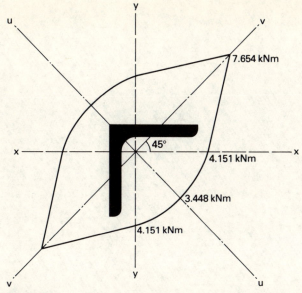

Fig. A.7

Appendix B

Steel section tables

350

(reproduced from BS 4 and BS 4848: Parts 2 and 4)

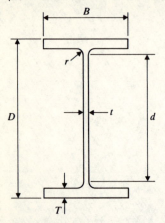

Universal beams

Designation		Depth of section	Width of section	Thickness		Root radius	Depth between fillets	Area of section
				Web	Flange			
Serial size	Mass per unit length	D	B	t	T	r	d	
(mm)	(kg/m)	(mm)	(mm)	(mm)	(mm)	(mm)	(mm)	(cm²)
914 × 419	388	920.5	420.5	21.5	36.6	24.1	799.0	494.5
	343	911.4	418.5	19.4	32.0	24.1	799.0	437.5
914 × 305	289	926.6	307.8	19.6	32.0	19.1	824.4	368.8
	253	918.5	305.5	17.3	27.9	19.1	824.4	322.8
	224	910.3	304.1	15.9	23.9	19.1	824.4	285.3
	201	903.0	303.4	15.2	20.2	19.1	824.4	256.4
838 × 292	226	850.9	293.8	16.1	26.8	17.8	761.7	288.7
	194	840.7	292.4	14.7	21.7	17.8	761.7	247.2
	176	834.9	291.6	14.0	18.8	17.8	761.7	224.1
762 × 267	197	769.6	268.0	15.6	25.4	16.5	685.8	250.8
	173	762.0	266.7	14.3	21.6	16.5	685.8	220.5
	147	753.9	265.3	12.9	17.5	16.5	685.7	188.1
686 × 254	170	692.9	255.8	14.5	23.7	15.2	615.0	216.6
	152	687.6	254.5	13.2	21.0	15.2	615.0	193.8
	140	683.5	253.7	12.4	19.0	15.2	615.0	178.6
	125	677.9	253.0	11.7	16.2	15.2	615.0	159.6
610 × 305	238	633.0	311.5	18.6	31.4	16.5	537.2	303.8
	179	617.5	307.0	14.1	23.6	16.5	537.2	227.9
	149	609.6	304.8	11.9	19.7	16.5	537.2	190.1

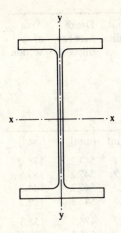

Second moment of area		Radius of gyration		Elastic modulus		Plastic modulus		Ratio $\dfrac{D}{T}$
x–x axis	y–y axis	x–x axis	y–y axis	x–x axis	y–y axis	x–x axis	y–y axis	
(cm⁴)	(cm⁴)	(cm)	(cm)	(cm³)	(cm³)	(cm³)	(cm³)	
718 742	45 407	38.13	9.58	15 616	2 160.0	17 657	3 339.0	25.2
625 282	39 150	37.81	9.46	13 722	1 871.0	15 474	2 890.0	28.5
504 594	15 610	36.99	6.51	10 891	1 014.0	12 583	1 603.0	29.0
436 610	13 318	36.78	6.42	9 507	871.9	10 947	1 372.0	32.9
375 924	11 223	36.30	6.27	8 259	738.1	9 522	1 162.0	38.1
325 529	9 427	35.63	6.06	7 210	621.4	9 522	982.5	44.7
339 747	11 353	34.30	6.27	7 986	772.9	9 157	1 211.0	31.8
279 450	9 069	33.63	6.06	6 648	620.4	7 648	974.4	38.7
246 029	7 792	33.13	5.90	5 894	534.4	6 809	841.5	44.4
239 894	8 174	30.93	5.71	6 234	610.0	7 167	958.7	30.3
205 177	6 846	30.51	5.57	5 385	513.4	6 197	807.3	35.3
168 966	5 468	29.97	5.53	4 483	412.3	5 174	649.0	43.0
170 147	6 621	28.03	5.53	4 911	517.7	5 624	810.3	29.3
150 319	5 782	27.85	5.46	4 372	454.5	4 997	710.0	32.7
136 276	5 179	27.62	5.38	3 988	408.2	4 560	637.8	36.0
118 003	4 379	27.19	5.24	3 481	346.1	3 996	542.0	41.9
207 571	15 838	26.14	7.22	6 559	1 017.0	7 456	1 574.0	20.2
151 631	11 412	25.79	7.08	4 911	743.3	5 521	1 144.0	26.1
124 860	9 300	25.61	6.99	4 090	610.3	4 572	936.8	31.0

Universal beams (*continued*)

Designation		Depth of section	Width of section	Thickness		Root radius	Depth between fillets	Area of section
Serial size	Mass per unit length			Web	Flange			
		D	B	t	T	r	d	
(mm)	(kg/m)	(mm)	(mm)	(mm)	(mm)	(mm)	(mm)	(cm²)
610 × 229	140	617.0	230.1	13.1	22.1	12.7	547.2	178.4
	125	611.9	229.0	11.9	19.6	12.7	547.2	159.6
	113	607.3	228.2	11.2	17.3	12.7	547.2	144.5
	101	602.2	227.6	10.6	14.8	12.7	547.2	129.2
533 × 210	122	544.6	211.9	12.8	21.3	12.7	476.5	155.8
	109	539.5	210.7	11.6	18.8	12.7	476.5	138.6
	101	536.7	210.1	10.9	17.4	12.7	476.5	129.3
	92	533.1	209.3	10.2	15.6	12.7	476.5	117.8
	82	528.3	208.7	9.6	13.2	12.7	476.5	104.4
457 × 191	98	467.4	192.8	11.4	19.6	10.2	407.9	125.3
	89	463.6	192.0	10.6	17.7	10.2	407.9	113.9
	82	460.2	191.3	9.9	16.0	10.2	407.9	104.5
	74	457.2	190.5	9.1	14.5	10.2	407.9	95.0
	67	453.6	189.9	8.5	12.7	10.2	407.9	85.4
457 × 152	82	465.1	153.5	10.7	18.9	10.2	406.9	104.5
	74	461.3	152.7	9.9	17.0	10.2	406.9	95.0
	67	457.2	151.9	9.1	15.0	10.2	406.9	85.4
	60	454.7	152.9	8.0	13.3	10.2	407.7	75.9
	52	449.8	152.4	7.6	10.9	10.2	407.7	66.5
406 × 178	74	412.8	179.7	9.7	16.0	10.2	360.5	95.0
	67	409.4	178.8	8.8	14.3	10.2	360.5	85.5
	60	406.4	177.8	7.8	12.8	10.2	360.5	76.0
	54	402.6	177.6	7.6	10.9	10.2	360.5	68.4
406 × 140	46	402.3	142.4	6.9	11.2	10.2	359.6	59.0
	39	397.3	141.8	6.3	8.6	10.2	359.6	49.4
356 × 171	67	364.0	173.2	9.1	15.7	10.2	312.2	85.4
	57	358.6	172.1	8.0	13.0	10.2	312.2	72.2
	51	355.6	171.5	7.3	11.5	10.2	312.2	64.6
	45	352.0	171.0	6.9	9.7	10.2	312.2	57.0
356 × 127	39	352.8	126.0	6.5	10.7	10.2	311.1	49.4
	33	348.5	125.4	5.9	8.5	10.2	311.1	41.8

Second moment of area		Radius of gyration		Elastic modulus		Plastic modulus		Ratio $\dfrac{D}{T}$
x–x axis	y–y axis	x–x axis	y–y axis	x–x axis	y–y axis	x–x axis	y–y axis	
(cm⁴)	(cm⁴)	(cm)	(cm)	(cm³)	(cm³)	(cm³)	(cm³)	
111 844	4 512	25.04	5.03	3 626.0	392.1	4 146.0	612.50	27.9
98 579	3 933	24.86	4.96	3 222.0	343.5	3 677.0	535.70	31.2
87 431	3 439	24.60	4.88	2 879.0	301.4	3 288.0	470.20	35.1
75 720	2 912	24.21	4.75	2 515.0	255.9	2 882.0	400.00	40.7
76 207	3 393	22.12	4.67	2 799.0	320.2	3 203.0	500.60	25.5
66 739	2 937	21.94	4.60	2 474.0	278.8	2 824.0	435.10	28.7
61 659	2 694	21.84	4.56	2 298.0	256.5	2 620.0	400.00	30.8
55 353	2 392	21.68	4.51	2 076.0	228.6	2 366.0	356.20	34.1
47 491	2 005	21.32	4.38	1 798.0	192.2	2 056.0	300.10	40.0
45 717	2 343	19.10	4.33	1 956.0	243.0	2 232.0	378.30	23.9
41 021	2 086	18.98	4.28	1 770.0	217.4	2 014.0	337.90	26.3
37 103	1 871	18.84	4.23	1 612.0	195.6	1 833.0	304.00	28.8
33 388	1 671	18.75	4.19	1 461.0	175.5	1 657.0	272.20	31.6
29 401	1 452	18.55	4.12	1 296.0	152.9	1 471.0	237.30	35.7
36 215	1 143	18.62	3.31	1 557.0	149.0	1 800.0	235.40	24.6
32 435	1 012	18.48	3.26	1 406.0	132.5	1 622.0	209.10	27.1
28 577	878	18.29	3.21	1 250.0	115.5	1 441.0	182.20	30.6
25 464	794	18.31	3.23	1 120.0	103.9	1 284.0	162.90	34.2
21 345	645	17.92	3.11	949.0	84.6	1 094.0	133.20	41.3
27 329	1 545	16.96	4.03	1 324.0	172.0	1 504.0	266.90	25.9
24 329	1 365	16.87	4.00	1 188.0	152.7	1 346.0	236.50	28.6
21 508	1 199	16.82	3.97	1 058.0	134.8	1 194.0	208.30	31.8
18 626	1 017	16.50	3.85	925.3	114.5	1 048.0	177.50	37.0
15 647	539	16.29	3.02	777.8	75.7	888.4	118.30	36.0
12 452	411	15.88	2.89	626.8	58.0	720.8	91.08	46.0
19 522	1 362	15.12	3.99	1 073.0	157.3	1 212.0	243.00	23.2
16 077	1 109	14.92	3.92	869.5	128.9	1 009.0	198.80	27.5
14 156	968	14.80	3.87	796.2	112.9	894.9	174.10	30.9
12 091	812	14.57	3.78	686.9	95.0	773.7	146.70	36.2
10 087	357	14.29	2.69	571.8	56.6	653.6	88.68	33.1
8 200	280	14.00	2.59	470.6	44.7	639.8	70.24	41.0

Universal beams (*continued*)

Designation		Depth of section	Width of section	Thickness		Root radius	Depth between fillets	Area of section
				Web	Flange			
Serial size	Mass per unit length	D	B	t	T	r	d	
(mm)	(kg/m)	(mm)	(mm)	(mm)	(mm)	(mm)	(mm)	(cm²)
305 × 165	54	310.9	166.8	7.7	13.7	8.9	265.6	68.4
	46	307.1	165.7	6.7	11.8	8.9	265.6	58.9
	40	303.8	165.1	6.1	10.2	8.9	265.6	51.5
305 × 127	48	310.4	125.2	8.9	14.0	8.9	264.6	60.8
	42	306.6	124.3	8.0	12.1	8.9	264.6	53.2
	37	303.8	123.5	7.2	10.7	8.9	264.6	47.5
305 × 102	33	312.7	102.4	6.6	10.8	7.6	275.8	41.8
	28	308.9	101.9	6.1	8.9	7.6	275.8	36.3
	25	304.8	101.6	5.8	6.8	7.6	275.8	31.4
254 × 146	43	259.6	147.3	7.3	12.7	7.6	218.9	55.1
	37	256.0	146.4	6.4	10.9	7.6	218.9	47.5
	31	251.5	146.1	6.1	8.6	7.6	218.9	40.0
254 × 102	28	260.4	102.1	6.4	10.0	7.6	225.0	36.2
	25	257.0	101.9	6.1	8.4	7.6	225.0	32.2
	22	254.0	101.6	5.8	6.8	7.6	225.0	28.4
203 × 133	30	206.8	133.8	6.3	9.6	7.6	172.3	38.0
	25	203.2	133.4	5.8	7.8	7.6	172.3	32.3

Second moment of area		Radius of gyration		Elastic modulus		Plastic modulus		Ratio $\dfrac{D}{T}$
x–x axis	y–y axis	x–x axis	y–y axis	x–x axis	y–y axis	x–x axis	y–y axis	
(cm^4)	(cm^4)	(cm)	(cm)	(cm^3)	(cm^3)	(cm^3)	(cm^3)	
11 710	1 061	13.09	3.94	753.3	127.3	844.8	195.30	22.7
9 948	897	13.00	3.90	647.9	108.3	722.7	165.80	26.0
8 523	763	12.86	3.85	561.2	92.4	624.5	141.50	29.9
9 504	460	12.50	2.75	612.4	73.5	706.1	115.70	22.2
8 143	388	12.37	2.70	531.2	62.5	610.5	98.24	25.4
7 162	337	12.28	2.67	471.5	54.6	540.5	85.66	28.4
6 487	193	12.46	2.15	415.0	37.8	479.9	59.85	29.0
5 421	157	12.22	2.08	351.0	30.8	407.2	48.92	34.8
4 387	120	11.82	1.96	287.9	23.6	337.8	37.98	44.6
6 558	677	10.91	3.51	505.3	92.0	568.2	141.20	20.4
5 556	571	10.82	3.47	434.0	78.1	485.3	119.60	23.4
4 439	449	10.53	3.35	353.1	61.5	395.6	94.52	29.1
4 008	178	10.52	2.22	307.9	34.9	353.4	54.84	26.0
3 408	148	10.29	2.14	265.2	29.0	305.6	45.82	30.8
2 867	120	10.04	2.05	225.7	23.6	261.9	37.55	37.2
2 887	384	8.72	3.18	279.3	57.4	313.3	88.05	21.5
2 356	310	8.54	3.10	231.9	46.4	259.4	71.39	26.0

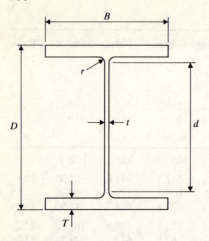

Universal columns

Designation		Depth of section	Width of section	Thickness		Root radius	Depth between fillets	Area of section
				Web	Flange			
Serial size	Mass per unit length	D	B	t	T	r	d	
(mm)	(kg/m)	(mm)	(mm)	(mm)	(mm)	(mm)	(mm)	(cm²)
356 × 406	634	474.7	424.1	47.6	77.0	15.2	290.1	808.1
	551	455.7	418.5	42.0	67.5	15.2	290.1	701.8
	467	436.6	412.4	35.9	58.0	15.2	290.1	595.5
	393	419.1	407.0	30.6	49.2	15.2	290.1	500.9
	340	406.4	403.0	26.5	42.9	15.2	290.1	432.7
	287	393.7	399.0	22.6	36.5	15.2	290.1	366.0
	235	381.0	395.0	18.5	30.2	15.2	290.1	299.8
Column core	477	427.0	424.4	48.0	53.2	15.2	290.1	607.2
356 × 368	202	374.7	374.4	16.8	27.0	15.2	290.1	257.9
	177	368.3	372.1	14.5	23.8	15.2	290.1	225.7
	153	362.0	370.2	12.6	20.7	15.2	290.1	195.2
	129	355.6	368.3	10.7	17.5	15.2	290.1	164.9
305 × 305	283	365.3	321.8	26.9	44.1	15.2	246.5	360.4
	240	352.6	317.9	23.0	37.7	15.2	246.5	305.6
	198	339.9	314.1	19.2	31.4	15.2	246.5	252.3
	158	327.2	310.6	15.7	25.0	15.2	246.5	201.2
	137	320.5	308.7	13.8	21.7	15.2	246.5	174.6
	118	314.5	306.8	11.9	18.7	15.2	246.5	149.8
	97	307.8	304.8	9.9	15.4	15.2	246.5	123.3

357

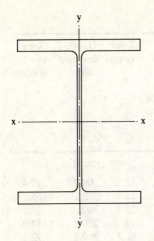

Second moment of area		Radius of gyration		Elastic modulus		Plastic modulus		Ratio $\dfrac{D}{T}$
x–x axis	y–y axis	x–x axis	y–y axis	x–x axis	y–y axis	x–x axis	y–y axis	
(cm⁴)	(cm⁴)	(cm)	(cm)	(cm³)	(cm³)	(cm³)	(cm³)	
275 140	98 211	18.5	11.00	11 592	4 632.0	14 247	7 114	6.2
227 023	82 665	18.0	10.90	9 964	3 951.0	12 078	6 058	6.7
183 118	67 905	17.5	10.70	8 388	3 292.0	10 009	5 033	7.5
146 765	55 410	17.1	10.50	7 004	2 723.0	8 229	4 157	8.5
122 474	46 816	16.8	10.40	6 027	2 324.0	6 994	3 541	9.5
99 994	38 714	16.5	10.30	5 080	1 940.0	5 818	2 952	10.8
79 110	31 008	16.2	10.20	4 153	1 570.0	4 689	2 384	12.6
172 391	68 056	16.8	10.60	8 075	3 207.0	9 700	4 979	8.0
66 307	23 632	16.0	9.57	3 540	1 262.0	3 977	1 917	13.9
57 153	20 470	15.9	9.52	3 104	1 100.0	3 457	1 668	15.5
48 525	17 469	15.8	9.46	2 681	943.8	2 964	1 430	17.5
40 246	14 555	15.6	9.39	2 264	790.4	2 482	1 196	20.3
78 777	24 545	14.80	8.25	4 314.0	1 525.0	5 101.0	2 337.0	8.3
64 177	20 239	14.50	8.14	3 641.0	1 273.0	4 245.0	1 947.0	9.3
50 832	16 230	14.20	8.02	2 991.0	1 034.0	3 436.0	1 576.0	10.8
38 740	12 524	13.90	7.89	2 368.0	806.3	2 680.0	1 228.0	13.1
32 838	10 672	13.70	7.82	2 049.0	691.4	2 298.0	1 052.0	14.7
27 601	9 006	13.60	7.75	1 755.0	587.0	1 953.0	891.7	16.8
22 202	7 268	13.40	7.68	1 442.0	476.9	1 589.0	723.5	20.0

Universal columns (*continued*)

Designation		Depth of section	Width of section	Thickness		Root radius	Depth between fillets	Area of section
				Web	Flange			
Serial size	Mass per unit length	D	B	t	T	r	d	
(mm)	(kg/m)	(mm)	(mm)	(mm)	(mm)	(mm)	(mm)	(cm^2)
254 × 254	167	289.1	264.5	19.2	31.7	12.7	200.2	212.4
	132	276.4	261.0	15.6	25.3	12.7	200.2	168.9
	107	266.7	258.3	13.0	20.5	12.7	200.2	136.6
	89	260.4	255.9	10.5	17.3	12.7	200.2	114.0
	73	254.0	254.0	8.6	14.2	12.7	200.2	92.9
203 × 203	86	222.3	208.8	13.0	20.5	10.2	160.8	110.1
	71	215.9	206.2	10.3	17.3	10.2	160.8	91.1
	60	209.6	205.2	9.3	14.2	10.2	160.8	75.8
	52	206.2	203.9	8.0	12.5	10.2	160.8	66.4
	46	203.2	203.2	7.3	11.0	10.2	160.8	58.8
152 × 152	37	161.8	154.4	8.1	11.5	7.6	123.4	47.4
	30	157.5	152.9	6.6	9.4	7.6	123.4	38.2
	23	152.4	152.4	6.1	6.8	7.6	123.4	29.8

Second moment of area		Radius of gyration		Elastic modulus		Plastic modulus		Ratio $\dfrac{D}{T}$
x–x axis	y–y axis	x–x axis	y–y axis	x–x axis	y–y axis	x–x axis	y–y axis	
(cm^4)	(cm^4)	(cm)	(cm)	(cm^3)	(cm^3)	(cm^3)	(cm^3)	
29 914	9 796	11.90	6.79	2 070.0	740.60	2 417.0	1 132.00	9.1
22 575	7 519	11.60	6.68	1 634.0	576.20	1 875.0	878.60	10.9
17 510	5 901	11.30	6.57	1 313.0	456.90	1 485.0	695.50	13.0
14 307	4 849	11.20	6.52	1 099.0	378.90	1 228.0	575.40	15.0
11 360	3 873	11.10	6.46	894.5	305.00	988.6	462.40	17.9
9 462	3 119	9.27	5.32	851.5	298.70	978.8	455.90	10.8
7 647	2 536	9.16	5.28	708.4	246.00	802.4	374.20	12.4
6 088	2 041	8.96	5.19	581.1	199.00	652.0	302.80	14.8
5 263	1 770	8.90	5.16	510.4	173.60	568.1	263.70	16.5
4 564	1 539	8.81	5.11	449.2	151.50	497.4	230.00	18.5
2 218	709	6.84	3.87	274.2	91.78	310.1	140.10	14.0
1 742	558	6.75	3.82	221.2	73.06	247.1	111.20	16.8
1 263	403	6.51	3.68	165.7	52.95	184.3	80.87	22.3

360

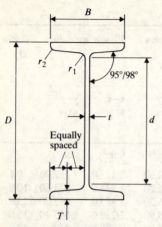

Joists with taper flanges

Designation		Depth of section	Width of section	Thickness		Radius		Depth between fillets	Area of section
				Web	Flange	Root	Toe		
Serial size	Mass per unit length	D	B	t	T	r_1	r_2		
(mm)	(kg/m)	(mm)	(mm)	(mm)	(mm)	(mm)	(mm)	(mm)	(cm²)
254 × 203	81.85	254.0	203.2	10.2	19.9	19.6	9.7	166.6	104.4
254 × 114	37.20	254.0	114.3	7.6	12.8	12.4	6.1	199.2	47.4
203 × 152	52.09	203.2	152.4	8.9	16.5	15.5	7.6	133.2	66.4
203 × 102*	25.33	203.2	101.6	5.8	10.4	9.4	3.2	161.0	32.3
178 × 102*	21.54	177.8	101.6	5.3	9.0	9.4	3.2	138.2	27.4
152 × 127	37.20	152.4	127.0	10.4	13.2	13.5	6.6	94.3	47.5
152 × 89*	17.09	152.4	88.9	4.9	8.3	7.9	2.4	117.7	21.8
152 × 76	17.86	152.4	76.2	5.8	9.6	9.4	4.6	111.9	22.8
127 × 114	29.76	127.0	114.3	10.2	11.5	9.9	4.8	79.4	37.3
127 × 114	26.79	127.0	114.3	7.4	11.4	9.9	5.0	79.5	34.1
127 × 76	16.37	127.0	76.2	5.6	9.6	9.4	4.6	86.5	21.0
127 × 76*	13.36	127.0	76.2	4.5	7.6	7.9	2.4	94.2	17.0
114 × 114	26.79	114.3	114.3	9.5	10.7	14.2	3.2	60.8	34.4
102 × 102	23.07	101.6	101.6	9.5	10.3	11.1	3.2	55.1	29.4
102 × 64*	9.65	101.6	63.5	4.1	6.6	6.9	2.4	73.2	12.3
102 × 44	7.44	101.6	44.4	4.3	6.1	6.9	3.3	74.7	9.5
89 × 89	19.35	88.9	88.9	9.5	9.9	11.1	3.2	44.1	24.9
76 × 76	14.67	76.2	80.0	8.9	8.4	9.4	4.6	38.0	19.1
76 × 76	12.65	76.2	76.2	5.1	8.4	9.4	4.6	37.9	16.3

*These sections have a 5° taper on the inside of the flanges; all others have an 8° taper.

Second moment of area		Radius of gyration		Elastic modulus		Plastic modulus		Ratio $\dfrac{D}{T}$
x–x axis	y–y axis	x–x axis	y–y axis	x–x axis	y–y axis	x–x axis	y–y axis	
(cm⁴)	(cm⁴)	(cm)	(cm)	(cm³)	(cm³)	(cm³)	(cm³)	
12 016.0	2 278.00	10.70	4.67	946.10	224.30	1 076.00	370.40	12.8
5 092.0	270.10	10.40	2.39	401.00	47.19	460.00	79.30	19.8
4 789.0	813.30	8.48	3.51	471.40	106.70	539.80	175.60	12.3
2 294.0	162.60	8.43	2.25	225.80	32.02	256.30	51.79	19.6
1 519.0	139.20	7.44	2.25	170.90	27.41	193.00	44.48	19.7
1 818.0	378.80	6.20	2.82	238.70	59.65	278.60	99.85	11.5
881.1	85.98	6.36	1.99	115.60	19.34	131.00	31.29	18.4
873.7	60.77	6.20	1.63	114.70	15.90	132.50	26.67	15.9
979.0	241.90	5.12	2.55	154.20	42.32	180.90	70.85	11.0
944.8	235.40	5.26	2.63	148.80	41.19	171.90	68.07	11.2
569.4	60.35	5.21	1.70	89.66	15.90	103.60	26.28	13.3
475.9	50.18	5.29	1.72	74.94	13.17	85.23	21.29	16.7
735.4	223.10	4.62	2.54	128.60	39.00	151.20	65.63	10.7
486.1	154.40	4.06	2.29	95.72	30.32	113.40	50.70	9.9
217.6	25.30	4.21	1.43	42.84	7.97	48.98	12.91	15.4
152.3	7.91	4.01	0.91	30.02	3.44	35.30	5.99	16.7
306.7	101.10	3.51	2.01	69.04	22.78	82.77	38.03	9.0
171.9	60.77	3.00	1.78	45.06	15.24	54.16	25.73	9.1
158.6	52.03	3.12	1.78	41.62	13.60	48.84	22.51	9.1

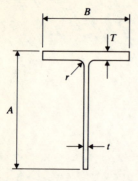

Structural tees cut from universal beams

Designation		Cut from universal beam (see table 5)		Width of section	Depth of section	Thickness		Root radius
						Web	Flange	
Serial size	Mass per unit length	Serial size	Mass per unit length	B	A	t	T	r
(mm)	(kg/m)	(mm)	(kg/m)	(mm)	(mm)	(mm)	(mm)	(mm)
305 × 457	127	914 × 305	253	305.5	459.2	17.3	27.9	19.1
	112		224	304.1	455.2	15.9	23.9	19.1
	101		201	303.4	451.5	15.2	20.2	19.1
292 × 419	113	838 × 292	226	293.8	425.5	16.1	26.8	17.8
	97		194	292.4	420.4	14.7	21.7	17.8
	88		176	291.6	417.4	14.0	18.8	17.8
267 × 381	99	762 × 267	197	268.0	384.8	15.6	25.4	16.5
	87		173	266.7	381.0	14.3	21.6	16.5
	74		147	265.3	376.5	12.9	17.5	16.5
254 × 343	85	686 × 254	170	255.8	346.5	14.5	23.7	15.2
	76		152	254.5	343.8	13.2	21.0	15.2
	70		140	253.7	341.8	12.4	19.0	15.2
	63		125	253.0	339.0	11.7	16.2	15.2
305 × 305	119	610 × 305	238	311.5	316.5	18.6	31.4	16.5
	90		179	307.0	308.7	14.1	23.6	16.5
	75		149	304.8	304.8	11.9	19.7	16.5
229 × 305	70	610 × 229	140	230.1	308.5	13.1	22.1	12.7
	63		125	229.0	305.9	11.9	19.6	12.7
	57		113	228.2	303.7	11.2	17.3	12.7
	51		101	227.6	301.1	10.6	14.8	12.7
210 × 267	61	533 × 210	122	211.9	272.3	12.8	21.3	12.7
	55		109	210.7	269.7	11.6	18.8	12.7
	51		101	210.1	268.4	10.9	17.4	12.7
	46		92	209.3	266.6	10.2	15.6	12.7
	41		82	208.7	264.2	9.6	13.2	12.7

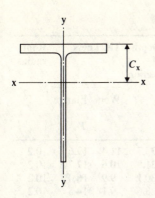

Area of section	Dimension C_x	Second moment of area		Radius of gyration		Elastic modulus		
		$x–x$ axis	$y–y$ axis	$x–x$ axis	$y–y$ axis	$x–x$ axis		$y–y$ axis
						Flange	Toe	
(cm^2)	(cm)	(cm^4)	(cm^4)	(cm)	(cm)	(cm^3)	(cm^3)	(cm^3)
161.4	12.00	32 696.0	6 659.00	14.20	6.42	2 722.0	964.2	435.9
142.6	12.10	20 036.0	5 611.00	14.30	6.27	2 392.0	869.8	369.1
128.2	12.50	26 439.0	4 713.00	14.40	6.06	2 108.0	810.8	310.7
144.4	10.80	24 660.0	5 676.00	13.10	6.27	2 277.0	777.6	386.5
123.6	11.10	21 381.0	4 535.00	13.20	6.06	1 928.0	690.9	310.2
112.1	11.40	19 589.0	3 896.00	13.20	5.90	1 723.0	644.8	267.2
125.4	9.90	17 528.0	4 087.00	11.80	5.71	1 770.0	613.3	305.0
110.2	9.99	15 495.0	3 423.00	11.90	5.57	1 551.0	551.3	256.7
94.0	10.20	13 328.0	2 734.00	11.90	5.39	1 308.0	484.5	206.1
108.3	8.67	12 036.0	3 310.00	10.50	5.53	1 388.0	463.4	258.8
96.9	8.60	10 738.0	2 891.00	10.50	5.46	1 249.0	416.5	227.2
89.3	8.65	9 939.0	2 589.00	10.50	5.38	1 149.0	389.3	204.1
79.8	8.87	8 997.0	2 190.00	10.60	5.24	1 015.0	359.5	173.1
151.9	7.11	12 292.0	7 919.00	9.00	7.22	1 730.0	500.9	508.4
114.0	6.65	8 949.0	5 706.00	8.86	7.08	1 346.0	369.5	371.7
95.1	6.43	7 365.0	4 650.00	8.80	6.99	1 145.0	306.3	305.1
89.2	7.61	7 745.0	2 256.00	9.32	5.03	1 018.0	333.3	196.1
79.8	7.55	6 911.0	1 966.00	9.31	4.96	915.8	299.8	171.8
72.2	7.60	6 295.0	1 720.00	9.34	4.88	827.8	276.6	150.7
64.6	7.80	5 710.0	1 456.00	9.40	4.75	731.8	255.9	127.9
77.9	6.67	5 182.0	1 696.00	8.16	4.67	776.8	252.1	160.1
69.3	6.60	4 593.0	1 469.00	8.14	4.60	696.2	225.4	139.4
64.6	6.57	4 282.0	1 347.00	8.14	4.56	651.7	211.3	128.3
58.9	6.56	3 905.0	1 196.00	8.14	4.51	594.9	194.4	114.3
52.2	6.73	3 517.0	1 003.00	8.21	4.38	522.2	178.7	96.1

Structural tees cut from universal beams (*continued*)

Designation		Cut from universal beam (see Table 5)		Width of section	Depth of section	Thickness		Root radius
Serial size	Mass per unit length	Serial size	Mass per unit length			Web	Flange	
				B	A	t	T	r
191 × 229	49	457 × 191	98	192.8	233.7	11.4	19.6	10.2
	45		89	192.0	231.8	10.6	17.7	10.2
	41		82	191.3	230.1	9.9	16.0	10.2
	37		74	190.5	228.6	9.1	14.5	10.2
	34		67	189.9	226.8	8.5	12.7	10.2
152 × 229	41	457 × 152	82	153.5	232.5	10.7	18.9	10.2
	37		74	152.7	230.6	9.9	17.0	10.2
	34		67	151.9	228.6	9.1	15.0	10.2
	30		60	152.9	227.3	8.0	13.3	10.2
	26		52	152.4	224.9	7.6	10.9	10.2
178 × 203	37	406 × 178	74	179.7	206.4	9.7	16.0	10.2
	34		67	178.8	204.7	8.8	14.3	10.2
	30		60	177.8	203.2	7.8	12.8	10.2
	27		54	177.6	201.3	7.6	10.9	10.2
140 × 203	23	406 × 140	46	142.4	201.2	6.9	11.2	10.2
	20		39	141.8	198.6	6.3	8.6	10.2
171 × 178	34	356 × 171	67	173.2	182.0	9.1	15.7	10.2
	29		57	172.1	179.3	8.0	13.0	10.2
	26		51	171.5	177.8	7.3	11.5	10.2
	23		45	171.0	176.0	6.9	9.7	10.2
127 × 178	20	356 × 127	39	126.0	176.4	6.5	10.7	10.2
	17		33	125.4	174.2	5.9	8.5	10.2
165 × 152	27	305 × 165	54	166.8	155.4	7.7	13.7	8.9
	23		46	165.7	153.5	6.7	11.8	8.9
	20		40	165.1	151.9	6.1	10.2	8.9
127 × 152	24	305 × 127	48	125.2	155.2	8.9	14.0	8.9
	21		42	124.3	153.3	8.0	12.1	8.9
	19		37	123.5	151.9	7.2	10.7	8.9
102 × 152	17	305 × 102	33	102.4	156.3	6.6	10.8	7.6
	14		28	101.9	154.4	6.1	8.9	7.6
	13		25	101.6	152.4	5.8	6.8	7.6
146 × 127	22	254 × 146	43	147.3	129.8	7.3	12.7	7.6
	19		37	146.4	128.0	6.4	10.9	7.6
	16		31	146.1	125.7	6.1	8.6	7.6
102 × 127	14	254 × 102	28	102.1	130.2	6.4	10.0	7.6
	13		25	101.9	128.5	6.1	8.4	7.6
	11		22	101.6	127.0	5.8	6.8	7.6
133 × 102	15	203 × 133	30	133.8	103.4	6.3	9.6	7.6
	13		25	133.4	101.6	5.8	7.8	7.6

Area of section	Dimension C_x	Second moment of area		Radius of gyration		Elastic modulus		
		x–x axis	y–y axis	x–x axis	y–y axis	x–x axis		y–y axis
						Flange	Toe	
62.6	5.55	2 978.0	1 172.00	6.90	4.33	536.5	167.2	121.5
57.0	5.49	2 701.0	1 043.00	6.89	4.28	491.5	152.7	108.7
52.3	5.48	2 482.0	935.70	6.89	4.23	453.0	141.5	97.8
47.5	5.42	2 246.0	835.60	6.88	4.19	414.5	128.8	87.7
42.7	5.47	2 037.0	725.90	6.90	4.12	372.7	118.3	76.5
52.5	6.03	2 608.0	571.60	7.07	3.31	432.6	151.4	74.5
47.5	5.98	2 364.0	505.90	7.06	3.26	395.0	138.4	66.3
42.7	5.99	2 128.0	438.90	7.06	3.21	355.5	126.1	57.8
38.0	5.82	1 870.0	397.20	7.02	3.23	321.4	110.6	52.0
33.2	6.03	1 667.0	322.30	7.08	3.11	276.3	101.3	42.3
47.5	4.80	1 758.0	772.40	6.08	4.03	366.1	111.0	86.0
42.7	4.73	1 574.0	682.30	6.07	4.00	332.6	100.0	76.3
38.0	4.62	1 383.0	599.40	6.03	3.97	299.5	88.1	67.4
34.2	4.81	1 282.0	508.30	6.12	3.85	266.6	83.7	57.2
29.5	5.05	1 130.0	269.40	6.19	3.02	223.9	75.0	37.8
24.7	5.27	968.3	205.60	6.26	2.89	183.6	66.4	29.0
42.7	4.01	1 158.0	681.20	5.21	3.99	289.1	81.6	78.6
36.1	3.95	979.1	544.50	5.21	3.92	248.1	70.0	64.4
32.3	3.92	877.9	483.90	5.21	3.87	223.7	63.4	56.5
28.5	4.02	791.8	406.10	5.27	3.78	196.9	58.3	47.5
24.7	4.41	719.6	178.30	5.40	2.69	163.2	54.4	28.3
20.9	4.52	618.4	140.10	5.44	2.59	136.7	47.9	22.4
34.2	3.19	636.4	530.60	4.31	3.94	199.5	51.5	63.6
29.5	3.09	541.0	448.50	4.29	3.90	175.3	44.1	54.1
25.8	3.07	476.2	381.40	4.30	3.85	155.3	39.3	46.2
30.4	3.91	653.6	230.00	4.64	2.75	167.1	56.3	36.7
26.6	3.85	567.8	194.10	4.62	2.70	147.5	49.5	31.2
23.7	3.80	504.4	168.60	4.61	2.67	132.6	44.3	27.3
20.9	4.14	486.7	96.67	4.83	2.15	117.4	42.4	18.9
18.2	4.23	426.7	78.39	4.85	2.08	101.0	38.0	15.4
15.7	4.48	375.7	60.01	4.89	1.96	83.9	34.9	11.8
27.6	2.67	349.1	338.70	3.56	3.51	130.9	33.9	46.0
23.7	2.58	296.7	285.60	3.54	3.47	115.2	29.0	39.0
20.0	2.68	263.2	224.50	3.63	3.35	98.1	26.6	30.7
18.1	3.25	278.8	89.10	3.93	2.22	85.7	28.6	17.5
16.1	3.35	252.6	73.87	3.96	2.14	75.3	26.6	14.5
14.2	3.49	227.2	59.97	4.00	2.05	65.2	24.7	11.8
19.0	2.10	152.6	191.90	2.83	3.18	72.8	18.5	28.7
16.1	2.12	133.7	154.80	2.88	3.10	63.0	16.6	23.2

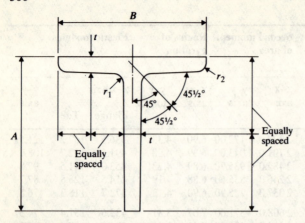

Rolled tees

Designation		Width of section	Depth of section	Thickness	Radius		Area of section
Serial size	Mass per unit length	B	A	t	Root r_1	Toe r_2	
(mm)	(kg/m)	(mm)	(mm)	(mm)	(mm)	(mm)	(cm²)
51 × 51	6.92	50.8	50.8	9.5	6.1	4.3	8.82
	4.76	50.8	50.8	6.4	6.1	4.3	6.06
44 × 44	4.11	44.4	44.4	6.4	5.8	3.8	5.24
	3.14	44.4	44.4	4.8	5.8	3.8	4.00

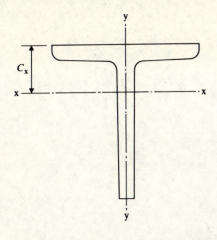

Dimension C_x	Second moment of area		Radius of gyration		Elastic modulus		
	x–x axis	y–y axis	x–x axis	y–y axis	x–x axis		y – y axis
					Flange	Toe	
(cm)	(cm⁴)	(cm⁴)	(cm)	(cm)	(cm³)	(cm³)	(cm³)
1.60	19.56	10.40	1.50	1.09	12.23	5.57	4.10
1.47	14.15	6.66	1.52	1.04	9.63	3.93	2.62
1.32	9.16	4.58	1.32	0.91	6.94	2.95	1.97
1.24	7.07	3.33	1.32	0.89	5.70	2.29	1.48

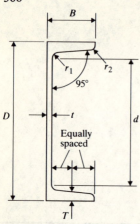

Channels

Designation		Depth of section	Width of section	Thickness		Radius		Depth between fillets	Area of section
Serial size	Mass per unit length	D	B	Web t	Flange T	Root r_1	Toe r_2		
(mm)	(kg/m)	(mm)	(mm)	(mm)	(mm)	(mm)	(mm)	(mm)	(cm²)
432 × 102	65.54	431.8	101.6	12.2	16.8	15.2	4.8	362.4	83.49
381 × 102	55.10	381.0	101.6	10.4	16.3	15.2	4.8	312.4	70.19
305 × 102	46.18	304.8	101.6	10.2	14.8	15.2	4.8	239.2	58.83
305 × 89	41.69	304.8	88.9	10.2	13.7	13.7	3.2	245.4	53.11
254 × 89	35.74	254.0	88.9	9.1	13.6	13.7	3.2	194.7	45.52
254 × 76	28.29	254.0	76.2	8.1	10.9	12.2	3.2	203.8	36.03
229 × 89	32.76	228.6	88.9	8.6	13.3	13.7	3.2	169.8	41.73
229 × 76	26.06	228.6	76.2	7.6	11.2	12.2	3.2	178.0	33.20
203 × 89	29.78	203.2	88.9	8.1	12.9	13.7	3.2	145.2	37.94
203 × 76	23.82	203.2	76.2	7.1	11.2	12.2	3.2	152.5	30.34
178 × 89	26.81	177.8	88.9	7.6	12.3	13.7	3.2	121.0	34.15
178 × 76	20.84	177.8	76.2	6.6	10.3	12.2	3.2	128.8	26.54
152 × 89	23.84	152.4	88.9	7.1	11.6	13.7	3.2	97.0	30.36
152 × 76	17.88	152.4	76.2	6.4	9.0	12.2	2.4	105.9	22.77
127 × 64	14.90	127.0	63.5	6.4	9.2	10.7	2.4	84.0	18.98
102 × 51	10.42	101.6	50.8	6.1	7.6	9.1	2.4	65.7	13.28
76 × 38	6.70	76.2	38.1	5.1	6.8	7.6	2.4	45.8	8.53

Note: Channels ordered to the standard thickness shall be practically accurate in profile. If the web thickness ordered is greater than the standard, the width of the flanges will be increased by the same amount as the increase in web thickness.

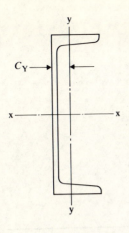

Dimension C_Y	Second moment of area		Radius of gyration		Elastic modulus		Plastic modulus		Ratio $\dfrac{D}{T}$
	x–x axis	y–y axis	x–x axis	y–y axis	x–x axis	y–y axis min.	x–x axis	y–y axis min.	
(cm)	(cm⁴)	(cm⁴)	(cm)	(cm)	(cm³)	(cm³)	(cm³)	(cm³)	
2.32	21 399	628.6	16.00	2.74	991.1	80.14	1 206.6	153.08	25.6
2.52	14 894	579.7	14.60	2.87	781.8	75.86	932.7	144.39	23.3
2.66	8 214	499.5	11.80	2.91	539.0	66.59	638.3	128.06	20.5
2.18	7 061	325.4	11.50	2.48	463.3	48.49	557.1	92.60	22.3
2.42	4 448	302.4	9.88	2.58	350.2	46.70	414.4	89.56	18.7
1.86	3 367	162.6	9.67	2.12	265.1	28.21	317.4	54.14	23.2
2.53	3 387	285.0	9.01	2.61	296.4	44.82	348.4	86.38	17.2
2.00	2 610	158.7	8.87	2.19	228.3	28.22	270.3	54.24	20.5
2.65	2 491	264.4	8.10	2.64	245.2	42.34	286.6	81.62	15.8
2.13	1 950	151.3	8.02	2.23	192.0	27.59	225.2	53.32	18.2
2.76	1 753	241.0	7.16	2.66	197.2	39.29	229.6	75.44	14.5
2.20	1 337	134.0	7.10	2.25	150.4	24.72	175.4	48.07	17.3
2.86	1 166.00	215.10	6.20	2.66	153.00	35.70	177.7	68.12	13.2
2.21	851.50	113.80	6.12	2.24	111.80	21.05	130.0	41.26	16.9
1.94	482.50	67.23	5.04	1.88	75.99	15.25	89.4	29.31	13.8
1.51	207.70	29.10	3.95	1.48	40.89	8.16	48.8	15.71	13.3
1.19	74.14	10.66	2.95	1.12	19.46	4.07	23.4	7.76	11.2

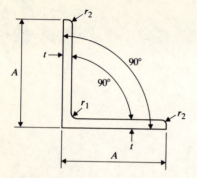

Equal angles

Designation		Leg length A	Thickness t	Radius		Mass/unit length	Area of section
Size	Thickness			Root r_1	Toe r_2		
(mm)	(mm)	(mm)	(mm)	(mm)	(mm)	(kg/m)	(cm²)
25 × 25	3		3			1.11	1.42
25 × 25	4	25	4	3.5	2.4	1.45	1.85
25 × 25	5		5			1.77	2.26
30 × 30	3		3			1.36	1.74
30 × 30	4	30	4	5	2.4	1.78	2.27
30 × 30	5		5			2.18	2.78
40 × 40	4		4			2.42	3.08
40 × 40	5	40	5	6	2.4	2.97	3.79
40 × 40	6		6			3.52	4.48
45 × 45	4		4			2.74	3.49
45 × 45	5	45	5	7	2.4	3.38	4.30
45 × 45	6		6			4.00	5.09
50 × 50	5		5			3.77	4.80
50 × 50	6	50	6	7	2.4	4.47	5.69
50 × 50	8		8			5.82	7.41
60 × 60	5		5			4.57	5.82
60 × 60	6	60	6	8	2.4	5.42	6.91
60 × 60	8		8			7.09	9.03
60 × 60	10		10			8.69	11.10
70 × 70	6		6			6.38	8.13
70 × 70	8	70	8	9	2.4	8.36	10.60
70 × 70	10		10			10.30	13.10
80 × 80	6		6			7.34	9.35
80 × 80	8	80	8	10	4.8	9.63	12.30
80 × 80	10		10			11.90	15.10

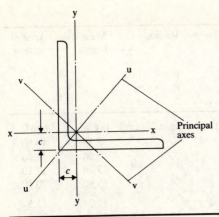

Principal axes

| Distance of centre of gravity c | Moment of inertia | | | Radius of gyration | | | Elastic modulus |
	About x–x, y–y	About u–u	About v–v	About x–x, y–y	About u–u	About v–v	About x–x, y–y
(cm)	(cm^4)	(cm^4)	(cm^4)	(cm)	(cm)	(cm)	(cm^3)
0.72	0.80	1.26	0.33	0.75	0.94	0.48	0.45
0.76	1.01	1.60	0.43	0.74	0.93	0.48	0.63
0.80	1.20	1.89	0.52	0.73	0.91	0.48	0.71
0.84	1.40	2.23	0.58	0.90	1.13	0.58	0.65
0.88	1.80	2.85	0.75	0.89	1.12	0.58	0.85
0.92	2.16	3.41	0.92	0.88	1.11	0.57	1.01
1.12	4.47	7.09	1.85	1.21	1.52	0.78	1.55
1.16	5.43	8.60	2.26	1.20	1.51	0.77	1.91
1.20	6.31	9.98	2.65	1.19	1.49	0.77	2.26
1.23	6.43	10.20	2.67	1.36	1.71	0.87	1.97
1.28	7.84	12.40	3.25	1.35	1.70	0.87	2.43
1.32	9.16	14.50	3.82	1.34	1.69	0.87	2.88
1.40	11.00	17.40	4.54	1.51	1.90	0.97	3.05
1.45	12.80	20.40	5.33	1.50	1.89	0.97	3.61
1.52	16.30	25.70	6.87	1.48	1.86	0.96	4.68
1.64	19.4	30.7	8.02	1.82	2.30	1.17	4.45
1.69	22.8	36.2	9.43	1.82	2.29	1.17	5.29
1.77	29.2	46.2	12.10	1.80	2.26	1.16	6.89
1.85	34.9	55.1	14.80	1.78	2.23	1.16	8.41
1.93	36.9	58.5	15.20	2.13	2.68	1.37	7.27
2.01	47.5	75.3	19.70	2.11	2.66	1.36	9.52
2.09	57.2	90.5	23.90	2.09	2.63	1.35	11.70
2.17	55.8	88.5	23.10	2.44	3.08	1.57	9.57
2.26	72.2	115.0	29.80	2.43	3.06	1.56	12.60
2.34	87.5	139.0	36.30	2.41	3.03	1.55	15.40

372

Equal angles (*continued*)

Designation		Leg length A	Thickness t	Radius		Mass/unit length	Area of section
Size	Thickness			Root r_1	Toe r_2		
(mm)	(mm)	(mm)	(mm)	(mm)	(mm)	(kg/m)	(cm^2)
90 × 90	6		6			8.30	10.60
90 × 90	8	90	8	11	4.8	10.90	13.90
90 × 90	10		10			13.40	17.10
90 × 90	12		12			15.90	20.30
100 × 100	8		8			12.2	15.5
100 × 100	12	100	12	12	4.8	17.8	22.7
100 × 100	15		15			21.9	27.9
120 × 120	8		8			14.7	18.7
120 × 120	10	120	10	13	4.8	18.2	23.2
120 × 120	12		12			21.6	27.5
120 × 120	15		15			26.6	33.9
150 × 150	10		10			23.0	29.3
150 × 150	12		12			27.3	34.8
150 × 150	15	150	15	16	4.8	33.8	43.0
150 × 150	18		18			10.1	51.0
200 × 200	16		16			48.5	61.8
200 × 200	18	200	18	18	4.8	54.2	69.1
200 × 200	20		20			59.9	76.3
200 × 200	24		24			17.1	90.6

Note 1: Some of the thickness given in this table are obtained by raising the rolls. (Practice in this respect is not uniform throughout the industry.) In such cases the legs will be slightly longer and the backs of the toes will be slightly rounded.

Note 2: Angles should be ordered by the flange length and thickness.

Note 3: Finished sections in which the angle between the legs is not less than 89° and not more than 91° shall be deemed to comply with the requirements of this British Standard.

Distance of centre of gravity c	Moment of inertia			Radius of gyration			Elastic modulus
	About x–x, y–y	About u–u	About v–v	About x–x, y–y	About u–u	About v–v	About x–x, y–y
(cm)	(cm⁴)	(cm⁴)	(cm⁴)	(cm)	(cm)	(cm)	(cm³)
2.41	80.3	127.0	33.30	2.76	3.47	1.78	12.20
2.50	104.0	166.0	43.10	2.74	3.45	1.76	16.10
2.58	127.0	201.0	52.80	2.72	3.42	1.76	19.80
2.66	148.0	234.0	62.00	2.70	3.40	1.75	23.30
2.74	145	230	59.8	3.06	3.85	1.96	19.9
2.90	207	328	85.7	3.02	3.80	1.94	29.1
3.02	249	393	104.0	2.98	3.75	1.93	35.6
3.23	255	405	105.0	3.69	4.65	2.37	29.1
3.31	313	497	129.0	3.67	4.63	2.36	36.0
3.40	368	584	151.0	3.65	4.60	2.35	4.27
3.51	445	705	185.0	3.62	4.56	2.33	5.24
4.03	624	991	258.0	4.62	5.82	2.97	56.9
4.12	737	1170	303.0	4.60	5.80	2.95	67.7
4.25	898	1430	370.0	4.57	5.76	2.93	83.5
4.37	1050	1670	435.0	4.54	5.71	2.92	98.7
5.52	2340	3720	959.0	6.16	7.76	3.94	162.0
5.60	2600	4130	1070.0	6.13	7.73	3.93	181.0
5.68	2850	4530	1170.0	6.11	7.70	3.92	199.0
5.84	3330	5280	1380.0	6.06	7.64	3.90	235.0

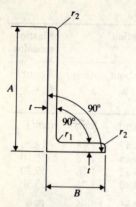

Unequal angles

Designation		Leg lengths		Thick-ness t	Radius		Mass unit length	Area of section	Distance of centre of gravity	
Size	Thick-ness	A	B		Root r_1	Toe r_2			c_x	c_y
(mm)	(mm)	(mm)	(mm)	(mm)	(mm)	(mm)	(kg/m)	(cm²)	(cm)	(cm)
40 × 25	4	40	25	4	4	2.4	1.91	2.46	1.36	0.62
60 × 30	5	60	30	5	6	2.4	3.37	4.29	2.15	0.68
60 × 30	6			6			3.99	5.08	2.20	0.72
65 × 50	5			5			4.35	5.54	1.99	1.25
65 × 50	6	65	50	6	6	2.4	5.16	6.58	2.04	1.29
65 × 50	8			8			6.75	8.60	2.11	1.37
75 × 50	6	75	50	6	7	2.4	5.65	7.19	2.44	1.21
75 × 50	8			8			7.39	9.41	2.52	1.29
80 × 60	6			6			6.37	8.11	2.47	1.48
80 × 60	7	80	60	7	8	4.8	7.36	9.38	2.51	1.52
80 × 60	8			8			8.34	10.60	2.55	1.56
100 × 65	7			7			8.77	11.20	3.23	1.51
100 × 65	8	100	65	8	10	4.8	9.94	12.70	3.27	1.55
100 × 65	10			10			12.30	15.60	3.36	1.63
100 × 75	8			8			10.60	13.50	3.10	1.87
100 × 75	10	100	75	10	10	4.8	13.00	16.60	3.19	1.95
100 × 75	12			12			15.40	19.70	3.27	2.03
125 × 75	8			8			12.20	15.50	4.14	1.68
125 × 75	10	125	75	10	11	4.8	15.00	19.10	4.23	1.76
125 × 75	12			12			17.80	22.70	4.31	1.84

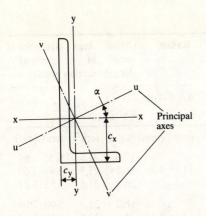

Moment of inertia				Radius of gyration				Angle α	Elastic modulus	
About x–x	About y–y	About u–u	About v–v	About x–x	About y–y	About u–u	About v–v		About x–x	About y–y
(cm^4)	(cm^4)	(cm^4)	(cm^4)	(cm)	(cm)	(cm)	(cm)	(tan α)	(cm^3)	(cm^3)
3.89	1.16	4.35	0.70	1.26	0.69	1.33	0.53	0.380	1.47	0.62
15.6	2.60	16.5	1.69	1.90	0.78	1.96	0.63	0.256	4.04	1.12
18.2	3.02	19.2	1.99	1.89	0.77	1.95	0.63	0.252	4.78	1.32
23.2	11.90	28.8	6.32	2.05	1.47	2.28	1.07	0.577	5.14	3.19
27.2	14.00	33.8	7.43	2.03	1.46	2.27	1.06	0.575	6.10	3.77
34.8	17.70	43.0	9.57	2.01	1.44	2.23	1.05	0.569	7.93	4.89
40.5	14.40	46.6	8.36	2.37	1.42	2.55	1.08	0.435	8.01	3.81
52.0	18.40	59.6	10.80	2.35	1.40	2.52	1.07	0.430	10.40	4.95
51.4	24.80	62.8	13.40	2.52	1.75	2.78	1.29	0.547	9.29	5.49
59.0	28.40	72.0	15.40	2.51	1.74	2.77	1.28	0.546	10.70	6.34
66.3	31.80	80.8	17.30	2.50	1.73	2.76	1.28	0.544	12.20	7.16
113.0	37.60	128.0	22.00	3.17	1.83	3.39	1.40	0.415	16.60	7.53
127.0	42.20	144.0	24.80	3.16	1.83	3.37	1.40	0.414	18.90	8.54
154.0	51.00	175.0	30.10	3.14	1.81	3.35	1.39	0.410	23.20	0.50
133.0	64.10	163.0	34.60	3.14	2.18	3.47	1.60	0.547	19.30	11.4
162.0	77.60	197.0	42.10	3.12	2.16	3.45	1.59	0.544	23.80	14.0
189.0	90.20	230.0	49.50	3.10	2.14	3.42	1.59	0.540	28.00	16.5
247.0	67.50	274.0	41.10	4.00	2.09	4.20	1.63	0.359	29.60	11.6
302.0	82.10	334.0	50.00	3.97	2.07	4.18	1.62	0.356	36.50	14.3
354.0	95.50	391.0	58.80	3.95	2.05	4.15	1.61	0.353	43.20	16.9

Unequal angles (*continued*)

Designation		Leg lengths		Thick-ness t	Radius		Mass unit length	Area of section	Distance of centre of gravity	
Size	Thick-ness	A	B		Root r_1	Toe r_2			c_x	c_y
(mm)	(mm)	(mm)	(mm)	(mm)	(mm)	(mm)	(kg/m)	(cm^2)	(cm)	(cm)
150 × 75 10				10			17.0	21.6	5.32	1.61
150 × 75 12		150	75	12	11	4.8	20.2	25.7	5.41	1.69
150 × 75 15				15			24.8	31.6	5.53	1.81
150 × 90 10				10			18.2	23.2	5.00	2.04
150 × 90 12		150	90	12	12	4.8	21.6	27.5	5.08	2.12
150 × 90 15				15			26.6	33.9	5.21	2.23
200 × 100 10				10			23.0	29.2	6.93	2.01
200 × 100 12		200	100	12	15	4.8	27.3	34.8	7.03	2.10
200 × 100 15				15			33.7	43.0	7.16	2.22
200 × 150 12				12			32.0	40.8	6.08	3.61
200 × 150 15		200	150	15	15	4.8	39.6	50.5	6.21	3.73
200 × 150 18				18			47.1	60.0	6.33	3.85

Moment of inertia				Radius of gyration				Angle α	Elastic modulus	
About x–x	About y–y	About u–u	About v–v	About x–x	About y–y	About u–u	About v–v		About x–x	About y–y
(cm^4)	(cm^4)	(cm^4)	(cm^4)	(cm)	(cm)	(cm)	(cm)	$(\tan \alpha)$	(cm^3)	(cm^3)
501	85.8	532	55.3	4.81	1.99	4.96	1.60	0.261	51.8	14.6
589	99.9	624	64.9	4.79	1.97	4.93	1.59	0.259	61.4	17.2
713	120.0	754	78.8	4.75	1.94	4.88	1.58	0.254	75.3	21.0
533	146.0	591	88.3	4.80	2.51	5.05	1.95	0.360	53.3	21.0
627	171.0	694	104.0	4.77	2.49	5.02	1.94	0.358	63.3	24.8
761	205.0	841	126.0	4.74	2.46	4.98	1.93	0.354	77.7	30.4
1220	210.0	1290	135.0	6.46	2.68	6.65	2.15	0.263	93.2	26.3
1440	247.0	1530	159.0	6.43	2.67	6.63	2.14	0.262	111.0	31.3
1758	299.0	1863	194.0	6.40	2.64	6.58	2.13	0.259	137.0	38.4
1652	803.0	2024	431.0	6.36	4.44	7.04	3.25	0.552	119.0	70.5
2022	979.0	2475	527.0	6.33	4.40	7.00	3.23	0.550	147.0	86.9
2376	1146.0	2902	620.0	6.29	4.37	6.95	3.21	0.548	174.0	103.0

Note 1: Some of the thicknesses given in this table are obtained by raising the rolls. (Practice in this respect is not uniform throughout the industry.) In such cases the legs will be slightly longer and the backs of the toes will be slightly rounded.

Note 2: Angles should be ordered by the flange length and thickness.

Note 3: Finished sections in which the angle between the legs is not less than 89° and not more than 91° shall be deemed to comply with the requirements of this British Standard.

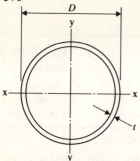

Circular hollow sections

Outside diameter D	Thickness t	Mass M	Sectional area A	Moment of inertia I	Radius of gyration r	Elastic modulus Z
(mm)	(mm)	(kg/m)	(cm²)	(cm⁴)	(cm)	(cm³)
21.3	3.2	1.43	1.82	0.77	0.650	0.72
26.9	3.2	1.87	2.38	1.70	0.846	1.27
33.7	2.6	1.99	2.54	3.09	1.100	1.84
	3.2	2.41	3.07	3.60	1.080	2.14
	4.0	2.93	3.73	4.19	1.060	2.49
42.4	2.6	2.55	3.25	6.46	1.410	3.05
	3.2	3.09	3.94	7.62	1.390	3.59
	4.0	3.79	4.83	8.99	1.360	4.24
48.3	3.2	3.56	4.53	11.60	1.600	4.80
	4.0	4.37	5.57	13.80	1.570	5.70
	5.0	5.34	6.80	16.20	1.540	6.69
60.3	3.2	4.51	5.74	23.50	2.020	7.78
	4.0	5.55	7.07	28.20	2.000	9.34
	5.0	6.82	8.69	33.50	1.960	11.10
76.1	3.2	5.75	7.33	48.80	2.580	12.80
	4.0	7.11	9.06	59.10	2.550	15.50
	5.0	8.77	11.20	70.90	2.520	18.60
88.9	3.2	6.76	8.62	79.20	3.030	17.80
	4.0	8.38	10.70	96.30	3.000	21.70
	5.0	10.30	13.20	116.00	2.970	26.20
114.3	3.6	9.83	12.50	192.00	3.920	33.60
	5.0	13.50	17.20	257.00	3.870	45.00
	6.3	16.80	21.40	313.00	3.820	54.70
139.7	5.0	16.60	21.20	481.00	4.770	68.80
	6.3	20.70	26.40	589.00	4.720	84.30
	8.0	26.00	33.10	720.00	4.660	103.00
	10.0	32.00	40.70	862.00	4.600	123.00

Plastic modulus S (cm³)	Torsional moment of inertia J (cm⁴)	Torsional modulus C (cm³)	Superficial area per m (m²)	Thickness t (mm)	Outside diameter D (mm)
1.06	1.54	1.44	0.067	3.2	21.3
1.81	3.41	2.53	0.085	3.2	26.9
2.52	6.19	3.67	0.106	2.6	33.7
2.99	7.21	4.28	0.106	3.2	
3.55	8.38	4.97	0.106	4.0	
4.12	12.90	6.10	0.133	2.6	42.4
4.93	15.20	7.19	0.133	3.2	
5.92	18.00	8.48	0.133	4.0	
6.52	23.20	9.59	0.152	3.2	48.3
7.87	27.50	11.40	0.152	4.0	
9.42	32.30	13.40	0.152	5.0	
10.40	46.90	15.60	0.189	3.2	60.3
12.70	56.30	18.70	0.189	4.0	
15.30	67.00	22.20	0.189	5.0	
17.00	97.60	25.60	0.239	3.2	76.1
20.80	118.00	31.00	0.239	4.0	
25.30	142.00	37.30	0.239	5.0	
23.50	158.00	35.60	0.279	3.2	88.9
28.90	193.00	43.30	0.279	4.0	
35.20	233.00	52.40	0.279	5.0	
44.10	384.00	67.20	0.359	3.6	114.3
59.80	514.00	89.90	0.359	5.0	
73.60	625.00	109.00	0.359	6.3	
90.80	961.00	138.00	0.439	5.0	139.7
112.00	1 177.00	169.00	0.439	6.3	
139.00	1 441.00	206.00	0.439	8.0	
169.00	1 724.00	247.00	0.439	10.0	

Circular hollow sections (*continued*)

Outside diameter D	Thickness t	Mass M	Sectional area A	Moment of inertia I	Radius of gyration r	Elastic modulus Z
(mm)	(mm)	(kg/m)	(cm^2)	(cm^4)	(cm)	(cm^3)
168.3	5.0	20.10	25.70	856.00	5.780	102.00
	6.3	25.20	32.10	1 053.00	5.730	125.00
	8.0	31.60	40.30	1 297.00	5.670	154.00
	10.0	39.00	49.70	1 564.00	5.610	186.00
193.7	5.4	25.10	31.90	1 417.00	6.660	146.00
	6.3	29.10	37.10	1 630.00	6.630	168.00
	8.0	36.60	46.70	2 016.00	6.570	208.00
	10.0	45.30	57.70	2 442.00	6.500	252.00
	12.5	55.90	71.20	2 934.00	6.420	303.00
	16.0	70.10	89.30	3 554.00	6.310	367.00
219.1	6.3	33.10	42.10	2 386.00	7.530	218.00
	8.0	41.60	53.10	2 960.00	7.470	270.00
	10.0	51.60	65.70	3 598.00	7.400	328.00
	12.5	63.70	81.10	4 345.00	7.320	397.00
	16.0	80.10	102.00	5 297.00	7.200	483.00
	20.0	98.20	125.00	6 261.00	7.070	572.00
244.5	6.3	37.00	47.10	3 346.00	8.420	274.00
	8.0	46.70	59.40	4 160.00	8.370	340.00
	10.0	57.80	73.70	5 073.00	8.300	415.00
	12.5	71.50	91.10	6 147.00	8.210	503.00
	16.0	90.20	115.00	7 533.00	8.100	616.00
	20.0	111.00	141.00	8 957.00	7.970	733.00
273.0	6.3	41.40	52.80	4 696.00	9.430	344.00
	8.0	52.30	66.60	5 852.00	9.370	429.00
	10.0	64.90	82.60	7 154.00	9.310	524.00
	12.5	80.30	102.00	8 697.00	9.220	637.00
	16.0	101.00	129.00	10 710.00	9.100	784.00
	20.0	125.00	159.00	12 800.00	8.970	938.00
	25.0	153.00	195.00	15 130.00	8.810	1 108.00
323.9	8.0	62.30	79.40	9 910.00	11.200	612.00
	10.0	77.40	98.60	12 160.00	11.100	751.00
	12.5	96.00	122.00	14 850.00	11.000	917.00
	16.0	121.00	155.00	18 390.00	10.900	1 136.00
	20.0	150.00	191.00	22 140.00	10.800	1 367.00
	25.0	184.00	235.00	26 400.00	10.600	1 630.00
355.6	8.0	68.60	87.40	13 200.00	12.300	742.00
	10.0	85.20	109.00	16 220.00	12.200	912.00
	12.5	106.00	135.00	19 850.00	12.100	1 117.00
	16.0	154.00	171.00	24 660.00	12.000	1 387.00
	20.0	166.00	211.00	29 790.00	11.900	1 676.00
	25.0	204.00	260.00	35 680.00	11.700	2 007.00

Plastic modulus S (cm^3)	Torsional moment of inertia J (cm^4)	Torsional modulus C (cm^3)	Superficial area per m (m^2)	Thickness t (mm)	Outside diameter D (mm)
133.00	1 712.00	203.00	0.529	5.0	168.3
165.00	2 107.00	250.00	0.529	6.3	
206.00	2 595.00	308.00	0.529	8.0	
251.00	3 128.00	372.00	0.529	10.0	
192.00	2 834.00	293.00	0.609	5.4	193.7
221.00	3 260.00	337.00	0.609	6.3	
276.00	4 031.00	416.00	0.609	8.0	
338.00	4 883.00	504.00	0.609	10.0	
411.00	5 869.00	606.00	0.609	12.5	
507.00	7 109.00	734.00	0.609	16.0	
285.00	4 772.00	436.00	0.688	6.3	219.1
357.00	5 919.00	540.00	0.688	8.0	
438.00	7 197.00	657.00	0.688	10.0	
534.00	8 689.00	793.00	0.688	12.5	
661.00	10 590.00	967.00	0.688	16.0	
795.00	12 520.00	1 143.00	0.688	20.0	
358.00	6 692.00	547.00	0.768	6.3	244.5
448.00	8 321.00	681.00	0.768	8.0	
550.00	10 150.00	830.00	0.768	10.0	
673.00	12 290.00	1 006.00	0.768	12.5	
837.00	15 070.00	1 232.00	0.768	16.0	
1 011.00	17 910.00	1 465.00	0.768	20.0	
448.00	9 392.00	688.00	0.858	6.3	273.0
562.00	11 700.00	857.00	0.858	8.0	
692.00	14 310.00	1 048.00	0.858	10.0	
849.00	17 390.00	1 274.00	0.858	12.5	
1 058.00	21 410.00	1 569.00	0.858	16.0	
1 283.00	25 600.00	1 875.00	0.858	20.0	
1 543.00	30 250.00	2 216.00	0.858	25.0	
799.00	19 820.00	1 224.00	1.020	8.0	323.9
986.00	24 320.00	1 501.00	1.020	10.0	
1 213.00	29 690.00	1 833.00	1.020	12.5	
1 518.00	36 780.00	2 271.00	1.020	16.0	
1 850.00	44 280.00	2 734.00	1.020	20.0	
2 239.00	52 800.00	3 260.00	1.020	25.0	
967.00	26 400.00	1 485.00	1.120	8.0	355.6
1 195.00	32 450.00	1 825.00	1.120	10.0	
1 472.00	39 700.00	2 233.00	1.120	12.5	
1 847.00	49 330.00	2 774.00	1.120	16.0	
2 255.00	59 580.00	3 351.00	1.120	20.0	
2 738.00	71 350.00	4 013.00	1.120	25.0	

Circular hollow sections (*continued*)

Outside diameter D	Thickness t	Mass M	Sectional area A	Moment of inertia I	Radius of gyration r	Elastic modulus Z
(mm)	(mm)	(kg/m)	(cm^2)	(cm^4)	(cm)	(cm^3)
406.4	10.0	97.80	125.00	24 480.00	14.000	1 205.00
	12.5	121.00	155.00	30 030.00	13.900	1 478.00
	16.0	154.00	196.00	37 450.00	13.800	1 843.00
	20.0	191.00	243.00	45 430.00	13.700	2 236.00
	25.0	235.00	300.00	54 700.00	13.500	2 692.00
	32.0	295.00	376.00	66 430.00	13.300	3 269.00
457.0	10.0	110.00	140.00	35 090.00	15.800	1 536.00
	12.5	137.00	175.00	43 140.00	15.700	1 888.00
	16.0	174.00	222.00	53 960.00	15.600	2 361.00
	20.0	216.00	275.00	65 680.00	15.500	2 874.00
	25.0	266.00	339.00	79 420.00	15.300	3 476.00
	32.0	335.00	427.00	97 010.00	15.100	4 246.00
	40.0	411.00	524.00	114 900.00	14.800	5 031.00

Plastic modulus S (cm³)	Torsional moment of inertia J (cm⁴)	Torsional modulus C (cm³)	Superficial area per m (m²)	Thickness t (mm)	Outside diameter D (mm)
1 572.00	48 950.00	2 409.00	1.280	10.0	406.4
1 940.00	60 060.00	2 956.00	1.280	12.5	
2 440.00	74 900.00	3 686.00	1.280	16.0	
2 989.00	90 860.00	4 472.00	1.280	20.0	
3 642.00	109 400.00	5 384.00	1.280	25.0	
4 497.00	132 900.00	6 539.00	1.280	32.0	
1 998.00	70 180.00	3 071.00	1.440	10.0	457.0
2 470.00	86 290.00	3 776.00	1.440	12.5	
3 113.00	107 900.00	4 723.00	1.440	16.0	
3 822.00	131 400.00	5 749.00	1.440	20.0	
4 671.00	158 800.00	6 951.00	1.440	25.0	
5 791.00	194 000.00	8 491.00	1.440	32.0	
6 977.00	229 900.00	10 060.00	1.440	40.0	

384

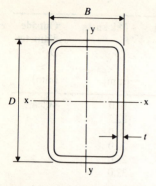

Rectangular hollow sections

Size $D \times B$	Thickness t	Mass M	Sectional area A	Moment of inertia I		Radius of gyration r r	
				x–x (cm⁴)	y–y (cm⁴)	x–x (cm)	y–y (cm)
(mm)	(mm)	(kg/m)	(cm²)	x–x (cm^4)	y–y (cm^4)	x–x (cm)	y–y (cm)
50 × 30	2.6	3.03	3.86	12.4	5.45	1.79	1.19
	3.2	3.66	4.66	14.5	6.31	1.77	1.16
60 × 40	3.2	4.66	5.94	28.3	14.80	2.18	1.58
	4.0	5.72	7.28	33.6	17.30	2.15	1.54
80 × 40	3.2	5.67	7.22	58.1	19.10	2.84	1.63
	4.0	6.97	8.88	69.6	22.60	2.80	1.59
90 × 50	3.6	7.46	9.50	99.8	39.10	3.24	2.03
	5.0	10.10	12.90	130.0	50.00	3.18	1.97
100 × 50	3.2	7.18	9.14	117.0	39.10	3.58	2.07
	4.0	8.86	11.30	142.0	46.70	3.55	2.03
	5.0	10.90	13.90	170.0	55.10	3.50	1.99
100 × 60	3.6	8.59	10.90	147.0	65.40	3.66	2.45
	5.0	11.70	14.90	192.0	84.70	3.60	2.39
	6.3	14.40	18.40	230.0	99.90	3.54	2.33
120 × 60	3.6	9.72	12.40	230.0	76.90	4.31	2.49
	5.0	13.30	16.90	304.0	99.90	4.24	2.43
	6.3	16.40	20.90	366.0	118.00	4.18	2.38
120 × 80	5.0	14.80	18.90	370.0	195.00	4.43	3.21
	6.3	18.40	23.40	447.0	234.00	4.37	3.16
	8.0	22.90	29.10	537.0	278.00	4.29	3.09
	10.0	27.90	35.50	628.0	320.00	4.20	3.00
150 × 100	5.0	18.70	23.90	747.0	396.00	5.59	4.07
	6.3	23.30	29.70	910.0	479.00	5.53	4.02
	8.0	29.10	37.10	1106.0	577.00	5.46	3.94
	10.0	35.70	45.50	1312.0	678.00	5.37	3.86

Elastic modulus Z		Plastic modulus S		Torsional moment of inertia J	Torsional modulus C	Super-ficial area per m	Thick-ness t	Size D × B
x–x (cm³)	y–y (cm³)	x–x (cm³)	y–y (cm³)	(cm⁴)	(cm³)	(m²)	(mm)	(mm)
4.96	3.63	6.21	4.30	12.1	5.90	0.154	2.6	50 × 30
5.82	4.21	7.39	5.08	14.2	6.81	0.153	3.2	
9.44	7.39	11.70	8.75	30.8	11.80	0.193	3.2	60 × 40
11.20	8.67	14.10	10.50	36.6	13.70	0.191	4.0	
14.50	9.56	18.30	11.10	46.1	16.10	0.233	3.2	80 × 40
17.40	11.30	22.20	13.40	55.1	18.90	0.231	4.0	
22.20	15.60	27.60	18.10	89.3	25.90	0.272	3.6	90 × 50
28.90	20.00	36.60	23.90	116.0	32.90	0.269	5.0	
23.50	15.60	29.20	17.90	93.3	26.40	0.293	3.2	100 × 50
28.40	18.70	35.70	21.70	113.0	31.40	0.291	4.0	
34.00	22.00	43.30	26.10	135.0	37.00	0.289	5.0	
29.30	21.80	36.00	25.10	142.0	35.60	0.312	3.6	100 × 60
38.50	28.20	48.10	33.30	187.0	45.90	0.309	5.0	
46.00	33.30	58.40	40.20	224.0	53.90	0.306	6.3	
38.30	25.60	47.60	29.20	183.0	43.30	0.352	3.6	120 × 60
50.70	33.30	63.90	38.80	242.0	56.00	0.349	5.0	
61.00	39.40	78.00	46.90	290.0	66.00	0.346	6.3	
61.70	48.80	75.40	56.70	401.0	77.90	0.389	5.0	120 × 80
74.60	58.40	92.30	69.10	486.0	93.00	0.386	6.3	
89.50	69.40	113.00	83.90	586.0	110.00	0.383	8.0	
105.00	80.00	134.00	90.40	688.0	126.00	0.379	10.0	
99.50	79.10	121.00	90.80	806.0	127.00	0.489	5.0	150 × 100
121.00	95.90	148.00	111.00	985.0	153.00	0.486	6.3	
147.00	115.00	183.00	137.00	1202.0	184.00	0.483	8.0	
175.00	136.00	220.00	164.00	1431.0	215.00	0.479	10.0	

Rectangular hollow sections (*continued*)

Size $D \times B$	Thickness t	Mass M	Sectional area A	Moment of inertia l		Radius of gyration r	
				x–x (cm⁴)	y–y (cm⁴)	x–x (cm)	y–y (cm)
(mm)	(mm)	(kg/m)	(cm²)	x–x (cm^4)	y–y (cm^4)	x–x (cm)	y–y (cm)
160×80	5.0	18.0	22.9	753	251	5.74	3.31
	6.3	22.3	28.5	917	302	5.68	3.26
	8.0	27.9	35.5	1113	361	5.60	3.19
	10.0	34.2	43.5	1318	419	5.50	3.10
200×100	5.0	22.7	28.9	1509	509	7.23	4.20
	6.3	28.3	36.0	1851	618	7.17	4.14
	8.0	35.4	45.1	2269	747	7.09	4.07
	10.0	43.6	55.5	2718	881	7.00	3.98
	12.5	53.4	68.0	3218	1022	6.88	3.88
	16.0	66.4	84.5	3808	1175	6.71	3.73
250×150	6.3	38.2	48.6	4178	1886	9.27	6.23
	8.0	48.0	61.1	5167	2317	9.19	6.16
	10.0	59.3	75.5	6259	2784	9.10	6.07
	12.5	73.0	93.0	7518	3310	8.99	5.97
	16.0	91.5	117.0	9089	3943	8.83	5.82
300×200	6.3	48.1	61.2	7880	4216	11.30	8.30
	8.0	60.5	77.1	9798	5219	11.30	8.23
	10.0	75.0	95.5	11940	6331	11.20	8.14
	12.5	92.6	118.0	14460	7619	11.10	8.04
	16.0	117.0	149.0	17700	9239	10.90	7.89
400×200	10.0	90.7	116.0	24140	8138	14.50	8.39
	12.5	112.0	143.0	29410	9820	14.30	8.29
	16.0	142.0	181.0	36300	11950	14.20	8.14
450×250	10.0	106.0	136.0	37180	14900	16.60	10.50
	12.5	132.0	168.0	45470	18100	16.50	10.40
	16.0	167.0	213.0	56420	22250	16.30	10.20

Elastic modulus Z		Plastic modulus S		Torsional moment inertia	Torsional of modulus C	Superficial area per m	Thickness t	Size $D \times B$
x–x (cm³)	y–y (cm³)	x–x (cm³)	y–y (cm³)	J (cm⁴)	(cm³)	(m²)	(mm)	(mm)
94.1	62.8	117	71.7	599	106	0.469	5.0	160 × 80
115.0	75.6	144	87.7	729	127	0.466	6.3	
139.0	90.2	177	107.0	882	151	0.463	8.0	
165.0	105.0	213	127.0	1041	175	0.459	10.0	
151.0	102.0	186	115.0	1202	172	0.589	5.0	200 × 100
185.0	124.0	231	141.0	1473	208	0.586	6.3	
227.0	149.0	286	174.0	1802	251	0.583	8.0	
272.0	176.0	346	209.0	2154	296	0.579	10.0	
322.0	204.0	417	249.0	2541	342	0.573	12.5	
381.0	235.0	505	297.0	2988	393	0.566	16.0	
334.0	252.0	405	284.0	4049	413	0.786	6.3	250 × 150
413.0	309.0	505	353.0	5014	506	0.783	8.0	
501.0	371.0	618	430.0	6082	606	0.779	10.0	
601.0	441.0	751	520.0	7317	717	0.773	12.5	
727.0	526.0	924	635.0	8863	851	0.766	16.0	
525.0	422.0	627	475.0	8468	681	0.986	6.3	300 × 200
653.0	522.0	785	593.0	10 549	840	0.983	8.0	
796.0	633.0	964	726.0	12 890	1016	0.979	10.0	
964.0	762.0	1179	886.0	15 654	1217	0.973	12.5	
1180.0	924.0	1462	1094.0	19 227	1469	0.966	16.0	
1207.0	814.0	1492	916.0	19 236	1377	1.180	10.0	400 × 200
1471.0	982.0	1831	1120.0	23 408	1657	1.170	12.5	
1815.0	1195.0	2285	1388.0	28 835	2011	1.170	16.0	
1653.0	1192.0	2013	1338.0	33 247	1986	1.380	10.0	450 × 250
2021.0	1448.0	2478	1642.0	40 668	2407	1.370	12.5	
2508.0	1780.0	3103	2047.0	50 478	2948	1.370	16.0	

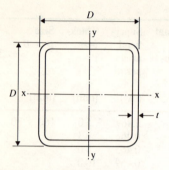

Square hollow sections

Size $D \times D$	Thickness t	Mass M	Sectional area A	Moment of inertia l	Radius of gyration r	Elastic modulus Z
(mm)	(mm)	(kg/m)	(cm²)	(cm⁴)	(cm)	(cm³)
20 × 20	2.0	1.12	1.42	0.76	0.73	0.76
	2.6	1.39	1.78	0.88	0.70	0.88
30 × 30	2.6	2.21	2.82	3.49	1.11	2.33
	3.2	2.65	3.38	4.00	1.09	2.67
40 × 40	2.6	3.03	3.86	8.94	1.52	4.47
	3.2	3.66	4.66	10.40	1.50	5.22
	4.0	4.46	5.68	12.10	1.46	6.07
50 × 50	3.2	4.66	5.94	21.60	1.91	8.62
	4.0	5.72	7.28	25.50	1.87	10.20
	5.0	6.97	8.88	29.60	1.83	11.90
60 × 60	3.2	5.67	7.22	38.70	2.31	12.90
	4.0	6.97	8.88	46.10	2.28	15.40
	5.0	8.54	10.90	54.40	2.24	18.10
70 × 70	3.6	7.46	9.50	69.50	2.70	19.90
	5.0	10.10	12.90	90.10	2.64	25.70
80 × 80	3.6	8.59	10.90	106.00	3.11	26.50
	5.0	11.70	14.90	139.00	3.05	34.70
	6.3	14.40	18.40	165.00	3.00	41.30
90 × 90	3.6	9.72	12.40	154.00	3.52	34.10
	5.0	13.30	16.90	202.00	3.46	45.00
	6.3	16.40	20.90	242.00	3.41	53.90
100 × 100	4.0	12.00	15.30	234.00	3.91	46.80
	5.0	14.80	18.90	283.00	3.87	56.60
	6.3	18.40	23.40	341.00	3.81	68.20
	8.0	22.90	29.10	408.00	3.74	81.50
	10.0	27.90	35.50	474.00	3.65	94.90

Plastic modulus S (cm^3)	Torsional moment of inertia J (cm^4)	Torsional modulus C (cm^3)	Superficial area per m (m^2)	Thickness t (mm)	Size $D \times D$ (mm)
0.95	1.22	1.07	0.076	2.0	20 × 20
1.15	1.44	1.23	0.074	2.6	
2.88	5.56	3.30	0.114	2.6	30 × 30
3.37	6.44	3.75	0.113	3.2	
5.39	14.00	6.41	0.154	2.6	40 × 40
6.40	16.50	7.43	0.153	3.2	
7.61	19.50	8.56	0.151	4.0	
10.40	33.80	12.40	0.193	3.2	50 × 50
12.50	40.40	14.50	0.191	4.0	
14.90	47.60	16.70	0.189	5.0	
15.30	60.10	18.60	0.233	3.2	60 × 60
18.60	72.40	22.00	0.231	4.0	
22.30	86.30	25.80	0.229	5.0	
23.60	108.00	28.70	0.272	3.6	70 × 70
31.20	142.00	36.80	0.269	5.0	
31.30	164.00	38.50	0.312	3.6	80 × 80
41.70	217.00	49.80	0.309	5.0	
50.50	261.00	58.80	0.306	6.3	
40.00	237.00	49.70	0.352	3.6	90 × 90
53.60	315.00	64.90	0.349	5.0	
65.30	381.00	77.10	0.346	6.3	
54.90	361.00	68.20	0.391	4.0	100 × 100
67.10	439.00	81.90	0.389	5.0	
82.00	533.00	97.90	0.386	6.3	
99.90	646.00	116.00	0.383	8.0	
119.00	761.00	134.00	0.379	10.0	

Square hollow sections (*continued*)

Size $D \times D$	Thickness t	Mass M	Sectional area A	Moment of inertia l	Radius of gyration r	Elastic modulus Z
(mm)	(mm)	(kg/m)	(cm²)	(cm⁴)	(cm)	(cm³)
120 × 120	5.0	18.00	22.90	503.00	4.69	83.80
	6.3	22.30	28.50	610.00	4.63	102.00
	8.0	27.90	35.50	738.00	4.56	123.00
	10.0	34.20	43.50	870.00	4.47	145.00
150 × 150	5.0	22.70	28.90	1009.00	5.91	135.00
	6.3	28.30	36.00	1236.00	5.86	165.00
	8.0	35.40	45.10	1510.00	5.78	201.00
	10.0	43.60	55.50	1803.00	5.70	240.00
	12.5	53.40	68.00	2125.00	5.59	283.00
	16.0	66.40	84.50	2500.00	5.44	333.00
180 × 180	6.3	34.2	43.6	2186	7.08	243
	8.0	43.0	54.7	2689	7.01	299
	10.0	53.0	67.5	3237	6.92	360
	12.5	65.2	83.0	3856	6.92	428
	16.0	81.4	104.0	4607	6.66	512
200 × 200	6.3	38.2	48.6	3033	7.90	303
	8.0	48.0	61.1	3744	7.83	374
	10.0	59.3	75.5	4525	7.74	452
	12.5	73.0	93.0	5419	7.63	542
	16.0	91.5	117.0	6524	7.48	652
250 × 250	6.3	48.1	61.2	6049	9.94	484
	8.0	60.5	77.1	7510	9.87	601
	10.0	75.0	95.5	9141	9.78	731
	12.5	92.6	118.0	11050	9.68	884
	16.0	117.0	149.0	13480	9.53	1078
300 × 300	10.0	90.7	116.0	16150	11.80	1077
	12.5	112.0	143.0	19630	11.70	1309
	16.0	142.0	181.0	24160	11.60	1610
350 × 350	10.0	106.0	136.0	26050	13.90	1489
	12.5	132.0	168.0	31810	13.80	1817
	16.0	167.0	213.0	39370	13.60	2250
400 × 400	10.0	122.0	156.0	39350	15.90	1968
	12.5	152.0	193.0	48190	15.80	2409

Plastic modulus S	Torsional moment of inertia J	Torsional modulus C	Superficial area per m	Thickness t	Size $D \times D$
(cm^3)	(cm^4)	(cm^3)	(m^2)	(mm)	(mm)
98.40	775.00	122.00	0.469	5.0	120 × 120
121.00	949.00	147.00	0.466	6.3	
149.00	1159.00	176.00	0.463	8.0	
178.00	1381.00	206.00	0.459	10.0	
157.00	1548.00	197.00	0.589	5.0	150 × 150
194.00	1907.00	240.00	0.586	6.3	
240.00	2348.00	291.00	0.583	8.0	
290.00	2829.00	345.00	0.579	10.0	
348.00	3372.00	403.00	0.573	12.5	
421.00	4029.00	468.00	0.566	16.0	
283	3357	355	0.706	6.3	180 × 180
352	4156	434	0.703	8.0	
429	5041	519	0.699	10.0	
519	6062	613	0.693	12.5	
634	7339	725	0.686	16.0	
353	4647	444	0.786	6.3	200 × 200
439	5770	545	0.783	8.0	
536	7020	655	0.799	10.0	
651	8479	779	0.773	12.5	
799	10 330	929	0.766	16.0	
559	9228	712	0.986	6.3	250 × 250
699	11 511	880	0.983	8.0	
858	14 086	1065	0.979	10.0	
1048	17 139	1279	0.973	12.5	
1298	21 109	1548	0.966	16.0	
1254	24 776	1575	1.180	10.0	300 × 300
1538	30 290	1905	1.170	12.5	
1916	37 566	2327	1.170	16.0	
1725	39 840	2186	1.380	10.0	350 × 350
2122	48 869	2655	1.370	12.5	
2655	60 901	3265	1.370	16.0	
2272	60 028	2896	1.580	10.0	400 × 400
2800	73 815	3530	1.570	12.5	

Appendix C

Useful computer programs

This appendix lists some of the computer programs used by the author during the normal course of his work. In all cases they have been written and tested by him, for use on the Commodore PET micro-computer with a storage capacity of not less than 8K bytes. In fact the listings given are those of the versions which are stored on disk for use on the newer version of the PET which has a storage of 32K bytes. Whilst he feels confident of the accuracy of the programs, the author cannot guarantee them in any way, but will be only too pleased to assist with any problems which may be encountered. The appendix is ordered as follows.

Each program is documented in four stages:

1. The listing is given as it is produced by the printer.
2. A brief explanation is given of the program contents; this is meant to give an experienced programmer an idea of which parts of the program can be omitted without endangering the operation of the program.
3. A sample problem is set.
4. Photographic plates of the actual display during the program's operation are presented as it is used to solve the problem which has been set.

The programs which are here presented are as follows:
R/C Beam check-design (cf. Ch. 2 p. 22)
Flanged R/C Beam check-design

C.1 R/C Beam check-design
C.1.1 Program listing

R/C BEAM - CHECK/DESIGN

READY.

```
5 PRINT"?":GOSUB1000
10 PRINT"ENTER CONCRETE AND STEEL STRENGTHS":INPUTFCU,FY
30 PRINT"ENTER DIMENSIONS B,D & D'":INPUTBR,DE,Y1:K=Y1/DE:C=2300+FY:X2=7*K/3
43 X1=16100*K/(16100-16*FY):X3=805/(1265+FY):X4=2300/(16*FY/7+2300)
46 K1=.45*(1-SQR(FCU)/52.5):K2=(2+(2-SQR(FCU)/17.5)↑2)/4/(3-SQR(FCU)/17.5)
47 PRINT"ENTER 1 FOR DESIGN; 2 FOR CHECK";:INPUTFL:IFFL=1THEN65
50 PRINT"ENTER TENSION & COMPRESSION STEEL AREAS":INPUTAT,AC
60 RT=AT/(BR*DE):RC=AC/(BR*DE):GOTO160
65 PRINT"ENTER THE DESIGN MOMENT":INPUTMU1:MU=MU1*1E+06
66 IFMU1<(DE↑2*K1*FCU*BR)/4/K2THEN72
67 PRINT"THIS SECTION MUST CONTAIN COMPRESSION":PRINT"STEEL."
68 PRINT"NEUTRAL AXIS <=".6*DE"MM":GOTO74
72 X=(DE-SQR(DE↑2-4*K2*MU1/K1/FCU/BR))/2/K2:IFX>DE/2THEN67
73 PRINT"THE NEUTRAL AXIS MUST BE <="X"MM"
74 PRINT"ENTER THE DEPTH TO THE NEUTRAL AXIS":INPUTXD:XD=XD/DE
75 MS=K1*FCU*BR*XD*DE*(DE-K2*XD*DE):MD=MU1-MS:IFXD<=X2THEN95
90 GOSUB720
91 GOTO110
95 IFXD<=X1THEN105
100 GOSUB710
101 GOTO110
105 GOSUB700
110 IFXD<=X4THEN120
115 GOSUB730
```

```
116 GOTO135
120 IFXD<=X3THEN130
125 GOSUB740
126 GOTO135
130 GOSUB750
135 AC=MD/F1/(DE-Y1):AT=(K1*FCU*BR*XD*DE+AC*F1)/F2:GOTO600
160 H=RT/1.15/K1:J=700*RC/K1:M=.5*(H*FY-J)/FCU
170 XD=M+SQR(M↑2+J*K/FCU):IFXD>X1GOTO220
190 GOSUB700
200 GOSUB750
210 GOTO600
220 B=FCU/FY:L=J*K/C:I=H/2-1150*RC/C/K1:XD=(I+SQR(I↑2+B*L))/B
240 IFXD>X20RXD>X3GOTO280
250 GOSUB710
260 GOSUB750
270 GOTO600
280 IFX2>X3GOTO340
290 F=2000*FY/K1:XD=H/B-F*RC/(C*FCU):D=B*C:E=700*RT/K1
300 IFXD>X3GOTO400
310 GOSUB720
320 GOSUB750
330 GOTO600
340 N=1150*RT-2875*RC:XD=(N+SQR(N↑2+B*C*(E+J*K)))/B/C
360 IFXD>X2GOTO400
370 GOSUB710
380 GOSUB740
390 GOTO600
400 A=(450*RT-1000*RC)/K1:XD=(A+SQR(A↑2+D*E))/D
```

```
420 IFXD>X4GOTO460
430 GOSUB720
440 GOSUB740
450 GOTO600
460 G=-(E+F*RC/C)*0.5/FCU:XD=G+SQR(G↑2+E/FCU)
480 GOSUB720
490 GOSUB730
600 PRINT"NEUTRAL AXIS DEPTH="XD*DE"MM":PRINT:PRINT:IFXD>=KTHEN603
601 PRINT"   ";:FORI=1TO30:PRINTTAB(13)" WARNING ";:PRINTTAB(13)" WARNING ":PRINT:NEXTI
602 PRINT"   THE NEUTRAL AXIS IS ABOVE THE LEVEL":PRINT"OF THE COMPRESSION STEE
L.";:GOTO650
603 IFAC<1THEN605
604 PRINT"COMPRESSION STEEL STRESS="F1"     N/SQ.MM."
605 PRINT"   TENSION STEEL STRESS="F2"     N/SQ.MM.";:PRINT:PRINT:IFAC<1THEN60
0
607 PRINT"COMPRESSION STEEL AREA ="AC"MM↑2"
608 PRINT"   TENSION STEEL AREA ="AT"MM↑2"
620 MU1=K1*FCU*BR*XD*DE*DE*(1-K2*XD)+F1*AC*(DE-Y1)
630 PRINT"ULTIMATE MOMENT OF RESISTANCE=     "MU1/1E+06"KNM"
635 IFFL<>1THEN650
640 PRINT"ANOTHER NEUTRAL AXIS DEPTH - YES OR NO":INPUTF#:IFF#="NO"THEN650
645 GOTO74
650 END
700 F1=700*(1-K/XD):RETURN
710 F1=(2300-700*K/XD)*FY/C:RETURN
720 F1=2000*FY/C:RETURN
730 F2=700*(1/XD-1):RETURN
740 F2=(700/XD+900)*FY/C:RETURN
```

396

```
750 F2=FY/1.15:RETURN
1000 PRINT"THIS PROGRAM WILL
1010 PRINT"DESIGN OR CHECK A
1020 PRINT"REINFORCED BEAM
1030 PRINT"WITH OR WITHOUT
1040 PRINT"COMPRESSION STEEL.
1050 PRINT"
1060 PRINT"
1070 PRINT"YOU WILL NEED TO
1080 PRINT"ENTER THE VALUES
1090 PRINT"OF STRENGTH IN
1100 PRINT"N/SQ.MM., AND THE
1110 PRINT"DIMENSIONS IN MM.
1120 PRINT"STEEL AREAS ARE
1130 PRINT"ENTERED IN SQ.MM.
1140 PRINT"AND MOMENTS IN
1160 PRINT"KNM.
1170 PRINT"
1180 RETURN
READY.
```

C.1.2 Explanation

Line 5

The first program line consists of two program statements separated by a colon. In PET Basic a colon signifies a statement separator. It allows more than one Basic command to reside at one line number thus saving on the storage required to accommodate the program.

The first statement is a PRINT statement, and the symbol inside the inverted commas tells the computer to clear the entire screen and return the cursor to the top L.H. corner of the screen before any information is printed.

The second statement directs the execution of the program to line 1000 and following. These lines completed the program RETURNS from line 1180 to line 10

Lines 1000–1180

This subroutine displays the program explanation on the screen and ensures that someone using the program for the first time knows what the program expects of him (or her). These lines are not germane to the execution of the calculation routines.

Lines 10–75

This is the data entry section of the program.

Line 90–490

This is the calculation section of the program.

Lines 600–750

This is the results output section of the program.

Some of the variables used are:

FCU $= f_{cu}$ the characteristic strength of concrete
FY $= f_y$ the characteristic strength of steel
F1 $= f_{yd1}$ the actual stress in the compression steel
F2 $= f_{yd2}$ the actual stress in the tension steel
AC $=$ the area of the compression steel
AT $=$ the area of the tension steel
XD $= x/d$ the neutral axis depth factor
MU1 $=$ the ultimate moment of resistance of the section

C.1.3 Sample problem

Design the steel needed in a section 250 mm wide by 500 mm deep, to carry a moment of 300 kNm if the concrete has a characteristic strength, $f_{cu} = 30 \, \text{N/mm}^2$; the steel is of $f_y = 410 \, \text{N/mm}^2$ and the depth to both steels from the concrete surface is 50 mm (i.e. to the centre of both steels)

C.1.4 Plates

Note: This program is based on the 'parabolic – rectangular' concrete stress block.

398

Load the program from cassette or disk, then:

Step I: Type RUN and press the 'RETURN' key.

Step II: The screen now displays the information shown in Fig. C.1 The computer is asking the operator to enter the values for f_{cu} and f_y for the concrete and steel being used. These are entered in N/mm² and are separated by a COMMA. It is most important that nothing other than a comma is used, otherwise the computer will misinterpret the information.

Step III: Type in 30, 410 and press the 'RETURN' key (Fig. C.2).

Step IV: This is followed by the information requested by the computer as shown in Fig. C.3. Note that the computer checks the necessity of compression steel. In this case it is necessary, and the computer tells the operator that the neutral axis must be less than 270 mm which corresponds to 0.6 times the effective depth. In this particular case the operator chooses the maximum neutral axis depth and enters 270 followed by the 'RETURN' key.

Step V: Immediately the 'RETURN' key is pressed the screen clears and the results shown in Fig. C.4 appear. At this stage, a new value for neutral axis may be tried, but this time the operator decides not to do so.

Step VI: Once the actual values of steel area are chosen, the designer may wish to check that they are satisfactory. Start the program again by typing RUN and pressing the 'RETURN' key. This time the value '2' is entered to obtain the check routine (Fig. C.5).

Step VII: Enter the values of steel areas as requested. Immediately the 'RETURN' key is pressed, the results of Fig. C.6 are displayed.

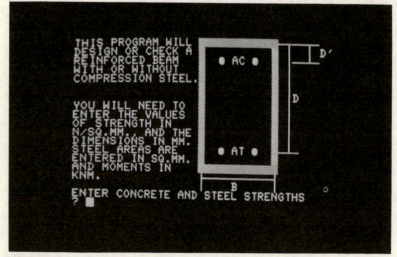

Fig. C.1

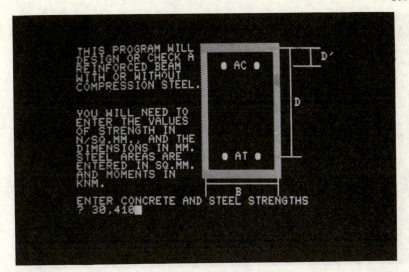

Fig. C.2

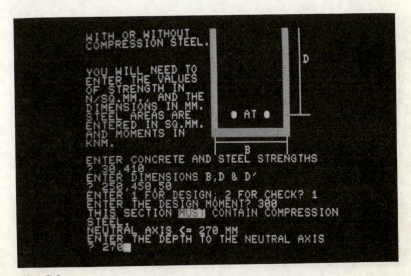

Fig. C.3

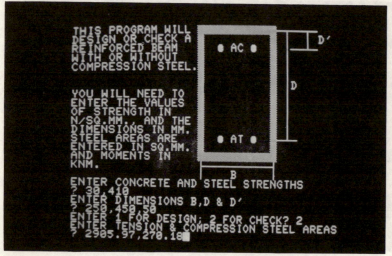

```
NEUTRAL AXIS DEPTH= 270 MM

COMPRESSION STEEL STRESS= 302.583026
                              N/SQ.MM.
    TENSION STEEL STRESS= 312.669127
                              N/SQ.MM.


COMPRESSION STEEL AREA = 265.027565 MM↑2
    TENSION STEEL AREA = 2866.84479 MM↑2


ULTIMATE MOMENT OF RESISTANCE=

                         300 KNM
ANOTHER NEUTRAL AXIS DEPTH - YES OR NO
? NO
```

Fig. C.4

```
THIS PROGRAM WILL
DESIGN OR CHECK A
REINFORCED BEAM
WITH OR WITHOUT
COMPRESSION STEEL.

YOU WILL NEED TO
ENTER THE VALUES
OF STRENGTH IN
N/SQ.MM., AND THE
DIMENSIONS IN MM.
STEEL AREAS ARE
ENTERED IN SQ.MM.
AND MOMENTS IN
KNM.

ENTER CONCRETE AND STEEL STRENGTHS
? 30,410
ENTER DIMENSIONS B,D & D'
? 250,450,50
ENTER 1 FOR DESIGN; 2 FOR CHECK? 2
ENTER TENSION & COMPRESSION STEEL AREAS
? 2905.97,270.18
```

Fig. C.5

```
NEUTRAL AXIS DEPTH= 272.175113 MM

COMPRESSION STEEL STRESS= 302.583026
                                N/SQ.MM.

   TENSION STEEL STRESS= 311.258556
                                N/SQ.MM.


COMPRESSION STEEL AREA = 270.18 MM↑2
   TENSION STEEL AREA = 2905.97 MM↑2

ULTIMATE MOMENT OF RESISTANCE=
                        301.975117 KNM


READY.
```

Fig. C.6

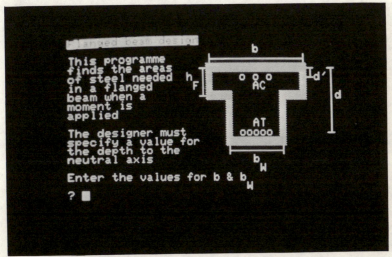

Fig. C.7

402

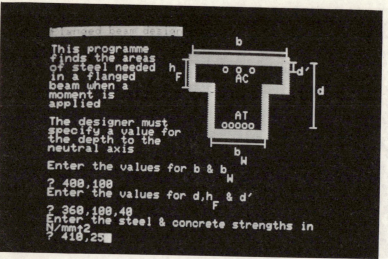

Fig. C.8

Fig. C.9

C.2 Flanged R/C Beam check design

C.2.1 Program listing (see page 404 and following pages for computer print outs)

C.2.2 Explanation
Line 4
This POKE statement puts the computer into LOWER case mode. Normally, the computer switches on in the graphics mode, viz. when the keys are pressed together with the 'SHIFT' key, special graphics symbols are output rather than lower case letters. This statement allows the lower case symbols to be used. Therefore, some of the listing will seem strange since the printer is not affected by the statement and simply lists in the graphics mode.

Lines 5–230
These are the program explanation lines

Lines 240–375
These are the data input lines

Lines 400–490
These are the results output lines

Lines 500–
These are the calculation routines

C.2.3 Sample problem
Design the steel needed in a section which has a flange of 400 by 100 mm deep and a web of size 100 mm thick and 300 mm deep measured from the underside of the flange. The section is required to take a moment of 150 kNm.

$f_{cu} = 25 \, \text{N/mm}^2$; $f_y = 410 \, \text{N/mm}^2$ cover to steel = 25 mm

C.2.4 Plates
Note: This program is based on the 'rectangular' concrete stress block.

Step I: Load the program and type RUN and press the 'RETURN' key. The information shown in Fig. C.7 is then displayed.

Step II: Enter the geometric values b and b_w d, h_F and d' as requested (Fig. C.8).

Step III: Enter the steel and concrete strengths f_{cu} and f_y in N/mm^2 (Fig. C.8).

Step IV: (Not shown) Enter '1' for DESIGN.

Step V: Not shown)
Enter 150 as the required moment in kNm.

Step VI: (Not shown)
Enter 173 as the chosen depth to the neutral axis.

Step VII: Immediately the 'RETURN' key is hit, the results shown in Fig. C.9 are displayed.

Step VIII: As in the last example, the designer may now proceed by running the check routine.

FLANGED R/C BEAM - CHECK/DESIGN

READY.

```
4 POKE59468,14
5 PRINT"?"
10 PRINT""
20 PRINT""
30 PRINT""
40 PRINT""
50 PRINT""
60 PRINT""
70 PRINT".C."
80 PRINT""
90 PRINT""
100 PRINT""
110 PRINT""
120 PRINT""
130 PRINT""
140 PRINT"      FLANGED BEAM DESIGN"
150 PRINT"      THIS PROGRAMME"
160 PRINT"FINDS THE AREAS"
170 PRINT"OF STEEL NEEDED"
180 PRINT"IN A FLANGED"
190 PRINT"BEAM WHEN A"
195 PRINT"MOMENT IS":PRINT"APPLIED"
200 PRINT:PRINT"THE DESIGNER MUST"
210 PRINT"SPECIFY A VALUE FOR"
```

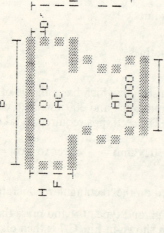

405

```
220 PRINT"THE DEPTH TO THE"
230 PRINT"NEUTRAL AXIS"
240 PRINT"ENTER THE VALUES FOR B & BW"
250 INPUTBR,BW
260 PRINT"ENTER THE VALUES FOR D,HF & D'":PRINT
270 INPUTDE,HF,DC
300 PRINT"ENTER THE STEEL & CONCRETE STRENGTHS IN N/MM↑2"
310 INPUTFY,FC
311 X1=805*DC/(805-.8*FY):X2=7*DC/3:X3=805*DE/(1265+FY):X4=805*DE/(805+.8*FY)
312 PRINT"ENTER DESIGN - 1; CHECK - 2"
314 INPUTCH:IFCH=2THEN1200
320 PRINT"ENTER THE VALUE OF MOMENT NEEDED IN KNM"
330 INPUTMU:MU=MU*1E+06
332 PRINT"ENTER THE DEPTH TO THE NEUTRAL AXIS"
334 INPUTX
340 IFX>HFTHEN700
350 MS=.2*FC*BR*X*(2*DE-X)
360 IFMSC=MUTHEN380
365 PRINT"M,? ="MS"THEREFORE":PRINT
370 PRINT"NO COMPRESSION STEEL IS NEEDED. CHOOSE A NEW VALUE FOR NEUTRAL AXIS D
EPTH"
375 GOTO334
380 MD=MU-MS
390 GOSUB1000
395 A1=MD/F1/(DE-DC):A2=MS/F2/(DE-X/2)+A1*F1/F2
397 CF=(.4*FC*BR*X+F1*A1)/1000
400 PRINT"MOMENT ="MU/1E+06"/N/ & ":PRINT"NEUTRAL AXIS DEPTH ="X
410 PRINT"REQUIRED AREA OF COMPRESSION STEEL":PRINT"A1"="A1"/↑2"
```

```
420 PRINT"⬛REQUIRED AREA OF TENSION STEEL":PRINT"  ="A2"↑/↑2"
430 PRINT"⬛⬛STEEL STRESSES⬛":PRINT"⬛COMPRESSION ="F1"N↑/↑2"
440 PRINT"⬛   TENSION ="F2"N↑/↑2"
450 PRINT"⬛⬛⬛CHECK-COMPRESSIVE AND TENSILE FORCES⬛"
460 PRINT"⬛⬛COMPRESSIVE FORCE ="CF"↑/N":PRINT"⬛   TENSILE FORCE ="F2*A2/1000"KN
470 PRINT"⬛⬛DO YOU WISH TO TRY A NEW VALUE OF":PRINT"NEUTRAL AXIS DEPTH? (YES OR
NO)
480 INPUTA$:IFA$="YES"THEN500
490 END
500 PRINT"ENTER THE NEUTRAL AXIS DEPTH"
510 INPUTX:GOTO340
700 MS=.2*FC*((BR-BW)*HF*(2*DE-HF)+BW*X*(2*DE-X))
710 IFMS<=MUTHEN730
715 PRINT"M,◆ ="MS"THEREFORE":PRINT
720 GOTO370
730 MD=MU-MS
740 GOSUB1000
745 A1=MD/F1/(DE-DC):A2=0.4*FC*((BR-BW)*HF+BW*X)/F2+A1*F1/F2
747 CF=(.4*FC*((BR-BW)*HF+BW*X)+F1*A1)/1000
750 GOTO400
790 FL=1
800 IFQ=0ANDP=1THEN830
810 IFXD>=HF/DETHEN850
820 Q=0:P=1:GOTO860
830 IFXD<=HF/DETHEN850
840 P=BW/BR:Q=(1-P)*HF/DE:GOTO860
850 GL=1
860 RETURN
```

```
1000 IFXC=X1THEN1040
1010 IFXC=X2THEN1030
1020 F1=2000*FY/(2300+FY):GOTO1050
1030 F1=(2300-700*DC/X)*FY/(2300+FY):GOTO1050
1040 F1=700*(1-DC/X)
1050 IFXC=X3THEN1090
1060 IFXC=X4THEN1080
1070 F2=700*(DE/X-1):RETURN
1080 F2=(700*DE/X+900)*FY/(2300+FY):RETURN
1090 F2=FY/1.15:RETURN
1200 PRINT"ENTER AREAS OF TENSION & COMPRESSION":PRINT"STEEL IN MM↑2"
1210 INPUTAT,AC
1220 RT=AT/BR/DE:RC=AC/BR/DE:K=DC/DE:H=RT/.46:J=1750*RC:M=.5*(H*FY-J)/FC
1240 Q=0:P=1:FL=0:GL=0
1250 XD=(M+SQR(M*M+P*J*K/FC+Q*Q/4-M*Q)-Q/2)/P
1260 IFFL=1THEN1290
1270 FL=1:GOSUB830
1280 IFGL=0THEN1250
1290 IFXD<=X1/DETHEN1650
1300 FL=0:GL=0
1310 C=2300+FY:B=FC/FY:I=H/2-2875*RC/C:L=J*K/C
1320 XD=((I+SQR(I*I+B*P*L+(B*Q))↑2/4-I*B*Q))/B-Q/2)/P
1330 IFFL=1THEN1360
1340 GOSUB790
1350 IFGL=0THEN1320
1360 IFXD<=X2/DEANDXD<=X3/DETHEN1650
1370 FL=0:GL=0
1380 IFX2>X3THEN1460
```

408

```
1390 F=5000*FY
1400 XD=(H/B-F*RC/Q/FC-Q)/P
1410 IFFL=1THEN1440
1420 GOSUB790
1430 IFGL=0THEN1400
1440 IFXD<=X3/DETHEN1650
1450 GOTO1520
1460 E=1750*RT:N=1150*RT-2875*RC
1461 PRINT"E="E
1470 XD=((N+SQR(N*N+B*C*P*(E+J*K)+(B*C*Q)↑2/4-N*B*C*D*Q))/B/C-Q/2)/P
1480 IFFL=0THEN1510
1490 GOSUB790
1500 IFGL=0THEN1470
1510 IFXD<=X2/DETHEN1650
1520 FL=0:GL=0:E=1750*RT
1530 A=1125*RT-2500*RC:D=B*C
1540 XD=((A+SQR(A*A+D*E*P-A*D*Q+(D*Q)↑2/4))/D-Q/2)/P
1550 IFFL=1THEN1580
1560 GOSUB790
1570 IFGL=0THEN1540
1580 IFXD<=X4/DETHEN1650
1590 FL=GL=0
1600 G=-(E+F*RC/Q)/2/FC
1610 XD=(G+SQR(G*G+E*P/FC+Q*Q/4-G*Q)-Q/2)/P
1620 IFFL=1THEN1650
1630 GOSUB790
1640 IFGL=0THEN1610
1650 X=XD*DE:GOSUB1000
```

```
1660 IFXC=HFTHEN1690
1670 MU=.4*FC*((BR-BW)*HF*(DE-HF/2)+BW*X*(DE-X/2))+F1*AC*(DE-DC):GOTO1700
1690 MU=.4*FC*BR*X*(DE-X/2)+F1*AC*(DE-DC)
1700 PRINT"ULTIMATE MOMENT OF RESISTANCE":PRINT ="MU/1E+06"KNM"
1702 PRINT"NEUTRAL AXIS DEPTH ="X"MM"
1703 PRINT"NEUTRAL AXIS DEPTH RATIO ="XD
1710 PRINT"STEELSTRESSES":PRINT"COMPRESSION ="F1"N/MM↑2":PRINT        TENSIC
="F2"N/MM↑2
1711 PRINT"DO YOU WISH TO RUN ANOTHER CHECK?":PRINT"ENTER YES OR NO":INPUTC$
1712 IFC$="YES"THEN1200
1720 END
```

Appendix D

Combination of bars table

The following table is used to choose bars to give the required area in a design calculation, when more than one bar size is allowed.

Example

If the required area = 750 mm²

on consulting the table the following sets of bars are the closest:

Bars	Actual area (mm²)
Two at 10-mm ⌀ and three at 16-mm ⌀	760.2
Four at 12-mm ⌀ and one at 20-mm ⌀	766.5
Four at 10-mm ⌀ and four at 12-mm ⌀	766.5
Four at 8-mm ⌀ and five at 12-mm ⌀	766.5

56.5	2 @ 6		150.7	3 @ 8
84.8	3 @ 6		157.0	2 @ 10
100.5	2 @ 8			
106.8	2 @ 6 & 1 @ 8		157.0	2 @ 6 & 2 @ 8
113.0	4 @ 6		163.3	4 @ 6 & 1 @ 8
128.8	1 @ 6 & 2 @ 8		179.0	2 @ 8 & 1 @ 10
135.0	2 @ 6 & 1 @ 10		185.3	1 @ 6 & 2 @ 10
141.3	5 @ 6		185.3	3 @ 6 & 2 @ 8

191.6	4 @ 6 & 1 @ 10
201.0	4 @ 8
207.3	1 @ 8 & 2 @ 10
207.3	2 @ 6 & 3 @ 8
213.6	2 @ 8 & 1 @ 12
213.6	2 @ 6 & 2 @ 10
213.6	4 @ 6 & 2 @ 8
226.1	2 @ 12
229.3	1 @ 6 & 4 @ 8
235.6	3 @ 10
241.9	3 @ 6 & 2 @ 10
241.9	5 @ 6 & 2 @ 8
251.3	5 @ 8
257.6	2 @ 8 & 2 @ 10
257.6	2 @ 6 & 4 @ 8
263.8	4 @ 6 & 3 @ 8
270.1	2 @ 10 & 1 @ 12
270.1	4 @ 6 & 2 @ 10
276.4	1 @ 8 & 2 @ 12
279.6	4 @ 8 & 1 @ 10
285.8	3 @ 6 & 4 @ 8
292.1	2 @ 6 & 3 @ 10
298.4	5 @ 6 & 2 @ 10
304.7	1 @ 10 & 2 @ 12
307.8	3 @ 8 & 2 @ 10
307.8	2 @ 6 & 5 @ 8
314.1	4 @ 10
314.1	4 @ 8 & 1 @ 12
314.1	4 @ 6 & 4 @ 8
326.7	2 @ 8 & 2 @ 12
336.1	2 @ 8 & 3 @ 10
339.2	3 @ 12
342.4	1 @ 6 & 4 @ 10
342.4	5 @ 6 & 4 @ 8
348.7	4 @ 6 & 3 @ 10
358.1	2 @ 10 & 1 @ 16
358.1	4 @ 8 & 2 @ 10
364.4	1 @ 8 & 4 @ 10
364.4	4 @ 6 & 5 @ 8
370.7	2 @ 6 & 4 @ 10
376.9	3 @ 8 & 2 @ 12
383.2	2 @ 10 & 2 @ 12
392.6	5 @ 10
398.9	3 @ 6 & 4 @ 10
402.1	2 @ 16
408.4	5 @ 8 & 2 @ 10
414.6	2 @ 8 & 4 @ 10
427.2	2 @ 12 & 1 @ 16
427.2	4 @ 10 & 1 @ 12
427.2	4 @ 8 & 2 @ 12
427.2	4 @ 6 & 4 @ 10
436.6	4 @ 8 & 3 @ 10
439.8	2 @ 8 & 3 @ 12
449.2	2 @ 6 & 5 @ 10
452.3	4 @ 12
455.5	5 @ 6 & 4 @ 10
461.8	3 @ 10 & 2 @ 12
464.9	3 @ 8 & 4 @ 10
477.5	5 @ 8 & 2 @ 12
480.6	1 @ 10 & 2 @ 16
493.2	2 @ 8 & 5 @ 10
496.3	2 @ 10 & 3 @ 12
502.6	1 @ 8 & 4 @ 12
505.7	4 @ 6 & 5 @ 10
515.2	1 @ 12 & 2 @ 16
515.2	4 @ 10 & 1 @ 16
515.2	4 @ 8 & 4 @ 10
530.9	1 @ 10 & 4 @ 12
540.3	2 @ 12 & 1 @ 20
540.3	4 @ 10 & 2 @ 12
540.3	4 @ 8 & 3 @ 12
552.9	2 @ 8 & 4 @ 12
559.2	2 @ 10 & 2 @ 16
565.4	5 @ 12
565.4	5 @ 8 & 4 @ 10
593.7	4 @ 8 & 5 @ 10
603.1	3 @ 16
603.1	3 @ 8 & 4 @ 12
609.4	2 @ 10 & 4 @ 12
618.8	5 @ 10 & 2 @ 12
628.3	2 @ 20
628.3	2 @ 12 & 2 @ 16
637.7	3 @ 10 & 2 @ 16
653.4	4 @ 12 & 1 @ 16
653.4	4 @ 10 & 3 @ 12
653.4	4 @ 8 & 4 @ 12
666.0	2 @ 8 & 5 @ 12
688.0	3 @ 10 & 4 @ 12
703.7	5 @ 8 & 4 @ 12
716.2	2 @ 16 & 1 @ 20

412

716.2	4 @ 10 & 2 @ 16
722.5	2 @ 10 & 5 @ 12
741.4	1 @ 12 & 2 @ 20
741.4	3 @ 12 & 2 @ 16
760.2	2 @ 10 & 3 @ 16
766.5	4 @ 12 & 1 @ 20
766.5	4 @ 10 & 4 @ 12
766.5	4 @ 8 & 5 @ 12
794.8	5 @ 10 & 2 @ 16
804.2	4 @ 16
829.3	1 @ 16 & 2 @ 20
829.3	2 @ 12 & 3 @ 16
845.0	5 @ 10 & 4 @ 12
854.5	2 @ 12 & 2 @ 20
854.5	4 @ 12 & 2 @ 16
879.6	4 @ 10 & 5 @ 12
882.7	1 @ 10 & 4 @ 16
892.9	2 @ 16 & 1 @ 25
917.3	1 @ 12 & 4 @ 16
917.3	4 @ 10 & 3 @ 16
942.4	3 @ 20
961.3	2 @ 10 & 4 @ 16
967.6	3 @ 12 & 2 @ 20
967.6	5 @ 12 & 2 @ 16
981.7	2 @ 25
1005.3	5 @ 16
1030.4	2 @ 16 & 2 @ 20
1030.4	2 @ 12 & 4 @ 16
1039.8	3 @ 10 & 4 @ 16
1055.5	4 @ 12 & 3 @ 16
1080.7	4 @ 12 & 2 @ 20
1118.4	4 @ 16 & 1 @ 20
1118.4	4 @ 10 & 4 @ 16
1119.1	2 @ 20 & 1 @ 25
1143.5	3 @ 12 & 4 @ 16
1162.3	2 @ 10 & 5 @ 16
1168.6	2 @ 12 & 3 @ 20
1182.8	1 @ 16 & 2 @ 25
1193.8	5 @ 12 & 2 @ 20
1196.9	5 @ 10 & 4 @ 16
1231.5	3 @ 16 & 2 @ 20
1231.5	2 @ 12 & 5 @ 16
1256.6	4 @ 20
1256.6	4 @ 12 & 4 @ 16
1295.1	4 @ 16 & 1 @ 25
1295.9	1 @ 20 & 2 @ 25
1319.4	4 @ 10 & 5 @ 16
1344.6	2 @ 16 & 3 @ 20
1369.7	1 @ 12 & 4 @ 20
1369.7	5 @ 12 & 4 @ 16
1383.8	2 @ 16 & 2 @ 25
1394.8	4 @ 12 & 3 @ 20
1432.5	2 @ 20 & 1 @ 32
1432.5	4 @ 16 & 2 @ 20
1457.6	1 @ 16 & 4 @ 20
1457.6	4 @ 12 & 5 @ 16
1472.6	3 @ 25
1482.8	2 @ 12 & 4 @ 20
1570.7	5 @ 20
1584.9	3 @ 16 & 2 @ 25
1595.9	3 @ 12 & 4 @ 20
1608.4	2 @ 32
1610.0	2 @ 20 & 2 @ 25
1633.6	5 @ 16 & 2 @ 20
1658.7	2 @ 16 & 4 @ 20
1709.0	4 @ 12 & 4 @ 20
1746.7	4 @ 16 & 3 @ 20
1747.5	4 @ 20 & 1 @ 25
1785.9	2 @ 25 & 1 @ 32
1785.9	4 @ 16 & 2 @ 25
1796.9	2 @ 12 & 5 @ 20
1822.1	5 @ 12 & 4 @ 20
1859.8	3 @ 16 & 4 @ 20
1874.7	2 @ 16 & 3 @ 25
1922.6	1 @ 20 & 2 @ 32
1924.2	3 @ 20 & 2 @ 25
1963.4	4 @ 25
1972.9	2 @ 16 & 5 @ 20
1987.0	5 @ 16 & 2 @ 25
2023.1	4 @ 12 & 5 @ 20
2060.8	4 @ 20 & 1 @ 32
2060.8	4 @ 16 & 4 @ 20
2099.3	1 @ 25 & 2 @ 32
2100.9	2 @ 20 & 3 @ 25
2164.5	1 @ 16 & 4 @ 25
2236.8	2 @ 20 & 2 @ 32
2238.3	2 @ 25 & 1 @ 40
2238.3	4 @ 20 & 2 @ 25

2261.9	5 @ 16 & 4 @ 20
2276.8	4 @ 16 & 3 @ 25
2277.6	1 @ 20 & 4 @ 25
2365.6	2 @ 16 & 4 @ 25
2375.0	4 @ 16 & 5 @ 20
2412.7	3 @ 32
2454.3	5 @ 25
2513.2	2 @ 40
2550.9	3 @ 20 & 2 @ 32
2552.5	5 @ 20 & 2 @ 25
2566.6	3 @ 16 & 4 @ 25
2590.2	2 @ 25 & 2 @ 32
2591.8	2 @ 20 & 4 @ 25
2729.2	4 @ 20 & 3 @ 25
2767.7	4 @ 25 & 1 @ 32
2767.7	4 @ 16 & 4 @ 25
2856.4	2 @ 16 & 5 @ 25
2865.1	2 @ 32 & 1 @ 40
2865.1	4 @ 20 & 2 @ 32
2905.9	3 @ 20 & 4 @ 25
2968.8	5 @ 16 & 4 @ 25
3004.1	1 @ 25 & 2 @ 40
3041.0	2 @ 20 & 3 @ 32
3081.1	3 @ 25 & 2 @ 32
3082.6	2 @ 20 & 5 @ 25
3179.2	5 @ 20 & 2 @ 32
3216.9	4 @ 32
3220.1	4 @ 25 & 1 @ 40
3220.1	4 @ 20 & 4 @ 25
3258.6	4 @ 16 & 5 @ 25
3317.5	1 @ 32 & 2 @ 40
3394.4	2 @ 25 & 3 @ 32
3495.0	2 @ 25 & 2 @ 40
3531.1	1 @ 20 & 4 @ 32
3534.2	5 @ 20 & 4 @ 25
3571.9	4 @ 25 & 2 @ 32
3669.3	4 @ 20 & 3 @ 32
3707.8	1 @ 25 & 4 @ 32
3711.0	4 @ 20 & 5 @ 25
3769.9	3 @ 40
3845.3	2 @ 20 & 4 @ 32
3985.8	3 @ 25 & 2 @ 40
4021.2	5 @ 32
4062.8	5 @ 25 & 2 @ 32
4121.7	2 @ 32 & 2 @ 40
4159.4	3 @ 20 & 4 @ 40
4198.7	2 @ 25 & 4 @ 32
4376.2	4 @ 25 & 3 @ 32
4473.6	4 @ 32 & 1 @ 40
4473.6	4 @ 20 & 4 @ 32
4476.7	4 @ 25 & 2 @ 40
4649.5	2 @ 20 & 5 @ 32
4689.6	3 @ 25 & 4 @ 32
4751.6	2 @ 25 & 3 @ 40
4787.7	5 @ 20 & 4 @ 32
4926.0	3 @ 32 & 2 @ 40
4967.6	5 @ 25 & 2 @ 40
5002.9	2 @ 25 & 5 @ 32
5026.5	4 @ 40
5180.4	4 @ 25 & 4 @ 32
5277.8	4 @ 20 & 5 @ 32
5378.4	2 @ 32 & 3 @ 40
5517.4	1 @ 25 & 4 @ 40
5671.3	5 @ 25 & 4 @ 32
5730.2	4 @ 32 & 2 @ 40
5733.4	4 @ 25 & 3 @ 40
5830.7	1 @ 32 & 4 @ 40
5984.7	4 @ 25 & 5 @ 32
6008.2	2 @ 25 & 4 @ 40
6283	5 @ 40
6499.1	3 @ 25 & 4 @ 40
6534.5	5 @ 32 & 2 @ 40
6635.0	2 @ 32 & 4 @ 40
6986.9	4 @ 32 & 3 @ 40
6990.0	4 @ 25 & 4 @ 40
7264.9	2 @ 25 & 5 @ 40
7439.2	3 @ 32 & 4 @ 40
7480.9	5 @ 25 & 4 @ 40
7891.6	2 @ 32 & 5 @ 40
8243.5	4 @ 32 & 4 @ 40
8246.6	4 @ 25 & 5 @ 40
9047.7	5 @ 32 & 4 @ 40
9500.1	4 @ 32 & 5 @ 40

Index

416

joint, rigid, 211

kern, 99

limit-state, 3
load sharing in slabs, 70
loads, 5
   combinations, 7
local buckling in the web of rolled hollow
      sections, 188

minimum ply thickness, 262
minimum steel areas, 29
modification factors, 57

neutral axis, 13
   stipulations of CP110, 20
nominal stress, 216

overturning, 130

passive pressure, 127
Perry–Robertson Formula, 177
pivotal method, 259
plinth, 116
ply thickness, 262
Poisson's ratio, 205
polar moment of inertia, 289
pressure, earth, 127
principle axes, 307
proof load, 220
prying, 213, 269
purlins, 312

radius of gyration, 142, 174, 200, 315
Rankine earth pressures, 127
reinforcement
   cover, 113
   strength of, 32
residual stress, 176
retaining walls, modes of failure, 124
rigid joints, 211
rivets, 211
rolled hollow section, 188, 169

safety factors, 7, 126
secondary moments due to slenderness, 196
serviceability limit-state, 3
shear
   mechanism of failure, 36
   mechanism of resistance, 38

shear, average, 149
shear, beams of small span, 45
shear, connections, 212
shear, in steel sections, 148
shear, punching, 78, 107, 110
shear, reinforcement, 39, 80
   stipulations of CP 110, 42, 80
shear, resistance of flanged sections, 46
shear, ultimate stress, 39
shear-lag, 33
slab bases, 203, 237
slab, effective width under a point load, 73
slab, secondary reinforcement, 76
slab, shear
   ordinary, 76
   punching, 78
   reinforcement, 80
slag, 277
slenderness ratio, 142, 152, 175, 179,
      196, 313
sliding, 130
slip factor, 220
span, effective, 12, 139
stability, 4
standard deviation, 6
steel stress
   allowable stress in sections, 142
   compression, 24
   residual, 176
   tension, 19
stress concentration, 211
strip foundation, 117

tee-sections, 32
throat thickness, 279
throat, weld, 278
truss analogy, 41
types of weld, 275

ultimate limit-state, 3
universal beam, 139

vibration, 4

web bearing, 139, 140, 150, 247
web buckling, 139, 140, 152, 259
   rolled hollow sections, 188, 259
web stiffeners, 158
weld throat, 278
welded connections, 274
welding, 211, 276
   allowable stresses, 279